Internet Resources for Professional Astronomers

This book presents a comprehensive guide to the impact of the internet on professional astronomy. Each chapter deals with a range of the electromagnetic spectrum, with authors first discussing the corresponding problems for each range, particularly those that can be treated by access to very large databases. They then describe how data and information can be found on the internet, and explain how to access and process this information. This is complemented by a detailed guide to astronomy on the internet, including several hundred links of interest to internet resources, arranged by topic. The book gives examples of the analysis of data from telescopes such as Hubble Space Telescope, with a step-by-step guide to using such data. Written by experts from all over the world, this book will be of interest to all astronomers, both professional and amateur, and provides a key to the high-tech work of modern astrophysics.

Internet Resources
for Professional Astronomy

Proceedings of the IX Canary Islands Winter School of Astrophysics

Edited by: Mark R. Kidger,
Ismael Pérez-Fournon and Francisco Sánchez

Instituto de Astrofiçica de Canarias

CAMBRIDGE
UNIVERSITY PRESS

PUBLISHED BY THE PRESS SYNDICATE OF THE UNIVERSITY OF CAMBRIDGE
The Pitt Building, Trumpington Street, Cambridge, United Kingdom

CAMBRIDGE UNIVERSITY PRESS
The Edinburgh Building, Cambridge CB2 2RU, UK   www.cup.cam.ac.uk
40 West 20th Street, New York, NY 10011–4211, USA   www.cup.org
10 Stamford Road, Oakleigh, Melbourne 3166, Australia
Ruiz de Alarcón 13, 28014 Madrid, Spain

© Cambridge University Press 1999

First published 1999

Printed in the United Kingdom at the University Press, Cambridge

A catalogue record for this book is available from the British Library

Library of Congress Cataloguing in Publication data
Canary Islands Winter School of Astrophysics (9th: 1996: Tenerife, Canary Islands)
Internet resources for professional astronomy: proceedings of the IX Canary Islands
Winter School of Astrophysics / edited by Mark R. Kidger, Ismael Pérez-Fournon
and Francisco Sánchez.
p.   cm.
Includes index
ISBN 0 521 66308 3
1. Astronomy–Computer network resources Congresses.
2. Internet (Computer network) Congresses.
I. Kidger, Mark R., II. Pérez-Fournon, I. III. Sánchez, Francisco. IV. Title
QB14.3.C36   1996
025.0652–dc2199–24121CIP

ISBN 0 521 66308 3 hardback

Contents

Contents

Preface

There is no doubt that the Internet has made a tremendous impact on many fields in the last few years, amongst others, on astronomy and its impact is only going to increase in the foreseeable future. When one combines the power of the Internet, with the ever-increasing trend to produce large databases and data archives, a new and extraordinarily powerful research tool suddenly becomes available. Access to the largest telescopes and the most important space-borne instruments is now no longer limited to scientists in a few highly advanced nations. Any scientist, any member of the public even, who has a PC and a modem, can now access data from instruments such as the Hubble Space Telescope, or the Very Large Array, or the new generation of 8 and 10-m telescopes springing up around the world (soon to be joined by the Spanish 10-m telescope here in the Canary Islands). It is this development which has provided the inspiration for the IX Canary Islands Winter School of Astrophysics, under the title "Astrophysics With Large Databases in the Internet Age".

A total of 50 students, from 20 countries, met in Tenerife, in the Museo de la Ciencias y el Cosmos (of the Cabildo de Tenerife) from 17-28th November 1997. Their aim, to learn about astronomical databases and their application, where to find them, and how to use them efficiently. The number of students was limited to permit them to receive ample computer access and individual tuition, which meant that almost two-thirds of all the applications to attend had to be rejected. Eight first-class lecturers, all experts in their fields and all with great practical experience of the problems involved in the study of modern astrophysical data, faced and out-faced what seemed to be the well-nigh impossible task of giving the students a solid grounding in this field in just two weeks.

The lectures, were organised, as is this book, with a series of introductory lectures about Internet resources for astrophysics, followed by a set of theoretical and practical lectures on the four major ranges of the electromagnetic spectrum. The theoretical lectures given by acknowledged experts, the practical lectures (including demonstrations of data reduction) by scientists with intimate knowledge of facilities such as ISO, ROSAT and the HST. Together they form an invaluable whole and a huge resource for students.

La Laguna   The Editors

Internet Services for Professional Astronomy

By HEINZ ANDERNACH[1]

[1]Depto. de Astronomía, IFUG, Universidad de Guanajuato, Guanajuato, C.P. 36000, Mexico
Email: `heinz@astro.ugto.mx`

A (subjective) overview of Internet resources relevant to professional astronomers is given. Special emphasis is put on databases of astronomical objects and servers providing general information, e.g. on astronomical catalogues, finding charts from sky surveys, bibliographies, directories, browsers through multi-wavelength observational archives, etc. Archives of specific observational data will be discussed in more detail in other chapters of this book, dealing with the corresponding part of the electromagnetic spectrum. About 200 different links are mentioned, and every attempt was made to make this report as up-to-date as possible. As the field is rapidly growing with improved network technology, it will be just a snapshot of the situation in mid-1998.

1. Introduction

During the five or so years since the advent of the *World Wide Web* (*WWW*) the number of servers offering information for astronomers has grown as explosively as that of other servers (cf. Adorf (1995)). Perhaps even more than other media, the Internet is flooding us with an excessive amount of information and it has become a challenge to distinguish signal from noise. This report is yet another attempt to do this.

A web address is usually referred to as "Universal Resource Locator" (*URL*), and starts with the characters "`http://`". For better readability I omit these characters here, since one of the most common web browsers (`netscape`) assumes these by default anyway, unless other strings like "`ftp://`" are specified. All *URL* links in this overview and the rest of the book are given in teletype font. It should be emphasized that, owing to the nature of the *WWW*, some of these *URL*s may change without notice. The "File Transfer Protocol" (FTP) will appear as `ftp` in this text, so as to match the corresponding Unix command.

A useful introduction to the basics of Internet, explaining electronic mail, telnet, `ftp`, bulletin boards, "Netiquette", Archie, Gopher, Veronica, WAIS can be found e.g. in Grothkopf (1995) or Thomas (1997). I shall not repeat these basics here, but rather concentrate on practical tools to obtain astronomical information from the Internet. Many search engines have been developed for browsing *WWW* servers by keywords, e.g. AltaVista (`www.altavista.digital.com`), Lycos (`www.lycos.com`), Yahoo (`www.yahoo.com`), Savvy (`www.cs.colostate.edu/~dreiling/smartform.html`), WebGlimpse (`donkey.CS.Arizona.EDU/webglimpse`).

A comprehensive list of *URL*s for "resource discovery" can be found in Adorf (1995) or at `searchenginewatch.com`, and see the *URL* `www.cnet.com/Content/Features/Dlife/Search/ss06.html` for some useful tips to define your search adequately. For specifically astronomical topics the flood of *URL*s returned is smaller at the AstroWeb server, or its mirrors:

- `fits.cv.nrao.edu/www/astronomy.html`
- `cdsweb.u-strasbg.fr/astroweb.html`
- `www.stsci.edu/astroweb/astronomy.html`
- `msowww.anu.edu.au/anton/astronomy.html`

FIGURE 1. A typical example of an astronomer in the Internet age.
(Drawing courtesy Georges Paturel)

- www.vilspa.esa.es/astroweb/astronomy.html

AstroWeb is a consortium of several astronomers who have been collecting astronomy-relevant links since 1995. However, AstroWeb does not actively skim the web for relevant sites regularly, although this has been done occasionally. It relies mainly on forms sent in by the authors or webmasters of those sites, and currently collects about 2500 links.

Recent reviews related to Internet resources in astronomy have been given by a number of authors including: Andernach, Hanisch & Murtagh (1994), Egret & Heck (1995), Murtagh, Grothkopf & Albrecht (1995), Egret & Albrecht (1995), Ochsenbein & Hearn (1997), Heck (1997) and Grothkopf *et al* (1998). Equally useful are the proceedings of the meetings on *Astronomical Data Analysis Software and Systems, I–VII*, held annually since 1991 and published as *ASP* Conference Series vols. 25, 52, 61, 77, 101, 125, and 145. For information and electronic proceedings on meetings III–VII, look at:

- cadcwww.dao.nrc.ca/ADASS/adass_proc/adass3/adasstoc/adasstoc.html
- www.stsci.edu/stsci/meetings/adassIV/toc.html
- iraf.noao.edu/ADASS/adass.html
- www.stsci.edu/meetings/adassVI
- www.stsci.edu/stsci/meetings/adassVII

I do not discuss services explicitly dedicated to amateurs, although there is no well-defined boundary between professional and amateur astronomy, and the latter can be of vital use to professionals, e.g. in the field of variable stars, comets or special solar system events. Indeed, stellar photometry – a field traditionally dominated by amateur observers – boasts a database of variable star observations (`www.aavso.org/internatl_db.html`) that is the envy of many professionals and a shining example to the professional world of cooperation, organization and service. A further proof of a fruitful interaction between professionals and amateurs is "The Amateur Sky Survey" (*TASS*) which plans to monitor the sky down to 14-16 mag and study variable stars, asteroids and comets (`www.tass-survey.org`; Gombert & Droege (1998)).

A cautionary note: by its very nature of describing "sites" on the "web", this work is much like a tourist guide with all its imperfections; hotels or restaurants may have closed or changed their chefs, new roads may have been opened, and beaches may have deteriorated or improved. As similar things happen constantly with web pages or *URL*s, take this work as a suggestion only. You will be the one to adapt it to your own needs, and maintain it as your own reference.

Although I "visited" virtually all links quoted in the present report, be prepared to find obsolete, incomplete or no information at all at some *URL*s. However, after having convinced yourself of missing, obsolete, or incorrect information offered at a given site, do not hesitate to contact the "webmaster" or manager of that site and make constructive suggestions for improvement (rather than merely complain). This is even more important if you find a real error in database services used by a wide community. Vice versa, in your own efforts to provide your web pages, try to avoid links to other documents which do not exist or merely claim to be "under construction", just imagine the time an interested user with a slow connection may waste in calling such a link.

2. Data Centres

The two largest, general purpose data centres for astronomy world-wide are the "Astronomical Data Center" (*ADC*; `adc.gsfc.nasa.gov/adc.html`) of *NASA*'s Astrophysics Data Facility (*ADF*), and the "Centre de Données astronomiques de Strasbourg" (*CDS*; `cdsweb.u-strasbg.fr`). They were the first institutions to systematically collect machine-readable versions of astronomical catalogues but have now widened their scope considerably. Several of their services are so diverse that they will be mentioned in different sections of this paper. At other, medium-sized, regional data centres (e.g. in Moscow, Tokyo, Beijing, Pune, etc.) some of the services of the two major centres are mirrored to reduce the network load on *ADC* and *CDS*.

The increasing number of space missions has led to the creation of mission-oriented data centres like:

- *IPAC* (`www.ipac.caltech.edu`)
- *STScI* (`www.stsci.edu`)
- *ST-ECF* (`ecf.hq.eso.org`)
- *MPE* (`www.rosat.mpe-garching.mpg.de`)
- *ESTEC* (`astro.estec.esa.nl`)
- *VILSPA* (`www.vilspa.esa.es`)
- *CADC* (`cadcwww.dao.nrc.ca`)

and many others. Several major ground-based observatories have evolved data centres offering access to their archives of past observations. Some important examples are:

- The European Southern Observatory (`archive.eso.org`),

- The Royal Greenwich Observatory Isaac Newton Group of Telescopes on La Palma (`archive.ast.cam.ac.uk/ingarch`),
- The National Radio Astronomy Observatory (*NRAO*) Very Large Array (`info.aoc.nrao.edu/doc/vladb/VLADB.html`),
- The Westerbork Synthesis Radio Telescope (*WSRT*) (`www.nfra.nl/scissor`),

Many bulky data sets, like e.g. the *IRAS* data products, the Hubble Space Telescope *HST*) Guide Star Catalogue (*GSC*), etc., have been widely distributed on CD-ROMs. For a comprehensive, though not complete, list of CD-ROMs in astronomy see the "Mediatheque" maintained at *CDS* (`cdsweb.u-strasbg.fr/mediatheque.htm`) and D. Golombek's contribution to these proceedings.

3. Astronomical Catalogues

We consider here "astronomical catalogues" as static, final compilations of data for a given set of cosmic objects. According to Jaschek (1989) they can be further subclassified into (1) observational catalogues, (2) compilation catalogues, and (3) critical compilation catalogues and bibliographic compilation catalogues. In class (1) we also include tables of observational data commonly published in research papers.

The *CDS* maintains the most complete set of astronomical catalogues in a publicly accessible archive at `cdsweb.u-strasbg.fr/Cats.html`. Currently ∼2700 catalogues and published data tables are stored, of which ∼2200 are available for downloading via `ftp`, and a subset of ∼1760 of them are searchable through the VizieR browser (see below). Catalogues can be located by author name or keyword. Since 1993 tables published in *Astronomy & Astrophysics* and its *Supplement Series* are stored and documented in a standard way at *CDS*. As the authors are only recommended, but not obliged, to deposit their tables at *CDS*, there is a small incompleteness even for recently published tabular material. *CDS* is also making serious efforts in completing its archive by converting older or missing published tables into electronic form using a scanner and "Optical Character Recognition" (*OCR*) software. However, catalogues prepared in such a way (as stated in the accompanying documentation file) should be treated with some caution, since *OCR* is never really free of errors and careful proof-reading is necessary to confirm its conformity with the original.

Access to most catalogues is offered via anonymous `ftp` to `cdsarc.u-strasbg.fr` in the subdirectory `/pub/cats`. This directory is further subdivided into nine sections of catalogues (from stellar to high-energy data), and the "J" directory for smaller tables from journals. These subdirectories are directly named after their published location, e.g., `/pub/cats/J/A+AS/90/327` has tabular data published in A&AS Vol. 90, p. 327.

There are other useful commands on the *CDS* node `simbad` which are also available without a "proper" *SIMBAD* account (§4.1). You can telnet to `simbad.u-strasbg.fr`, login as `info`, give `<CR>` as password. This account allows one to query the "Dictionary of Nomenclature of Celestial Objects" and provides comments on the inclusion (or not) in *SIMBAD* of object names with a certain acronym. Other useful commands are e.g. `findcat`, allowing one to locate electronic catalogues by author or keyword, `findacro` to resolve acronyms of object designations (§9), `findgsc` to search the Guide Star Catalogue (*GSC* 1.1), and `findpmm` to search in the *USNO*-A1.0 catalogue of ∼5 10^8 objects (§5.2). The commands `simbib` and `simref` are useful to interrogate the *SIMBAD* bibliography remotely (by author's names or words in paper titles) or resolve the 19-digit "refcodes" (§6). The syntax of all commands can be checked by typing `command -help`. Users with frequent need for these utilities may install these commands on their own machine, by retrieving the file `cdsarc.u-strasbg.fr/pub/cats/cdsclient.tar.Z` (∼40 kb). This allows them to access the above information instantaneously from the command line.

NASA-ADC and *CDS* maintain mirror copies of their catalogue collections of *CDS* (see the *URL* `adc.gsfc.nasa.gov/adc/adc_archive_access.html`). A major fraction of the *CDS* and *NASA-ADC* catalogue collection is also available at the Japanese "Astronomical Data Analysis Center" (*ADAC*; `adac.mtk.nao.ac.jp`), at the Indian "Inter-University Centre for Astronomy & Astrophysics" (*IUCAA*; `www.iucaa.ernet.in/adc.shtml`, click on "Facilities"), and at the Beijing Astronomical Data Center (`www.bao.ac.cn/bdc`) in China.

NASA-ADC has issued a series of four CD-ROMs with collections of the "most popular" (i.e. most frequently requested) catalogues in their `ftp` archive. Some of these CDs (see `adc.gsfc.nasa.gov/adc/adc_other_media.html`) come with a simple software to browse the catalogues. However, the CD-ROM versions of these catalogues are static, and errors found in them (see errata at `ftp://adc.gsfc.nasa.gov/pub/adc/errors`), are corrected only in the `ftp` archive.

Many tables or catalogues published in journals of the American Astronomical Society (*AAS*), like the *Astrophysical Journal* (ApJ), its *Supplement* (ApJS), the *Astronomical Journal* (AJ), and also the *Publications of the Astronomical Society of the Pacific* (PASP), have no longer appeared in print since 1994. Instead, the articles often present a few sample lines or pages only, and the reader is referred to the *AAS* CD-ROM to retrieve the full data. As these CD-ROMs are published only twice a year, in parallel with the journal subscription, this would imply that the reader has to wait up to six months or more to be able to see the data. At the time of writing neither the printed papers nor the electronic ApJ and AJ mention that the CD-ROM data are also available via anonymous `ftp` (at `ftp://ftp.aas.org/cdrom`) before or at the time of the publication on paper. The `ftp` service appears much more practical than the physical CD-ROMs for active researchers (i.e. probably the majority of readers), and perhaps for this reason *AAS* has decided that CD-ROMs will not be issued beyond Vol. 9 (Dec. 1997; see also `www.journals.uchicago.edu/AAS/cdrom`). It would be desirable if readers of electronic *AAS* journals were able to access the newly published tabular data in their entirety and more directly in the future. With the trend to publish tabular data only in electronic form, there is also hope that authors may be released from the task of marking up large tables in TeX or LaTeX , since the main use of these tables will be their integration in larger databases, their processing with TeX -incompatible programming languages, or browsing on the user's computer screen, where TeX formatting symbols would only be disturbing (cf. also §11).

Most of the *AAS* CD-ROM data are also ingested in the catalogue archives of *CDS* and *ADC*. However, many other catalogues, not necessarily available from these two centres, can be found within various other services, e.g. in *CATS, DIRA2, EINLINE, HEASARC, STARCAT* (see §4.2), or the *LANL/SISSA* server (§6.2), etc. In particular, the present author has spent much effort to collect (or recover via *OCR*) many datasets published as tables in journals (see `cats.sao.ru/~cats/doc/Andernach.html` and `cats.sao.ru/~cats/doc/Ander_noR.html`). Almost 800 have been included so far. About half of these were kindly provided upon request by the the authors of the tables. The others were prepared either via scanning and *OCR* by the present author, applying careful proof-reading, or they were found on the *LANL/SISSA* preprint server and converted from TeX or even PostScript (*PS*) to *ASCII* format by the present author. There is no master database that would indicate on which of these servers a certain catalogue is available.

4. Retrieving Information on Objects

We may distinguish several facets to the task of retrieving information on a given astronomical object. To find out what has been published on that object, a good start is made by consulting the various object databases (§4.1) and looking at the bibliographical references returned by them. Some basic data on the object will also be offered, but to retrieve further published measurements a detailed search in appropriate catalogue collections will be necessary (§4.2). Eventually one may be interested in retrieving and working with raw or reduced data on that object such as spectra, images, time series, etc., which may only be found in archives of ground- or space-based observatories (§4.3).

Only objects outside the solar system will be discussed in this section. Good sites for information on solar system objects are *NASA*'s "Planetary Data System" (PDS) at JPL (Planetary and Space Science (1996); `pds.jpl.nasa.gov` and `ssd.jpl.nasa.gov`), the "Lunar and Planetary Institute" (LPI; `cass.jsc.nasa.gov/lpi.html`), the "Solar Physics" and "Planetary Sciences" links at the NSSDC (`nssdc.gsfc.nasa.gov/solar` and `.../planetary`), and the "Minor Planet Center" (`cfa-www.harvard.edu/cfa/ps/mpc.html`). The "Students for the Exploration and Development of Space" homepage also offers useful links on solar system information (`seds.lpl.arizona.edu`).

4.1. *Object Databases*

Object databases are understood here as those which gather both bibliographical references and measured quantities on Galactic and/or extragalactic objects. There are three prime ones: *SIMBAD*, *NED* and *LEDA*; the latter two are limited to extragalactic objects only. A comparison of their extragalactic content has been given by Andernach (1995). All three involve an astronomical "object-name resolver", which accepts and returns identifiers; they also permit retrieval of all objects within a stated radius around coordinates in various different systems or equinoxes. For large lists of objects the databases can also support batch jobs, which are prepared according to specific formats and submitted via email. The results can either be mailed back or be retrieved by the user via anonymous `ftp`. While *SIMBAD* and *NED* allow some limited choice of output format, *LEDA* is the only one that delivers the result in well-aligned tables with one object per line.

SIMBAD (Set of Identifications, Measurements, and Bibliography for Astronomical Data) is a database of astronomical objects outside the solar system, produced and maintained by *CDS*. Presently *SIMBAD* contains 1.54 million objects under 4.4 million identifying names, cross-indexed to over 2200 catalogues. It provides links to 95 700 different bibliographical references, collected for stars systematically since 1950. Presently over 90 journals are perused for *SIMBAD*, which is the most complete database for Galactic objects (stars, HII regions, planetary nebulae, etc.), but since 1983 it has included galaxies and other extragalactic objects as well. *SIMBAD* is not quite self-explanatory; its user's guide can be retrieved from `ftp://cdsarc.u-strasbg.fr/pub/simbad/guide13.ps.gz`, or may be consulted interactively at `simbad.u-strasbg.fr/guide/guide.html`.

Access to *SIMBAD* requires a password, and applications may be sent by email to `question@simbad.u-strasbg.fr`. By special agreement, access is free for astronomers affiliated to institutions in Europe, USA and Japan, while users from other countries are charged for access. The `telnet` address of *SIMBAD* is `simbad.u-strasbg.fr` and its web address in Europe is `simbad.u-strasbg.fr/Simbad`. It has a mirror site in USA, at `simbad.harvard.edu`.

The "*NASA/IPAC* Extragalactic Database" (*NED*, Helou *et al* (1995)) currently contains positions, basic data, and over 1 275 000 names for 767 000 extragalactic objects, nearly 880 000 bibliographic references to 33 000 published papers, and 37 000 notes from catalogues and other publications, as well as over 1 200 000 photometric measurements, and 500 000 position measurements. *NED* includes 15 500 abstracts of articles of extragalactic interest that have appeared in A&A(S), AJ, ApJ(S), MNRAS, and PASP since 1988, and from several other journals in more recent years. Although *NED* is far more complete in extragalactic objects than is *SIMBAD*, it is definitely worthwhile consulting *SIMBAD* to cover the extragalactic literature for the five years in its archives before *NED* commenced in 1988. Samples of objects may be extracted from *NED* through filters set by parameters like position in the sky, redshift, object type, and many others. *NED* has started to provide digital images, including finding charts from the "Digitized Sky Survey" (*DSS*) for some 120 000 of their objects. A unique feature of *NED* is that the photometric data for a given object may be displayed in a plot of the "Spectral Energy Distribution" (*SED*). *NED* is accessible without charge at `nedwww.ipac.caltech.edu` or via telnet to `ned.ipac.caltech.edu` (login as `ned`).

The "Lyon–Meudon Extragalactic Database" (*LEDA*), created in 1983 and maintained at Lyon Observatory, offers free access to the main (up to 66) astrophysical parameters for about 165 000 galaxies in the "nearby" Universe (i.e. typically z<0.3). All raw data as compiled from literature are available, from which mean homogenized parameters are calculated according to reduction procedures refined every year (Paturel *et al* (1997)). Finding charts of galaxies, at almost any scale, with or without stars from the *GSC*, can be created and ~74 000 images of part of these galaxies can be obtained in *PS* format. These images were taken with a video camera from the *POSS*-I for identification purposes only and are of lower quality than those from the Digitized Sky Survey (§5). However, they have been used successfully by the *LEDA* team to improve positions and shape parameters of the catalogued galaxies. *LEDA* also incorporates the galaxies (20 000 up to now) which are being detected in the ongoing "Deep Near-IR Survey of the Southern Sky" (*DENIS*; `www.strw.leidenuniv.nl/denis` or `denisexg.obspm.fr/denis/denis.html`). A flexible query language allows the user to define and extract galaxy samples by complex criteria. *LEDA* can be accessed at `www-obs.univ-lyon1.fr/leda` or via telnet to `lmc.univ-lyon1.fr` (login as `leda`). The homogenized part of *LEDA*'s data, together with simple interrogation software, is being released on "PGCROMs"'s every four years (1992, 1996, 2000...).

4.2. *Pseudo-Databases: Searchable Collections of Catalogues*

It should be kept in mind that object databases like *SIMBAD*, *NED* and *LEDA*, generally do not include the full information contained in the *CDS/ADC* collections of catalogues and tables. This is especially true for older tables which may have become available in electronic form only recently. The catalogues and tables frequently contain data columns not (yet) included in the object databases. Thus the table collections should be considered as a valuable complement to the databases. Different sites support different levels of search of those collections.

Probably the largest number of individual catalogues (~1560) that can be browsed from one site is that offered by `VizieR` at *CDS* (`vizier.u-strasbg.fr`). You may select the catalogues by type of object, wavelength range, name of space mission, etc. An advantage of `VizieR` is that the result comes with hyperlinks (if available) to *SIMBAD* or other relevant databases, allowing more detailed inquiries on the retrieved objects. A drawback is that a search on many of them at the same time requires selecting them individually by clicking on a button. An interface allowing searches through many or all of them is under construction.

The *DIRA2* service ("Distributed Information Retrieval from Astronomical files") at `www.ira.bo.cnr.it/dira/gb` is maintained by the *ASTRONET* Database Working Group in Bologna, Italy. It provides access to data from astronomical catalogues (see the manual at `www.pd.astro.it/prova/prova.html`). The *DIRA2* database contains about 270 original catalogues of Galactic and extragalactic data written in a *DIRA*-specific *ASCII* format. The output of the searches are *ASCII* or *FITS* files that can be used in other application programs. *DIRA2* allows one to plot objects in an area of sky taken from various catalogues onto the screen with various symbols of the user's choice. Sorting as well as selecting and cross-identification of objects from different catalogues is possible, but there is no easy way to search through many catalogues at a time. The software is publicly available for various platforms.

The "CATalogs supporting System" (*CATS*; `cats.sao.ru`) has been developed at the Special Astrophysical Observatory (*SAO*) in Russia. Apart from dozens of the larger general astronomical catalogues it offers the largest collection of radio source catalogues searchable with a single command (see chapter on Radio Astronomy by H. Andernach in these proceedings). A search through the entire catalogue collection "in one shot" is straightforward.

A set of about 100 catalogues, dominated by X-ray source catalogues and mission logs, can be browsed at the "High Energy Astrophysics Science Archive Research Center" (*HEASARC*) at `heasarc.gsfc.nasa.gov/W3Browse`. The same collection is available at the "Leicester Database and Archive Service" (*LEDAS*; `ledas-www.star.le.ac.uk`).

A similar service, offered via telnet and without a web interface, is the "Einstein On-line Service" (*EOLS*, or *EINLINE*) at the Harvard-Smithsonian Center for Astrophysics (*CfA*). It was designed to manage X-ray data from the *EINSTEIN* satellite, but it also served in 1993/94 as a testbed for the integration of radio source catalogues. Although *EOLS* is still operational with altogether 157 searchable catalogues and observing logs, lack of funding since 1995 has prevented any improvement of the software and interface or the integration of new catalogues.

NASA's "Astrophysics Data System" (*ADS*) offers a "Catalog Service" at `adscat.harvard.edu/catalog_service.html`. With the exception of the Minnesota Plate Scanning project (*APS*, cf. §5.1), all available 130 catalogues are stored at the Smithsonian Astrophysical Observatory (*SAO*). For a complete list, request "catalogues by name" from the catalogue service. Not all available catalogues can be searched simultaneously. The service has not been updated for several years and will eventually be merged with `VizieR` at *CDS*.

A growing number of catalogues is available in *dat*OZ (see `146.83.9.18/datoz_t.html`) at the University of Chile as described in Ortiz (1998). It offers visualization and cross-correlation tools. The *STARCAT* interface at *ESO* (`arch-http.hq.eso.org/starcat.html`) with only 65 astronomical catalogues is still available, but has become obsolete. *ASTROCAT* at *CADC* (`cadcwww.dao.nrc.ca/astrocat`) offers about 14 catalogues.

4.3. *Archives of Observational Data*

As these will be discussed in more detail in the various chapters of this book dedicated to Internet resources in different parts of the electromagnetic spectrum, I list only a few *URLs* from which the user may start to dig for data of his/her interest.

Abstracts of sections of the book by Egret & Albrecht (1995) are available on-line at `cdsweb.u-strasbg.fr/data-online.html` and provide links to several archives. The Astro-Web consortium offers a list of currently 129 records for "Data and Archive Centers" at `www.cv.nrao.edu/fits/www/yp_center.html`. *STScI* (`archive.stsci.edu`) has been designated by *NASA* as a multi-mission archive centre, focussing on optical and UV mission data sets. It now plays a role similar to *HEASARC* for high-energy data (see below) and *IPAC* (`www.ipac.caltech.edu`) for infra-red data. The latter are associated in the "Astrophysics Data Centers Coordinating Council" (*ADCCC*; `hea-www.harvard.edu/adccc`).

The Canadian Astronomy Data Center (*CADC*; cadcwww.dao.nrc.ca) includes archives of the *CFHT*, *JCMT* and *UKIRT* telescopes on Hawaii. The *ESO* and *ST-ECF* Science Archive Facility at archive.eso.org offers access to data obtained with the "New Technology Telescope" (*NTT*), and to the catalogue of spectroscopic plates obtained at *ESO* telescopes before 1984. The "Hubble Data Archive" at archive.stsci.edu includes the *HST* archive, the International Ultraviolet Explorer (*IUE*) archive and the *VLA FIRST* survey. The La Palma database at archive.ast.cam.ac.uk/ingarch contains most observations obtained with the Isaac Newton group of telescopes of *RGO*. At the National Optical Astronomy Observatory (*NOAO*; www.noao.edu/archives.html) most data from *CTIO* and *KPNO* telescopes are now being saved, and are available by special permission from the Director. Data from several high-energy satellite missions may be retrieved from *HEASARC* (heasarc.gsfc.nasa.gov), and from *LEDAS*. For results of the Space Astrometry Mission *HIPPARCOS* see astro.estec.esa.nl/Hipparcos/hipparcos.html or cdsweb.u-strasbg.fr.

Images from many surveys of the whole or most of the sky can be retrieved from the SkyView facility (skyview.gsfc.nasa.gov) at Goddard Space Flight Center (*GSFC*). Documentation of these surveys is available at skyview.gsfc.nasa.gov/cgi-bin/survey.pl. Overlays of these surveys with either contours from another survey or objects from a large set of object catalogues may be requested interactively.

Until late 1997 there was no tool which unified the access to the multitude of existing observatory archives. A serious approach to this goal has been made within the "AstroBrowse" project (sol.stsci.edu/~hanisch/astrobrowse_links.html). This resulted in the "Starcast" facility faxafloi.stsci.edu:4547/starcast allowing one to find (and retrieve) relevant data (photometric, imaging, spectral or time series) on a given object or for a given region of sky, in any range of the electromagnetic spectrum from ground- or space-based observatories. The *HEASARC* "AstroBrowser" (legacy.gsfc.nasa.gov/ab) provides an even wider scope, including astronomical catalogues and VizieR at the same time. The "Multimission Archive at *STScI*" (*MAST*; archive.stsci.edu/mast.html) currently combines the archive of *HST*, *IUE*, Copernicus, *EUVE*, *UIT*, plus that of the *FIRST* radio survey and the *DSS* images. Its interface with astronomical catalogues allows one to query e.g. which high-redshift QSOs have been observed with *HST*, or which Seyfert galaxies with *IUE*.

An "Astronomy Digital Image Library" (*ADIL*) is available from the National Center for Supercomputing Applications (*NCSA*) at imagelib.ncsa.uiuc.edu/imagelib.

5. Digital Optical Sky Surveys, Finding Charts, and Plate Catalogues

The first Palomar Observatory Sky Survey (*POSS*-I) was taken on glass plates in red (E) and blue (O) colour from 1950 to 1958. While glass copies are less commonly available, the printed version of *POSS*-I provided the first reference atlas of the whole sky north of $-30°$ declination down to ~20 mag. Together with its southern extensions, provided 20–30 years later by *ESO* in B, and by the UK Schmidt Telescope (*UKST*) in B$_J$, it was the basic tool for optical identification of non-optical objects. However, reliable optical identification required a positional accuracy on the order of a few arcsec. A common, but not too reliable, tool for this were transparent overlays with star positions taken from the *SAO* star catalogue. Major limitations are the low density of stars at high Galactic latitudes and differences in the scale and projection between the transparency and the Palomar print or plate. For higher accuracy than a few arcsec the use of a plate measuring machine was required for triangulation of fainter stars closer to the object in question, but such machines were only available at a few observatories.

Curiously, it was mainly the necessity, around 1983, to prepare the *HST* Guide Star Catalog (*GSC*; www-gsss.stsci.edu/gsc/gsc.html), which led to a whole new Palomar Sky survey (the "quick-V") with shorter exposures in the V band, which was then fully digitized at *STScI* with 1.7″ pixel size to extract guide stars for the *HST*. Later the deeper red plates of *POSS*-I were also scanned at the same resolution to provide an image database of the whole northern sky. For the first time almost the whole sky was available with absolute accuracy of ∼1″, but owing to the sheer volume the pixel data were accessible only to local users at *STScI* during the first years. By the time that they had been prepared for release on 102 CD-ROMs and sold by the Astronomical Society of the Pacific (*ASP*), the Internet and the *WWW* had advanced to a point where small extractions of these pixel data could be accessed remotely. Other observatories also employed plate-scanning machines to scan *POSS*, *ESO* and *UKST* surveys at even finer pixel sizes, and some catalogues were prepared that contained several 10^8 objects detected on these plates. Such catalogues usually include a classification of the object into stars, galaxies or "junk" (objects which fit into neither class and may be artefacts). However, such classifications have a limited reliability. They should not be taken for granted, and it is wise to check the object by visual inspection on the plates (or prints or films), or at least on the digitized image. It is important to distinguish between these different media: the glass plates may show objects of up to ∼1 mag fainter than are visible on the paper prints. Thus the pixel data, being digitized from the glass plates, may show fainter objects than those visible on the prints. Moreover, they offer absolute positional accuracy of better than 1″. On the other hand, the pixel size of the standard *DSS* (1.7″) represents an overriding limitation in deciding on the morphology of faint (i.e. small) objects. Eventually we have the "finding charts" which are merely sketches of all the objects extracted from the image, plotted to scale, but with artificial symbols representing the object's magnitude, shape, orientation, etc. (usually crosses or full ellipses for stars, and open ellipses for non-stellar objects). They should not be taken as a true image of the sky, but rather as an indication of the presence of an optical object at a given position, or as an accurate orientation indicator for observers (see Fig. 2 for an example).

The following presents a quick guide to the varied data products which can be freely accessed now. The Royal Observatory Edinburgh (*ROE*) offers comprehensive information on the status of ongoing optical sky surveys at www.roe.ac.uk/ukstu/ukst.html (go to the "Survey Progress" link). Other places to watch for such information are "Spectrum" (the *RGO/ROE* Newsletter), the *ESO* Messenger (www.eso.org/gen-fac/pubs/messenger), the *STScI* Newsletter (www.stsci.edu/stsci/newsletters/newsletters.html), the Anglo-Australian Observatory (*AAO*) Newsletter (www.aao.gov.au/news.html) and the Newsletter of the "Working Group on Sky Surveys" (formerly "WG on Wide Field Imaging") of *IAU* Commission 9 (chaired by Noah Brosch, email noah@stsci.edu, a *URL* at http://www-gsss.stsci.edu/iauwg/welcome.html is in preparation). See also the chapter by D. Golombek for a summary of plate digitizations available at the *STScI*.

5.1. *Digitized Sky Survey Images*

Digitizations of the *POSS*-E plates in the northern sky and *UKST*-B$_J$ plates in the southern sky are available from SkyView at *URL* skyview.gsfc.nasa.gov. For the vast majority of astronomical institutions without a CD-ROM juke box, this web service allows much easier access to the *DSS* than working locally with the set of 102 CD-ROMs of the *ASP*. Batch requests for large lists of *DSS* extractions can be formulated from the command line using server *URLs* based on perl scripts (see skyview.gsfc.nasa.gov/batchpage.html or www.ast.cam.ac.uk/~rgm/first/collab/first_batch.html). Several other sites offer the standard *DSS* through their servers, e.g. the *CADC* at cadcwww.dao.nrc.ca/dss/dss.html,

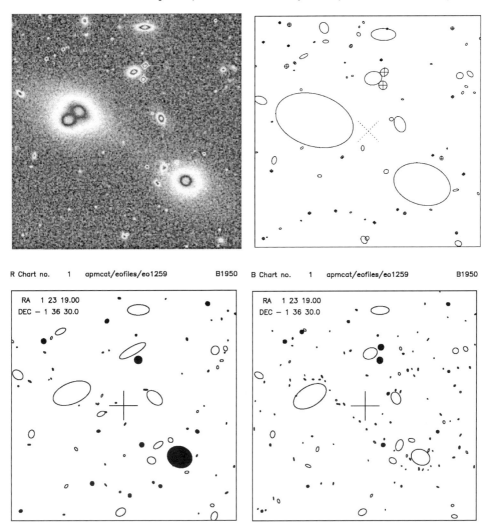

FIGURE 2. Digitized Sky Survey (*DSS*) images versus finding charts from object catalogues, for an 8′×8′ region in the core of the galaxy cluster Abell 194. Upper left: *DSS* image of *POSS*-I red plate from `SkyView`; upper right: *COSMOS* finding chart from B$_J$ plate (open ellipses are galaxies, star-like objects are crossed); lower panels: *APM* charts from *POSS*-I, red (R) and blue (B) plate (filled symbols are star-like, open ones are galaxy-like, and the central cross is 1 arcmin wide). Note the different classification of objects in *APM* and *COSMOS* and in the R and B scans of *APM*. Several multiple objects, clearly separated on the *DSS*, are blended into single, often elongated objects in the *APM* or *COSMOS* catalogues.

ESO at `archive.eso.org/dss/dss`, or `Skyeye` at Bologna (`www.ira.bo.cnr.it/skyeye/gb`), the University of Leicester in UK (`ledas-www.star.le.ac.uk/DSSimage`), and NAO in Japan at `dss.mtk.nao.ac.jp`. Each of these offer a slightly different "look and feel", e.g. *CADC* offers absolute coordinates labels around the charts (but only in J2000), Sky-Eye offers simultaneous extraction of objects from *DIRA2* catalogues (§4.2) and an easy batch request for charts via email, etc. A comparison of performance and speed between these *DSS* servers can be found at `star-www.rl.ac.uk/~acc/archives/archives.html` (see Charles & Shayler (1998)).

Meanwhile almost all of the "second-epoch survey" (SES) plates have been taken: the northern *POSS*-II (`www.eso.org/research/data-man/poss2`) at Palomar, and the southern *UKST SES-R* survey at *AAO* (`www.roe.ac.uk/ukstu/ukst.html`). Most of these have been digitized. Early in 1998 the *STScI* server (`archive.stsci.edu/dss/dss_form.html`) was the only one offering the second-epoch surveys (*POSS*-II or *UKST* R), if available, and otherwise *POSS*-I or *UKST*-B$_J$ surveys.

A digitization of the *POSS*-I E- and O-survey plates was also performed with the Automated Plate Scanner (*APS*) at the Astronomy Department of the University of Minnesota. Only those plates (E and O) with $|b| >20°$ have been scanned. The *APS* uses a flying laser spot to record the transmission of each plate only above ∼65% of the background transmittance as determined in an initial low-resolution scan. This compromise was needed to achieve the small pixel size of 0.3″ for the final images which contain only signal above background, i.e. their background is black. They can be retrieved in *FITS* format from `aps.umn.edu/homepage.aps.html`.

5.2. *Object Catalogues and Finding Charts from* DSS

The *HST* Guide Star Catalog (*GSC*) was the first all-sky catalogue of optical objects extracted from plate digitizations. For declinations north of +3° the Palomar "Quick-V" (epoch 1982) plates of 20 min exposure were used. For the south the 50–75 min exposures of the *SERC*-B$_J$ survey (epoch ∼1975) and its equatorial extension (epoch ∼1982) were used (see `www-gsss.stsci.edu/gsc/gsc12/description.html`). The *GSC* contains ∼19 10^6 objects in the range 6–15 mag. Most of them are stars, but an estimated 5 10^6 galaxies are present as well. The positional accuracy has been improved to better than 0.4″ in version 1.2. Note, however, that this catalogue is not magnitude-limited, but that the selection of stars has been carried out so as to provide a uniform *surface density* of guide stars over the sky.

The *Automated Plate Measuring Machine* (*APM*) is located at the Institute of Astronomy, Cambridge, UK and has been used to prepare object catalogues from Sky Survey plates at high Galactic latitudes ($|b| >20°$), see e.g. Lewis & Irwin (1996). Both colours of the *POSS*-I survey plates were scanned and the objects cross-identified, so that colour information is available for a matched object catalogue of well over 100 million objects down to m=21.5 in blue (O) and m=20 in red (E). For the southern sky the glass plates of the *UKST*-B$_J$ and later the *UKST SES-R* survey have been scanned, with limiting magnitudes of 22.5 in B$_J$ and 21 in R. All plates were scanned at 0.5″ scan interval and a scanning resolution of 1″. The pixel data of the scans are not available, and no copies of the entire catalogue are distributed. Both the northern hemisphere catalogue ($\delta > -3°$), and the southern hemisphere catalogue based on *UKST* B$_J$ and *SES-R* plates (∼50% complete) are available for routine interrogation at `www.ast.cam.ac.uk/~apmcat`. The *URL* `www.aao.gov.au/local/www/apmcatbin/forms` offers a standalone client program in C (`apmcat.c`) which allows queries for large sets of finding charts and object lists from the command line. The catalogues can also be accessed from a captive account (`telnet 131.111.68.56`, login as `catalogues` and follow the instructions).

COSMOS (COordinates, Sizes, Magnitudes, Orientations, and Shapes) is a plate scanning machine at the Royal Observatory Edinburgh, which was used to scan the whole southern sky ($\delta < +2.5°$) from the IIIa-J and Short Red Surveys, and led to an object catalogue of several hundred million objects (Drinkwater, Barnes & Ellison (1995)). Public access to the catalogue is provided through the Anglo-Australian Observatory (*AAO*) at `www.aao.gov.au/local/www/surveys/cosmos`, and through the Naval Research Laboratory (NRL) at `xweb.nrl.navy.mil/www_rsearch/RS_form.html`. The *AAO* facility requires the user to register and obtain a password. During a `telnet` session the user may either input coordinates on the fly, or have them read from a file previously transferred (via `ftp`) to the public *AAO* account, and create charts and/or object lists on various different output media. The user has to transfer the output back to the local account via `ftp`. This disadvantage is balanced by the possibility of extracting large amounts of charts for big cross-identification projects. Charts may be requested in stamp-size format resulting in PostScript files containing many charts per page.

The US Naval Observatory (*USNO*) has scanned the *POSS*-I E- and O-plates (for plate centres with $\delta \geq -30°$) and the *ESO*-R and *SERC*-J plates (centred at $\delta \leq -35°$) with the "Precision Measuring Machine" (*PMM*). A scan separation of $0.9''$ was used (i.e. finer than that of the *STScI* scans for *DSS*), and object fitting on these images resulted in the *USNO*-A1.0 catalogue of 488 006 860 objects down to the very plate limit (limiting mag O=21, E=20, J=22, F=21). Objects were accepted only if present to within $2''$ on both E- and O-plates, which implies an efficient rejection of plate faults, but also risks losing real, faint objects with extreme colours. This catalogue is available both as a set of 10 CD-ROMs and interactively at `psyche.usno.navy.mil/pmm`. Client programs at *CDS* (§3) and *ESO* (`archive.eso.org/skycat/servers/usnoa`) allow extraction of object lists of small parts of the sky very rapidly from the command line. There are plans to produce a *USNO*-B catalogue, which will combine *POSS*-I and *POSS*-II in the north, *UKST*-B$_J$, *ESO*-R, and *AAO*-R in the south, and will attempt to add proper motions and star/galaxy separation fields to the catalogue.

The images from the *APS* scans of *POSS*-I (see above) have been used to prepare a catalogue of $\sim 10^9$ stellar objects and 10^6 galaxies detected on *both* E and O plates. Extractions from this catalogue may be drawn from `aps.umn.edu/homepage.aps.html`, or from the *ADS* catalogue service at `adscat.harvard.edu/catalog_service.html`.

5.3. *Catalogues of Direct Plates*

While the object catalogues mentioned above were drawn from homogeneous sky surveys, there are almost 2 million wide-field photographic plates archived at individual observatories around the world (see the Newsletters of the *IAU* Commission 9 "Working Group on Sky Surveys"). To allow the location and retrieval of such plates for possible inspection by eye or with scanning machines, the "Wide-Field Plate Database" (*WFPDB*) (`www.wfpa.acad.bg`) is being compiled and maintained at the Institute of Astronomy of the Bulgarian Academy of Sciences. Described in Tsvetkov *et al* (1998), the *WFPDB* offers metadata for currently $\sim$330 000 plates from 57 catalogues (see `www.wfpa.acad.bg/~list`) collected from more than 30 observatories. Since August 1997 the *WFPDB* may be searched as catalogue `VI/90` via the *CDS* `VizieR` catalogue browser (`vizier.u-strasb.fr/cgi-bin/VizieR`).

5.4. *An Orientation Tool for the Galactic Plane*

The "Milky Way Concordance" is a graphical tool to create charts with objects from catalogues covering the Galactic Plane, as described by Barnes & Myers (1997), available at the *URL* `cfa-www.harvard.edu/~peterb/concord`. Currently 17 catalogues including H II regions, planetary and reflection nebulae, dark clouds, and supernova remnants are available to create colour-coded finding charts of user-specified regions.

5.5. *Future Surveys*

The All Sky Automated Survey (*ASAS*; `sirius.astrouw.edu.pl/~gp/asas/asas.html`) is a project of the Warsaw University Astronomical Observatory (Pojmański 1997). Its final goal is the photometric monitoring of $\sim 10^7$ stars brighter than 14 mag all over the sky from various sites distributed over the world. The first results on variable stars found in ∼100 square degrees have become available at the above *URL*.

While current Digitized Sky Surveys are all based on photographic material digitized off-line after observing, the future generation of optical Sky Surveys will be digital from the outset, like the "Sloan Digital Sky Survey" (*SDSS*; `www-sdss.fnal.gov:8000`, Margon (1998)). The *SDSS* will generate deep (r′=23.5 mag) images in five colours (u′, g′, r′, i′, and z′) of π steradians in the Northern Galactic Cap ($|b| > +30°$). The *SDSS* will be performed in drift-scan mode over a period of five years. A dedicated 2.5 m telescope at Apache Point Observatory (NM, USA; `www.apo.nmsu.edu`), equipped with a mosaic of 5×6 imaging CCD detectors of 2048^2 pixels will allow a uniquely large 3° field of view. Selected from the imaging survey, 10^6 galaxies (complete to r′ ∼18 mag) and 10^5 quasars (to r′ ∼19 mag) will be observed spectroscopically. The entire dataset produced during the course of the survey will be tens of terabytes in size. The *SDSS* Science Archive (`tarkus.pha.jhu.edu/scienceArchive`) will eventually contain several 10^8 objects in five colours, with measured attributes, and associated spectral and calibration data. Observations began in May 1998, and the data will be made available to the public after the completion of the survey.

6. Bibliographical Services

Long before the Internet age, abstracts of the astronomical literature were published annually in the *Astronomischer Jahresbericht* by the Astronomisches Rechen-Institut in Heidelberg (`www.ari.uni-heidelberg.de/publikationen/ajb`). The series started in 1899, one year after the first issue of *Science Abstracts* was published, the precursor of *INSPEC* (§6.1; cf. `www.iee.org.uk/publish/inspec/inspec.html`). Abstracts of many papers which originally did not have an English abstract were given in German. Since 1969 its successor, the *Astronomy & Astrophysics Abstracts* (*AAA*), published twice a year, have been THE reference work for astronomical bibliography. The slight drawback that it appears about 8 months after the end of its period of literature coverage is compensated by its impressive completeness of "grey literature", including conference proceedings, newsletters and observatory publications. Until about 1993, browsing these books was about the only means for bibliographic searches "without charge" (except for the cost of the books). In 1993 *NASA*'s "Astrophysics Data System" (*ADS*) Abstract Service with initially 160 000 abstracts became accessible via telnet. After a few months of negotiation about public accessibility outside the US, the service was eventually put on the *WWW* in early 1994, with abstracts freely accessible to remote users world-wide. Shortly thereafter they turned into (and continue to be) the most popular bibliographic service in astronomy (see Kurtz *et al* (1996), Eichhorn *et al* (1998)).

On the announcement during *IAU* General Assembly XIII (Kyoto, Japan, Aug. 1997) that *AAA* is likely to stop publication at the end of 1998, some Astronomy librarians compared the completeness of *AAA* with that of *ADS* and *INSPEC* (§6.1). The results show that, in particular, information about conference proceedings and observatory reports is missing from *ADS* and *INSPEC*. After the demise of *AAA*, *ADS* would be the *de facto* bibliography of astronomy literature, and there is a danger that it will not be as complete as *AAA* (see `www.eso.org/libraries/iau97/libreport.html` for a discussion). It would indeed be to the benefit of all astronomers if some day all abstracts from *Astronomischer Jahresbericht* and *AAA* (covering 100 years!) became available on the Internet (see §6.1 for *ARIBIB*).

Many bibliographic services in astronomy use a 19-digit reference code or "refcode" (Schmitz *et al* (1995), see e.g. `cdsweb.u-strasbg.fr/simbad/refcode.html`). They have the advantage of being unique, understandable to the human eye, and may be used directly to resolve the full reference and to see their abstracts on the web. Lists of refcodes are also maintained by *NED* and *ADS* (`adsabs.harvard.edu/abs_doc/journal_abbr.html`). Note, however, that *ADS* calls them "bibcodes", and that for less common bibliographic sources occasionally these may differ from *CDS* refcodes. Unique bibcodes do not exist as yet for proceedings volumes and monographs, but work is under way in this area.

6.1. *Abstract and Article Servers*

NASA's *ADS* Abstract Service (`adsabs.harvard.edu/abstract_service.html`) is offered at the Center for Astrophysics (*CfA*) of the Smithsonian Astrophysical Observatory (*SAO*). It goes back to several 10^5 abstracts prepared by *NASA*'s " Scientific and Technical Information" Group (*STI*) since 1975. Note that the latter abstracts may not be identical with the published ones and that complete coverage of the journals is not guaranteed. Since 1995 most of the abstracts are being received directly from the journal editors, and coverage is therefore much more complete. The service now contains abstracts from four major areas which need to be searched separately: Astronomy ($\sim$380 000 abstracts), Instrumentation, Physics & Geophysics and *LANL/SISSA* `astro-ph` preprints (§6.2). The preprints expire 6 months after their entry date. The four databases combined offer over 1.1 million references. The service is also useful to browse contents of recent journals using the `BIBCODE QUERY` or `TOC QUERY` (§6.5) links. Its popularity is enormous: it was accessed by $\sim$10 000 users per month, and about 5 million references per month were returned in response to these queries in late 1997. It has mirror sites in Japan (`ads.nao.ac.jp`) and France (`cdsads.u-strasbg.fr/ads_abstracts.html`).

The *ADS* provides very sophisticated search facilities, allowing one to filter by author, by title word(s) or words in the abstract, and even by object name, albeit with the silent help of *NED* and *SIMBAD*. The searches can be tuned with various weighting schemes and the resulting list of abstracts will be sorted in decreasing order of relevance (see `adsdoc.harvard.edu/abs_doc/abs_help.html` for an extensive manual). Each reference comes with links (if available) to items like (C) citations available (references that cite that article), (D) data tables stored at *CDS* or *ADC*, (E) electronic versions of the full paper (for users at subscribing institutions), (G) scanned version of the full paper, (R) references cited by that article, etc. Links between papers (citations and references) are gradually being completed for older papers. Citations are included for papers published since 1981 and were purchased from the "Institute for Scientific Information" (*ISI*), see below. When the recognition of the full text from the scanned images has been completed in a few years (see below), *ADS* plans to build its own R and C links.

The *ADS* also employed page scanners to convert printed pages of back issues of major astronomical journals into images ("bitmaps") accessible from the web. Early in 1995 the *ADS* started offering this "Article Service" at `adsabs.harvard.edu/article_service.html`. Images of over 60 000 scanned articles are now on line, and over 250 000 pages were retrieved monthly in 1997. The intention is to prepare page scans back to volume 1 for all major journals. Note, however, that these are images of printed pages and not *ASCII* versions of the full text. Eventually the full article database will be about 500 Gbyte of data. The images of printed pages may eventually be converted into *ASCII* text via *OCR*, but currently this exercise is estimated to take a full year of *CPU* time (excluding the subsequent effort in proof-reading and correcting the *OCR* result). No full-text search features are available as yet from *ADS*, not even for recent articles.

ARIBIB is a bibliographic database with information given in the printed bibliography "Astronomy and Astrophysics Abstracts" (*AAA*) by the Astrononomisches Rechteninstitut (*ARI*) in Heidelberg, Germany. Currently, at `www.ari.uni-heidelberg.de/aribib` references to the literature of 1983–1997 (Vols. 33–68 of *AAA*) are freely available, but may be restricted to subscribers of the printed *AAA* in future. No abstracts are available from *ARIBIB*. The *ARI* intends to prepare references of older literature in a machine-readable format, by scanning earlier volumes of *AAA* and "Astronomischer Jahresbericht".

The UnCover database of authors and titles of scientific papers (see §6.5 for details) may also be used for keyword searches, although it does not offer abstracts.

There are numerous well-established commercial bibliographic database services which charge for access. The use of these systems in astronomy has been reviewed by Davenhall (1993) and Michold *et al* (1995), and Thomas (1997) gives a more general overview. Typically these databases cover a range of scientific and engineering subjects and none of them is specifically astronomical. This has the disadvantage that more obscure astronomical journals are not covered, but the advantage that astronomical papers in non-astronomical publications will be included. The Institution of Electrical Engineers (*IEE*) in the UK produces *INSPEC* (`www.iee.org.uk/publish/inspec/inspec.html`) which is the main English-language commercial bibliographic database covering physics (including astrophysics), electrical engineering, electronics, computing, control and information technology. It currently lists some 5 million papers and reports, with over 300 000 new entries being added per year, and it covers the main astronomical journals. An abstract is usually included for each entry. Another important bibliographic database is the Science Citation Index (*SCI*) produced by the Institute for Scientific Information Inc. (*ISI*; `www.isinet.com`) in USA. The *SCI* contains details drawn from over 7500 journals and (*via* the *Index to Scientific and Technical Proceedings, ISTP*) over 4200 conferences per year. While the *SCI* does not contain abstracts, it offers cross-references to all the citations in each paper that is included, a unique and extremely valuable feature. Often commercial bibliographic services are not accessed directly, but rather through a third-party vendor. Typically such a vendor will make a number of bibliographic databases available, having homogenized their appearance and customized their contents to a greater or lesser extent. There is a number of such vendors; one example is the "Scientific and Technical Information Network" (STN; `www.cas.org/stn.html`) which includes links to about 200 databases.

6.2. *Preprint Servers*

The availability of electronic means has reduced the delay between acceptance and publication of papers in refereed journals from 6–10 down to 3–6 months (or even 1 month in case of letters or short communications), largely because authors prepare their own manuscripts electronically in the formatting requirements of the journals. Nevertheless, for conference proceedings the figure remains between 10 and 20 months. Such a delay did (and still does) crucially affect certain types of research. Thus, for several decades, the "remedy" to this delay has been a frequent exchange of preprints among astronomers, and until very recently, the preprint shelves used to be the most frequented areas in libraries. This situation has gradually changed since April 1992, when both *SISSA* (International School for Advanced Studies, Trieste, Italy) and *LANL* (Los Alamos National Laboratory, USA) started to keep mirror archives of electronically submitted preprints (`xxx.lanl.gov`). In the first years of their existence, preprints were dominant in the fields of theoretical cosmology and particle physics, and about 35 `astro-ph` preprints were submitted monthly in mid-1993. The popularity of this service has increased impressively since then: over 60 000 *daily* accesses to `xxx.lanl.gov` from 6 000 different hosts (see Fig. 3), and about 300 preprints submitted per month in 1997/8 only for `astro-ph` (and ~1800 altogether), with a fair balance between all parts of observational and theoretical astrophysics; mirror sites have been installed in 12 other countries (Australia, Brazil,

China, France, Germany, Israel, Japan, Russia, South Korea, Spain, Taiwan, and the UK, see xxx.lanl.gov/servers.html). A further mirror site is under construction in India. References to electronic preprints from *LANL/SISSA* are made more and more frequently in refereed journals, and the *LANL/SISSA* server also provides links to citations to, and references from, their electronic preprints to other preprints of the same collection.

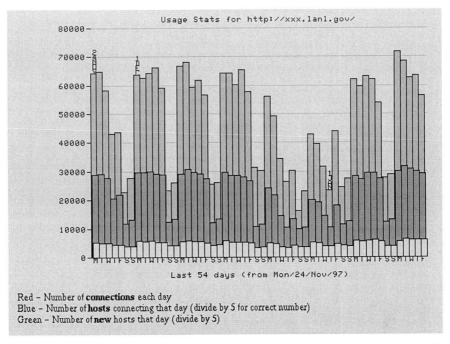

FIGURE 3. Daily accesses to the *LANL* preprint server. The regular 2-day dips are weekends, while the wider dips are due to Thanksgiving, Christmas and New Year 1997/98.

You may subscribe to receive a *daily* list of new preprints. To avoid this flood of emails you could rather send an email to astro-ph@xxx.lanl.gov just after the end of month MM of year YY, with subject "list YYMM" , e.g. "list 9710", to receive a full list of preprints archived in October 1997. To get the description of all the functionalities of the preprint server (e.g. how to submit or update your own preprints), send an email to the above address with "Subj: get bighelp.txt". If you want to check for the presence of a certain preprint, send an email with "Subj: find keyword" , where keyword is an author or part of the title of the paper. You need to specify a given year if you want to search further back than the default of one year from the current date. Once you have located the preprint(s) that you are interested in, the most efficient (least time-consuming) method to obtain a copy is to request it by its sequence number YYMMNNN, via email to the same address with Subj: get YYMMNNN. It comes in a self-extracting uuencoded file which you need to save in a file, say XYZ, strip off the mail header, and just execute it with the command csh XYZ. A common shortcoming is that authors sometimes do not include all necessary style files with their papers. Generally these can be obtained separately from the server by email with Subj: get whatever.sty. Another solution to this problem is to use the web interface at xxx.lanl.gov/ps/astro-ph/YYMMNNN, where (unlike the email or ftp service) the complete *PS* files of papers are indeed available (and are in fact created on the fly upon request).

Submission of a preprint to the *LANL/SISSA* server is today the most efficient way of world-wide "distribution" without expenses for paper or postage, since the preprint will be available to the entire community within 24 hours from receipt, provided it passes some technical checks of file integrity and processability. Note, however, that electronic preprint servers contain papers in different stages of publication: accepted, submitted to a refereed journal or to appear in conference proceedings. Occasionally the authors do not explain the status of the paper, and the preprint server may have been the only site to which the paper has ever been submitted. Of course, a reference made to a preprint of work that was never published should be regarded with caution in case it was subsequently rejected by a journal. Note also that only a fraction of all papers published in refereed journals (perhaps 30–50% now) is available from *LANL/SISSA* prior to publication. Many other preprints are offered from web pages of individual researchers or from institutional pages. A substantial collection of links to other sources of preprints is provided at `www.ucm.es/info/Astrof/biblio.html#preprints`.

The International Centre for Theoretical Physics in Trieste, Italy, provides the "One-Shot World-Wide Preprints Search" at `www.ictp.trieste.it/indexes/preprints.html`.

You can also check a list of "Papers Submitted to Selected Astrononomical Publications" at `www.noao.edu/apj/ypages/yp.html`. A list of papers submitted to the *Astronomical Journal* can be viewed at `www.astro.washington.edu/astroj/lemon/yp.html`. Note that these are *submitted* and not necessarily *accepted* articles. The pages come with links to email the authors, which is very helpful if you wish to ask some of them for a copy. There is also a page with "Titles and abstracts of ApJ Letters accepted but not yet published" at `cfa-www.harvard.edu/aas/apjl_abstracts.shtml`.

A "Distributed Database of Online Astronomy Preprints and Documents" is currently in the early stages of development. While not functional as yet, there is a web page describing the project. People wanting to monitor progress, or make suggestions, are invited to look at the *URL* `doright.stsci.edu/Epreps`.

6.3. *Preprint Lists from NRAO and STScI*

Although the *LANL/SISSA* server has become very popular recently, there are still a large number of preprints being distributed *only* on paper. The ideal places to get reasonably complete listings of these are the following.

The STEPsheet ("Space Telescope Exhibited Preprints") is a list of all preprints received during the last two weeks at the *STScI* and is prepared by its librarian, Sarah Stevens-Rayburn. It is delivered by email, and subscription requests should be sent to `library@stsci.edu`. Since about 1996 the number of items per month has dropped below the monthly number of electronic preprints submitted to `astro-ph`. Note that the STEPsheet preprints themselves are not distributed by the *STScI* librarian and must be requested from the individual authors. The full current *STScI* database contains everything received in the last several years, along with all papers received since 1982 and not yet published. Both databases are searchable at `sesame.stsci.edu/library.html`.

The RAPsheet ("Radio Astronomy Preprints") is a listing of all preprints received in the Charlottesville library of the National Radio Astronomy Observatory (*NRAO*) in the preceding two weeks (contact: `library@nrao.edu`). Interested persons should request copies of preprints from the authors. The tables of contents of all incoming journals and meeting proceedings are perused in order to find published references and to update the records. A database of preprints received since 1986, along with their added references, and including unpublished ones since 1978, is also searchable at `libwww.aoc.nrao.edu/aoclib/rapsheet.html`.

6.4. *Electronic Journals*

Most of the major astronomical journals are now available over the Internet. While they tended to be freely available for a test period of one to two years, most of them now require a license, which can be obtained for free only if the host institute of the user is subscribed to the printed version. Nevertheless, for most journals the tables of contents are accessible on the *WWW* for free (see §6.5). A very useful compilation of links to electronic journals and other bibliographical services is provided at `www.ucm.es/info/Astrof/biblio.html`.

Under the "*AAS* Electronic Journal Project" (`www.aas.org/Epubs/eapjl/eapjl.html`; cf. Boyce (1997)) the Astrophysical Journal (ApJ) had been made available electronically since 1996 (`www.journals.uchicago.edu/ApJ/journal`). As of April 1997, access to the full text of the ApJ Electronic Edition is available only to institutional and individual subscribers. What you *can* do without subscribing, is to browse the contents pages of even the latest issues. The policy of *AAS* is to sell a license for the whole set of both back and current issues. Licensees will have to keep paying in order to see any issues of the journal. In return *AAS* will continue to maintain electronic links between references, a facility which the database of scanned back journals at the *ADS* article service does not offer.

The electronic *Astronomical Journal* (`www.journals.uchicago.edu/AJ/journal`) came on-line in January 1998. *Astronomy & Astrophysics* (A&A) and its *Supplement Series* (A&AS) are available at `link.springer.de/link/service/journals/00230/tocs.htm`, and at `www.edpsciences.com/docinfos/AAS/OnlineAAS.html`, respectively. Abstracts of both journals can be viewed at `cdsweb.u-strasbg.fr/Abstract.html`.

The journal *New Astronomy*, initially designed to appear electronic-only released its first issue in September 1996 (see `www.elsevier.nl/locate/newast`). One may subscribe to a service alerting about new articles appearing in *New Astronomy* by sending email to `newast-e@elsevier.nl` with Subj: `subscribe newast-c`.

Several other journals are available electronically, e.g. the *Publications of the Astronomical Society of the Pacific* (PASP; `www.journals.uchicago.edu/PASP/journal`); the *Publications of the Astronomical Society of Australia* (PASA) has started an experimental web server, beginning from Vol. 14 (1997), at `www.atnf.csiro.au/pasa`; Pis'ma Astronomicheskii Zhurnal offers English abstracts at `hea.iki.rssi.ru/pazh`, and Bulletin of the American Astronomical Society (BAAS) at `www.aas.org/publications/baas/baas.html`. The *IAU* "Informational Bulletin of Variable Stars" (IBVS) has been scanned back to its volume 1 (1961) and is available at `www.konkoly.hu/IBVS/IBVS.html`. One volume (# 45) of the Bulletin d'Information of *CDS* (BICDS) is at `cdsweb.u-strasbg.fr/Bull.html`.

The *Journal of Astronomical Data* (JAD), announced during the 22nd *IAU* General Assembly in 1994 as the future journal for the publication of bulky data sets on CD-ROM, has produced three volumes on CD-ROM since October 1995, available at cost from `www.twinpress.nl/jad.htm`.

More and more electronic journals are linking their references directly to the *ADS* Abstract Service, thus working toward a virtual library on the Internet.

6.5. *Tables of Contents*

The *ADS* abstract service (§6.1) offers to browse specific volumes of journals via the `TOC QUERY` button on *URL* `adswww.harvard.edu/abstract_service.html`. For those journals not accessible by the `TOC QUERY` you can use the `BIBCODE QUERY`, e.g. to browse the *Bull. Astron. Soc. India*, search for bibcode `1997BASI...25`.

UnCover at the "Colorado Alliance of Research Libraries" (CARL) is a database of tables of contents of over 17 000 multidisciplinary journals. It can be accessed at *URL* `uncweb.carl.org`, or via telnet to `pac.carl.org` or `database.carl.org`. It contains article titles and authors only, and offers keyword searches for both. Data ingest started in September 1988. In late 1997 it included more than 7 000 000 articles, and 4 000 articles were added daily. About sixty astronomy-related journals are present, including ApJ, A&A, AJ, ApJS, MNRAS, PASP, Nature, Science, and many others which are less widely distributed. One of the most useful features is that the contents pages of individual journals can be viewed by volume and issue. This provides an important independent resource to monitor the current astronomical literature, especially if one's library cannot afford to subscribe to more than the major journals. Copies of all retrieved articles can be ordered (by FAX only) for a charge indicated by the database. UnCover also provides links to access other library databases and to browse the library catalogues of several North American libraries.

For some of the journals not covered by bibliographic services the Publishers' web pages may offer tables of contents, like e.g. for the *Monthly Notices of the Royal Astronomical Society* (MNRAS) at `www.blackwell-science.com/~cgilib/jnlpage.bin?Journal=MNRAS &File=MNRAS&Page=contents`. To browse *Chinese Astronomy & Astrophysics* click on "Contents Services" at `www.elsevier.com/inca/publications/store/5/8/5/`. For *Astrophysics & Space Science* see `www.wkap.nl/jrnltoc.htm/0004-640X`. The National Academy of Sciences of the USA offers the contents pages and full texts of its Proceedings and colloquia at `www.pnas.org`. The American Physical Society has its *Reviews of Modern Physics* at `rmp.aps.org`. The *Icarus* journal offers its tables of contents and lists of submitted papers at the *URL* `astrosun.tn.cornell.edu/Icarus/`. Tabular and other data from papers published in Icarus from 1994–97 were published on the *AAS* CD-ROMs.

6.6. *"Grey Literature": Newsletters, Observatory Publications, etc.*

A compilation of links to various astronomical newsletters is provided by P. Eenens at `www.astro.ugto.mx/~eenens/hot/othernews.html`. In 1994/95, Cathy Van Atta at *NOAO* (now retired) and a few others prepared a list of astronomical Newsletters. S. Stevens-Rayburn (*STScI*) has put this list on *URL* `sesame.stsci.edu/lib/NEWSLETTER.htm` and invites volunteers to complete and update the information. A catalogue of over 4 000 individual Observatory Publications, ranging from the 18th century to the present, has been prepared by Brenda Corbin and is available from within the *USNO* Library Online Catalog "Urania" (click on "Library Resources" at `www.usno.navy.mil/library/lib.html`).

A list of *IAU* Colloquia prepared by *STScI* librarian S. Stevens-Rayburn is offered at `sesame.stsci.edu/lib/other.html`. A Union List of Astronomy Serials II (*ULAS* II) compiled by Judy L. Bausch can be searched at `sesame.stsci.edu/lib/union.html`. It provides information on ∼2300 (primarily) non-commercial publications of observatories and institutions concerned with research in astronomy. For each item, it lists the holding records of 42 contributing libraries, representing the most comprehensive astronomical collections in North America, with selected holdings from China, Europe, India, and South America as well.

A database of book reviews in astronomy from 1987 to the present has been prepared by Marlene Cummins and is available at `www.astro.utoronto.ca/reviews1.html`.

6.7. *Library Holdings*

The card catalogues of hundreds of libraries from all disciplines (including the Library of Congress, which can be accessed on the *WWW* at `lcweb2.loc.gov/catalog`) are available over the net, and the number is continuously growing. The Libweb directory at `sunsite.berkeley.edu/Libweb` lists library catalogues which are accessible on the *WWW*. Libweb is frequently updated and currently provides addresses of over 1700 libraries from 70 countries. A more complete listing of astronomy-related libraries can be found at

`www.stsci.edu/astroweb/cat-library.html`, the "Libraries Resources" section of the AstroWeb. The *STScI* library holdings are searchable at `stlibrary.stsci.edu/html`, and
those of *ESO* at `www.eso.org/libraries/webcat.html`.

7. Directories and Yellow-Page Services

Occasionally, if not frequently, one needs to search for the e-mail of an astronomer
somewhere in the world. There are many ways to find out and several may have to
be tried. One is the "*RGO* email directory". It is maintained by C.R. Benn and
R. Martin until the closure of the *RGO* in 1998, and to be taken over by the Starlink project (`star-www.rl.ac.uk`) in 1999. The guide is made up of three parts. One is
a list of ~13 000 personal emails, another one offers phone/FAX numbers and emails
or *URL*s of ~950 astronomical research institutes, and a third one has postal addresses
for ~650 institutions. You should make sure that your departmental secretary knows
about the latter two! All lists are updated frequently and available for consultation at
`www.ast.cam.ac.uk/astrosearch.html`. The impatient and frequent user of these directories should draw a local copy from time to time (three files at `ftp.ast.cam.ac.uk/guide`
`/astro*.lis`), or ask your system manager to install a site-wide command to interrogate
these lists and draw updated copies from time to time. To request inclusion in this
directory or communicate updated addresses, send a message to `email@ast.cam.ac.uk`.

Be sure that you never (ab)use such lists of thousands of addresses to send your
announcements to the entire list. You will most likely offend the majority of the recipients
who are not interested in your message, which may be even regarded as "spamming" (see
e.g. `cdsweb.u-strasbg.fr/~heck/spams.htm` for a collection of links to defend users from
unsolicited email). However, these email directories may be useful for the legitimate
task of selecting a well-defined subset of researchers as a distribution list for specific
announcements.

The *RGO* email guide depends on personal and institutional input for its updates and
turns out to be fairly incomplete especially for North American astronomers. Complete
addresses for *AAS* members can be found in the *AAS* Membership Directory which
appears in print annually and is distributed to *AAS* members only. The *AAS* membership
directory has been put on-line and made searchable at `directory.aas.org`, but it cannot
be downloaded entirely.

Since 1995 the *IAU* membership directory has been accessible at the LSW Heidelberg
web site, but in early 1998 it moved to `www.iau.org/members.html`. The address database
is managed by the *IAU* office in Paris (`www.iau.org`), and requests for updates have to
be sent there by email (`iau@iap.fr`). From these the *IAU* office prepares an updated
database every few months which is then put on-line. Hopefully this web address for the
IAU membership directory will remain stable in the future, and not change every three
years with the election of a new General Secretary at each *IAU* General Assembly.

The European Astronomical Society (*EAS*) offers its membership directory under *URL*
`www.iap.fr/eas/directory.html`. It provides links to membership directories of several
national astronomical societies in Europe (see `www.iap.fr/eas/societies.html`).

The "Star*sFamily" of directories is maintained at *CDS* (Heck (1997)) and divided into
three parts. "StarWorlds" (`cdsweb.u-strasbg.fr/~heck/sfworlds.htm`) offers addresses
and many practical details for ~6 000 organizations, institutions, associations, companies
related to astronomy and space sciences from about 100 countries, including about 5 000
direct links to their homepages. "*StarHeads*" (`cdsweb.u-strasbg.fr/~heck/sfheads.htm`)
is a compilation of links to personal *WWW* homepages of about 4500 astronomers and
related scientists. For "*StarBits*" see §9.

Another way to search for email addresses is a search engine (mirrored at various sites) which can accessed via `telnet bruno.cs.colorado.edu`, login as user `netfind`. You should give either first, last or login name of the person you look for, plus keywords containing the institution and/or the city, state, or country where the person works.

There are more, rather "informal" ways of tracing emails of astronomers. One is to check whether they have contributed preprints to the *SISSA/LANL* server (§6.2) recently. If so, the address from which they sent it will be listed in the search result returned to you. Another way is the command `finger xx@node.domain`, where `xx` is either the family name or a best guess of a login name of the individual you seek. However, some nodes prefer "privacy" and disable this command. One last resort is to send an email to `postmaster` at the node where you believe the person is or used to be.

8. Meetings and Jobs

Since about 1990, the librarian at the Canada-France-Hawaii Telescope (*CFHT*, Hawaii), E. Bryson, has maintained a list of forthcoming astronomical meetings, including those back to Sept. 1996, at the *URL* `cadcwww.dao.nrc.ca/meetings/meetings.html`. One may subscribe to receive updates of this list by request to `library@cfht.hawaii.edu`. It is now the most complete reference in the world for future astronomy meetings. Organizers of meetings should send their announcements to `library@cfht.hawaii.edu` to guarantee immediate and world-wide diffusion. Official meetings of the *IAU* and some other meetings of interest to astronomers are announced in the *IAU* Information Bulletin (see `www.iau.org/bulletin.html`). A list of all past *IAU* Symposia is available at `www.iau.org/pastsym.html`.

Probably the most complete collection of job advertisements in astronomy is the "*AAS* Job Register" at `www.aas.org/JobRegister/aasjobs.html`. The European Astronomical Society (*EAS*) maintains a Job Register at `www.iap.fr/eas/jobs.html`. STARJOBS is an electronic notice board maintained at the Rutherford Appleton Laboratory. The service is co-sponsored by the *EAS* and includes announcements of all European astronomical jobs notified to the Starlink astronomical computing project (`telnet` to `star.rl.ac.uk`, login as `starjobs`). One can also copy (via `ftp`) the complete list of jobs as the file `starlink-ftp.rl.ac.uk:/pub/news/star_jobs`. For employment opportunities in the Space Industry consult the *URL* `www.spacejobs.com`. The AstroWeb offers links to job offers at `www.cv.nrao.edu/fits/www/yp_jobs.html`.

9. Dictionaries and Thesauri

The *Second Dictionary of the Nomenclature of Celestial Objects* (see Lortet, Borde & Ochsenbein (1994)) can be consulted at `vizier.u-strasbg.fr/cgi-bin/Dic` or via telnet to `simbad.u-strasbg.fr` (login as `info`, give `<CR>` as password, then issue the command `info cati XXX` to inquire about the acronym `XXX`). Authors of survey-type source lists are strongly encouraged to check that designations of their objects do not clash with previous namings and are commensurate with *IAU* recommendations on nomenclature `cdsweb.u-strasbg.fr/iau-spec.html`. In order to guarantee that designations of an ongoing survey will not clash with other names, authors or PIs of such surveys should consider to pre-register an acronym for their survey some time before publication at the *URL* given above.

Independently, a list of "Astronomical Catalog Designations" has been prepared by *INSPEC* (see `www.iee.org.uk/publish/inspec/astro_ob.html`). It is less complete than the *CDS* version and deviates in places from the *IAU* recommendations.

The dictionary *"StarBits"*, maintained by A. Heck (Strasbourg), offers ~120 000 abbreviations, acronyms, contractions and symbols from astronomy and space sciences and related fields. It is accessible at `cdsweb.u-strasbg.fr/~heck/sfbits.htm`. Astronomers are invited to consult this dictionary to avoid assigning an acronym that has been used previously.

On behalf of the *IAU* several librarians of large astronomical institutions prepared "The Astronomy Thesaurus" of astronomical terms (Shobbrook & Shobbrook (1993), and later its "Multi-Lingual Supplement" (Shobbrook & Shobbrook (1995)) in five different languages (English, French, German, Italian, and Spanish). It is freely available at `www.aao.gov.au/library/thesaurus`. It may be useful in many respects, e.g., to translate astronomical terms, to aid authors in better selection of keywords for their papers, and to help librarians improve the classification of publications. The Thesaurus is available via anonymous `ftp` for DOS, MAC and Unix systems (`www.aao.gov.au/lib/thesaurus.html`); it has not been updated for some years, but M. Cummins (`astlibr@astro.utoronto.ca`) is currently in charge of it and appreciates comments about its future.

As an aside I mention the "Electronic Dictionary of Space Sciences" by J. Kleczek and H. Kleczková, who have collected several 10 000 of words, synonyms and expressions from astronomy, space sciences, space technology, earth- and atmospheric sciences and related mathematics, physics, and engineering fields in five languages (English, French, German, Spanish, and Portuguese). An electronic version is available at cost (see `www.twinpress.nl/edss.htm`).

The "Oxford English Dictionary" (*OED*; `www1.oed.com/proto/`) is currently being revised to include a far more comprehensive set of astronomical terms than before (see the *OED* Newsletter at `www1.oup.co.uk/reference/`; Mahoney (1998)).

10. Miscellaneous

10.1. *Astronomical Software*

For an introduction to publicly available astronomical software and numerical libraries see Feigelson & Murtagh (1992). The "Astronomical Software and Documentation Service" (*ASDS*) at `asds.stsci.edu/asds` contains links to the major astronomical software packages and documentation. It allows one to search for keywords in all the documentation files available. The *Statistical Consulting Center for Astronomy* at Penn State University (`www.stat.psu.edu/scca/homepage.html`) offers advice and answers to frequently asked questions about statistical applications. The "StatCodes" (of statistical software for astronomy and related fields, at `www.astro.psu.edu/statcodes`) are now also in *ASDS*.

10.2. *Observatory and Telescope Manuals*

The librarian at *CFHT* (Hawaii), E. Bryson, has collected observatory and telescope manuals in electronic form from all over the world. Several dozen such documents are now available through the *ASDS* (§10.1) at `asds.stsci.edu/asds`. The documents may be searched by keywords.

10.3. *IAU Circulars, Minor Planets, ATEL, and Ephemerides*

Information about time-critical phenomena has been distributed by the *IAU* "Central Bureau for Astronomical Telegrams" (*CBAT*; `cfa-www.harvard.edu/cfa/ps/cbat.html`) via "*IAU* Circulars" since October 1922. The *IAU* "Minor Planet Center" (*MPC*; `cfa-www.harvard.edu/cfa/ps/mpc.html`) is responsible for the collection and dissemination of astrometric observations and orbits for minor planets and comets, via the "Minor Planet Electronic Circulars" (*MPEC*; `cfa-www.harvard.edu/cfa/ps/services/MPEC.html`), distributed on paper as the "Minor Planet Circulars" since before 1947. Through collaboration with *CBAT* and *MPC*, the *ADS* Abstract Service includes electronic circulars of these two sites. The title, author, and object names are freely available through the *ADS*, and the whole text of each circular is indexed by *ADS* so that it is found on searches. The references returned from the *ADS* include an on-line link to the full circular at *CBAT* and *MPC*.

In December 1997 the "Astronomer's Telegram" (*ATEL*; Rutledge (1998)) was released. It is a web-based publication system for short notices on time-critical information and is available at `fire.berkeley.edu:8080/`). Submission of telegrams is restricted to professional astronomers and requires special permission. The potential authors are given a special authentication code to be used at the time of submission, which is entirely automated and unmoderated, without any human intervention. Readers can freely access the telegrams or ask to be on a mailing list to receive telegrams within minutes of their submission. During the first six months of its existence about five telegrams per month were received.

Ephemerides and orbital elements of comets and minor planets can be consulted e.g. at `cfa-www.harvard.edu/iau/Ephemerides/index.html` or at JPL's HORIZONS system (`ssd.jpl.nasa.gov/horizons.html`). A free, interactive program for fancy calculations of ephemeris, visibility curves from any site on Earth, and graphical displays of finding charts based on the *HST* Guide Star Catalogue, orbits of solar system bodies, views of Earth and Moon, and much more is available (for X-Windows systems with Motif) for download at `iraf.noao.edu/~ecdowney/xephem.html`.

The "Astronomy Calculator", available at `w3.one.net/~rback/frames.html`, aspires to provide general information about the phases of the moon, lunar eclipse, next annual meteor shower and planets.

10.4. *Atomic Data*

A multitude of links to databases containing atomic and molecular data can be found at `cfa-www.harvard.edu`, in `.../amp/data/otherdb.html` or `.../~esmond/amdata.html`.

The "Opacity Project" (OP) at `vizier.u-strasbg.fr/OP.html` offers extensive atomic data required to estimate stellar envelope opacities.

The original implementation of the "Vienna Atomic Line Data Base" (*VALD*; Piskunov *et al* (1995)) is available at `www.ast.univie.ac.at/~weiss/vald.html`. The *VALD* manual (at `plasma-gate.weizmann.ac.il/VALD.html`) is part of a summary of "Databases for Atomic and Plasma Physics" (*DBfAPP*) of the Weizmann Institute (see `plasma-gate.weizmann.ac.il/DBfAPP.html`). An updated and improved interface for *VALD* is under construction (at `www.astro.uu.se/vald`), though actual data traffic will continue to be handled via e-mail. This page will include all necessary links including documentation, registration form, requests forms and examples. Send inquiries to F. Kupka (email: `valdadm@astro.univie.ac.at`).

10.5. *Libraries*

Two distribution lists are available specifically for astronomy librarians to share relevant information on widely varying subjects. Astrolib (started in 1988) with ~200 members is managed by E. Bouton (`library@nrao.edu`), librarian at the National Radio Astronomy Observatory (*NRAO*). The European Group of Astronomy Librarians (EGAL), is managed by I. Howard (`howard@ast.cam.ac.uk`), librarian at the Royal Greenwich Observatory (*RGO*). J. Regan (`library@mso.anu.edu.au`) is currently trying to set up a group in Asia and the Pacific Rim. People wishing to post announcements to libraries including physics and mathematics departments should get in touch with the moderator of the email distribution list of *PAMnet*, a network of Physics, Astronomy and Mathematics librarians. Send your inquiry to `david.e.stern@yale.edu`.

U. Grothkopf, *ESO* librarian, maintains a list of names, addresses, phone and fax numbers, email address and homepage *URLs* of astronomy librarians and libraries world-wide. A useful search engine (`www.eso.org/libraries/astro-addresses.html`) can find librarians even from incomplete information. Librarians not on the list are encouraged to send information to `esolib@eso.org`.

10.6. *Astronomy Education on the Internet*

Given that the Internet offers almost unlimited possibilities for interactive courses, more and more of these can be found on the web, see e.g. Benacchio, Brolis, & Saviane (1998). The "AstroEd" page, at `www-hpcc.astro.washington.edu/scied/astro/astroindex.html`, provides links to on-line astronomy education resources and some on-line courses. The National Research Council's project "Resources for Involving Scientists in Education" (*RISE*) features a web site at `www.nas.edu/rise/examp.htm`. An interesting example is given by the University of Oregon (`www.zebu.uoregon.edu`), where an astronomy book is being developed in "hypertext". A further advantage of these "books" is their potential of being kept up-to-date by a group of professionals.

The European Southern Observatory (*ESO*) organized an educational programme called "Astronomy On-line" in December 1997. Its web pages are still available at `www.eso.org/astronomyonline/` (Albrecht *et al* (1998)).

10.7. *Others*

D. Verner's compilation of people mentioned in acknowledgments of papers in major astronomical journals up to 1996 can be accessed from `www.pa.uky.edu/~verner/aai.html`.

11. Issues for the Future

The Internet has been with us for only about a decade. Users of the *WWW* should be aware that there is still more information, literature, data, etc., existing only in printed form, than is available on the Internet. While the possibilities of information and data retrieval have advanced at a tremendous pace in recent years, there is an infinite number of possible improvements. I shall mention only a few very subjective ones as an example here.

The increasing presence of commercial companies on the Internet is both an enrichment and a plague, the latter because more and more frequently unsolicited emails are being sent to global distribution lists with commercial offers. While this is annoying, and measures should be taken against it (see `cdsweb.u-strasbg.fr/~heck/spams.htm`), I do not think that it is a reason for astronomers to refrain from being listed in email guides or from making these guides available among colleagues. The damage to easy communication among scientists would be too severe.

The transition from printing large tables on paper to publishing them in electronic form (§3) either in the *ADC* and *CDS* archives (or on the *AAS* CD-ROMs), raises the question about the future of marking-up tables for printing. For many years authors have been obliged to convert their data tables to LaTeX format. Ironically, *AAS* requests a charge of US\$ 50 for the service to convert the data tables back to plain *ASCII* format for publication on their CD-ROM, except for tables marked up with the :AAS-TeX macros (see `ftp://ftp.aas.org/cdrom/guidelines.html`). It may be anticipated that the "publication" of tables in electronic form will eventually release authors from this task. However, special non-*ASCII* symbols, like e.g. Greek letters, will require to be transliterated to *ASCII* characters in the electronic version.

Unfortunately the journals in astronomy do not yet oblige authors to provide their tabular data to a data centre, as a requisite for publication. An agreement between all major journals and the data centres *ADC* and *CDS* is highly desirable, not only for the sake of the completeness of their electronic archives of tabular and catalogue data, but also to remedy the following problem. The clearing house of the *IAU* Task Group on Astronomical Designations of *IAU* Commission 5 (`cdsweb.u-strasbg.fr/iau-spec.html`) has frequently come across unconventional namings of astronomical objects causing confusion and redundancy of names in object databases like *NED* and *SIMBAD*. My experience in the Task Group was that the standard refereeing system of journals does not help to avoid this problem. Ideally, these tables should be run through an automatic cross-checking routine prior to publication or acceptance. For this purpose they should have at least two identifiers (a name and a coordinate) and could then be compared with databases like *SIMBAD*, *NED* or *LEDA*, in order to check the consistency of names and coordinates, and perhaps even part of the measured quantities. Of course this is useful only if the objects were known previously.

12. Concluding Remarks

The Internet and World Wide Web have added just another medium for fast access to large amounts of information. It can save researchers lots of time in retrieving the required information and allows access to unexplored data which are worth many research projects in their own right. However, the flood of information on the web has become so large that now, when searching for a given piece of information, we are about to spend more time in browsing the web than we used to need searching in the library a decade ago (when the amount of available information was substantially less).

In the early years of networking we were happy when we could get electronic copies of astronomical catalogues without the delays through shipments of tapes. Now we are so flooded with them that in the rush of using many of them at the same time we sometimes forget that each one of them is telling us a different story. We must still read their detailed documentation if we want to derive reliable results from the available data. We have gone a good part of the way already to the point where all past issues of the major astronomical journals will be available electronically on the web. However, network saturation still keeps us from skimming a journal in the way that we could in the library.

A compromise has yet to be found between a rigorous refereeing system of web pages (as proposed by some) and the absolute liberty we currently "enjoy" in offering our own information and expressing our interests on the web. Many people have tried in recent years to offer guides to certain parts of astrophysical information, and the present article is just another example. The challenge for the future is how to protect ourselves from too much redundant or superseded information. While preparing this paper I came across many web pages which at first sight looked promisingly complete. However, when the last update (if given at all) was more than about a year ago, I usually refrained from quoting it here, because of the danger that it would not be maintained any more, or that it would offer too many outdated links. Perhaps a step towards reducing this danger could be a web browser that automatically recognized the date of latest update of a web document and would allow to set filters on that date in a search for relevant links. Actually, the AltaVista search engine (`www.altavista.digital.com`) allows a range of "last modified" dates to be used in advanced searches.

I am grateful to the School organizers for the opportunity to give these lectures and for their financial support. My special thanks go to Clive Davenhall, Elizabeth Griffin, and André Fletcher for their careful reading of the manuscript. Useful information or comments on the text were provided by P. Boyce, P. Eenens, G. Eichhorn, D. Fullagar, G. Giovannini, D. Golombek, C. S. Grant, U. Grothkopf, R.J. Hanisch, M. Irwin, F. Kupka, S. Kurtz, N. Loiseau, A. Macdonald, F. Murtagh, F. Ochsenbein, G. Paturel, M. Schmitz, S. A. Trushkin, M. Tsvetkov, and M. J. West. My apologies to those I forgot to mention here.

REFERENCES

ADORF H.-M. 1995, *Vistas Astron.* **39**, 243–253

ALBRECHT, R., WEST, R., MADSEN, C., & NAUMANN, M., 1998, Astronomical Data Analysis and Software Systems – VII, eds. Albrecht, R., Hook, R. N., & Bushouse, H. A., ASP Conf. Ser. **145**, 248–251

ANDERNACH H., HANISCH R. J., & MURTAGH F. 1994, *PASP* **106**, 1190–1216

ANDERNACH H. 1995 *ApL&C* **31**, 1–8

BARNES, P. J., & MYERS, P. C. 1997, ApJS **109**, 461–471

BENACCHIO, L., BROLIS, M., SAVIANE, I. 1998, *Astronomical Data Analysis Software and Systems – VII*, ASP Conf. Ser., **145**, p. 244–247

BOYCE, P. B. 1997, Ap&SS, **247**, 55–92

CHARLES, A., & SHAYLER, D. 1998, *Starlink Bulletin* **20**, 20–21, April 1998

DAVENHALL, A. C. 1993, An Astronomer's Guide to On-line Bibliographic Databases and Information Services, Starlink User Note 174.1; available at `star-www.rl.ac.uk/docs /sun174.htx/sun174.html`

DRINKWATER, M. J., BARNES, D. G., & ELLISON, S. L. 1995, PASA **12**, 248–257

EGRET, D., & HECK, A. (EDS.) 1995, Weaving the Astronomy Web *Vistas Astron.* **39**, 1–127, cf. also `cdsweb.u-strasbg.fr/waw/proceedings.html`

EGRET, D., & ALBRECHT, M. A. (EDS.) 1995, *Information & On-line Data in Astronomy*, Astrophysics and Space Science Library, vol. 203, Kluwer Acad. Publishers, cf. also `cdsweb.u-strasbg.fr/data-online.html`

EICHHORN, G., ACCOMAZZI A., GRANT C. S., KURTZ M. J., & MURRAY, S. S. 1998, *Astronomical Data Analysis Software and Systems – VII*, ASP Conf. Ser., ASP, San Francisco, **145**, 378–381

FEIGELSON, E. & MURTAGH, F. 1992, PASP, **104**, 574–581

GOMBERT, G. & DROEGE, T. 1998, *Sky & Telescope*, **95**, 43–46

GROTHKOPF, U. 1995, *Vistas Astron.* **39**, 137–143

GROTHKOPF, U., ANDERNACH, H., STEVENS-RAYBURN, S., & GOMEZ, M. (EDS.) 1998, *Library and Information Services in Astronomy III (LISA-III)*, ASP Conf. Ser. , **153**, ASP, San Francisco

HECK, A. 1994, *Vistas Astron.* **38**, 401–418

HECK, A. (ED.) 1997, *Electronic Publishing for Physics and Astronomy*, Ap&SS, special issue **247**, Kluwer Acad. Publ.

HECK, A. 1997, Ap&SS, **247**, 211–220

HELOU, G., MADORE, B.F., SCHMITZ, M., WU, X., CORWIN, H.G. JR., LAGUE, C., BENNETT, J., & SUN, H. 1995, in *Information & On-Line Data in Astronomy*, Egret, D. & Albrecht, M. A., Dordrecht, Kluwer Acad. Publ., p. 95–113

JASCHEK, C., 1989 *Astronomical Data*, Cambridge University Press, Cambridge, UK

KURTZ, M. J., EICHHORN, G., MURRAY, S. S., GRANT, C. S., & ACCOMAZZI, A. 1996, in *Data and Knowledge in a Changing World: The Information Revolution: Impact on Science and Technology*, Dubois, J.-E., Gershon, N. (eds.), Springer Verlag, pp. 123–130

LEWIS, G., & IRWIN, M. 1996, Spectrum, Newsletter of the Royal Observatories, **12**, p. 22–24

LORTET, M.-C., BORDE, S., & OCHSENBEIN, F. 1994, A&AS **107**, 193–218

MAHONEY, T. J. 1998, in Library and Information Services in Astronomy III (LISA–III), eds. U. Grothkopf *et al*, ASP Conf. Ser. **153**, 218–223, ASP, San Francisco

MARGON, B. 1998, *Phil. Trans. Roy. Soc. London A*, in press; `astro-ph/9805314`

MICHOLD, U., CUMMINS, M., WATSON, J. M., HOLMQUIST, J., & SHOBBROOK, R. 1995, in *Information & On-line Data in Astronomy*, Egret, D. & Albrecht, M. (eds.), Dordrecht, Kluwer Acad. Publ., 207–228

MURTAGH, F., GROTHKOPF, U. & ALBRECHT, M. (EDS.) 1995, *Library and Information Services in Astronomy II (LISA–II)*, Vistas Astron. **39**, 127–286

OCHSENBEIN F. & HEARN A. (EDS.) 1997, Proc. *International Cooperation in Dissemination of the Astronomical Data*, Baltic Astronomy **6**, No. 2, p. 165–365

ORTIZ, P. F. 1998, *Astronomical Data Analysis Software and Systems – VII*, ASP Conf. Ser., **153**, 371–374, ASP, San Francisco

PATUREL G., ANDERNACH H., BOTTINELLI L., DI NELLA H., DURAND N., GARNIER R., GOUGUENHEIM L., LANOIX P., MARTHINET M.-C., PETIT C., ROUSSEAU J., THEUREAU G., & VAUGLIN I. 1997, A&AS **124**, 109–122

PISKUNOV, N. E., KUPKA, F., RYABCHIKOVA, T. A., WEISS, W. W. & JEFFERY, C. S. 1995, A&AS, **112**, 525–535

PLANETARY AND SPACE SCIENCE 1996, Special issue about the Planetary Data System: Vol. 44, no. 1, January 1996

POJMAŃSKI, G. 1997, *Acta Astron.* **47**, 467–481

RUTLEDGE, R. E. 1998, *PASP* **110**, 754–756

SCHMITZ, M., HELOU, G., DUBOIS, P., LAGUE, C., MADORE, B., CORWIN, H. G. JR., & LESTEVEN, S. 1995, in *Information & On-line Data in Astronomy*, Egret, D. & Albrecht, M. (eds.), Kluwer Acad. Publ., 347–351

SHOBBROOK, R. M., & SHOBBROOK, R. R. 1993, *The Astronomy Thesaurus*, Version 1.1, Anglo-Australian Observatory, Epping, Australia, 115 pp.

SHOBBROOK, R. R., & SHOBBROOK, R. M. 1995, *The Multi-Lingual Supplement to The Astronomy Thesaurus*, Version 2.1, Anglo-Australian Observatory, Epping, Australia, 95 pp.

THOMAS, B. J. 1997 *The Internet for Scientists and Engineers*, SPIE Press, PM 51

TSVETKOV, M. K., STAVREV, K. Y., TSVETKOVA, K. P., SEMKOV, E. H., MUTAFOV, A. S., & MICHAILOV, M.-E. 1998,
New Horizons From Multi-Wavelength Sky Surveys, McLean, B. J., Golombek, D. A., Hayes, J. J. E., & Payne, H. E. (eds.), IAU Symp. **179**, 462–464, Kluwer Acad. Publishers

Radio Astronomy

By G E O R G E   M I L E Y

Sterrewacht (Leiden Observatory), P.O. Box 9513, 2300 RA LEIDEN, The Netherlands

An introduction to the theoretical background to radio astronomy is given, explaining the history of its development and of the observing techniques which are most commonly used. This introduction is then used to analyse the application of radio astronomy to the study of extragalactic radio sources and the theory of their emitted radiation. Finally, we look at how radio sources are used in cosmology to study the history and evolution of the Universe – particularly such techniques as source counts, the luminosity function and the study of clusters of galaxies.

1. Introduction

I shall give you an introduction to the basics of radio astronomy. My goal is to set the scene for Heinz Andernach who will discuss radio astronomy in the context of the theme of this school, Astrophysics with Large Databases in the Internet Age. In deference to this topic, I feel compelled to mention the Internet at least once during these lectures, but I do so with a message of caution. We all know that during the last few years he Internet has become a powerful research tool and you will see many fine examples of how it can be exploited to do important astronomy. However, it is also addictive and like many addictive substances it can be misused. You should ensure that your use of the Internet is driven by the need to tackle an astronomical problem. Otherwise the Internet could even ruin your career. So my advice is to restrict your surfing to the beautiful beaches of Tenerife!

Radio astronomy is used to investigate an enormous diversity and range of phenomena and objects. Radio telescopes are used to probe stars and pulsars, galactic and extragalactic masers, and distant galaxies. Subjects of study include the structure and internal dynamics of galaxies, star formation, galaxy formation, general relativity and cosmology. Each of these topics warrants a lecture course in itself. In the five hours available to me I can only touch on a few of them. I shall introduce you to the tools of the trade and then discuss luminous extragalactic radio sources, emphasising some of the diagnostics commonly used to study them. Extragalactic radio sources are amongst the largest, the most energetic and the most distant known objects in the Universe.

2. Historical Highlights

Advances in astronomy and the resultant changes in our view of the Universe have been one of the most significant achievements of twentieth century civilisation. Extending astronomical observations beyond the narrow optical spectral window was the most important technological revolution in astronomy since the invention of the telescope. Radio astronomy was the first of these *new* windows on the Universe and still one of the most important. Due to the relatively few spectral lines observable at radio wavelengths, the range of physical diagnostics in radio astronomy are not as rich as in some other regions of the electromagnetic spectrum, but the angular resolution achievable at radio wavelengths is unique.

The development of radio astronomy is an excellent illustration of the process by which modern science advances, namely through a mutual interaction between fundamental knowledge, technological development and the willingness of society to fund basic research. Modern radio astronomy was made possible by

(*a*) technology developed for defence (e.g. radar during the Second World War, accurate clocks),

(*b*) the rise of computers,

(*c*) consumer electronics (e.g. video tape recorders).

This Winter School is a direct result of the enormous advances in information technology in the last few years. Because of their *roots* as electrical engineers and because of the enormous demands of radio- astronomical techniques on computing power, radio astronomy has been at the forefront of applying computing techniques to astronomy. Many of the discoveries of radio astronomy have been serendipitous and therefore were unplanned.

Now, some important landmarks in the history of radio astronomy:

2.1. *Breaking out of our optical prison*

1888: Hertz produced radio waves in laboratory

1894: Lodge unsuccessfully tried to detect radio waves from Sun.

1931: Jansky fortuitously (serendipitously) detected extraterrestrial radio waves.

2.2. *Our Galaxy and Exotic Objects within it*

1937: Reber made first radio maps of Milky Way

1944: van de Hulst predicted that HI should radiate 21-cm hyperfine line.

1951: 21 cm line detected by groups in the Netherlands, USA and Australia.

1950s: Mapping of our galaxy in HI and Continuum

1965: Discovery of Galactic OH Masers

1968: Little Green Men: Pulsars

2.3. *Radio astronomers become historians*

1943: Discovery of galaxies with broad emission lines by Carl Seyfert "Seyfert Galaxies"

1946: Discovery of discrete radio source Cygnus A (Hey, Parsons, Phillips)

1948: Several other discrete radio sources detected

1948: Radio Interferometer developed by Ryle (Cambridge) and Christiansen (Sydney).

1949: Bolton, Stanley, and Slee, by measuring positions to $10'$ accuracy, show that 2 of the brightest radio sources are associated with galaxies Virgo A = M87 ($\sim$15 Mpc) and Centaurus A = NGC 5128 ($\sim$ 3 Mpc)

1950: Alfvän and Herlofson attribute radiation from discrete radio sources to the synchrotron process (previously discussed by Schott in 1912)

1951: Graham Smith determined the position of Cygnus A to 1" accuracy. Baade and Minkowski identified Cygnus A with a peculiar galaxy having z = 0.06 ($\sim$250 Mpc!). This showed that Cygnus A was $> 10^6$ more powerful radio emitters than Milky Way "Radio Galaxies"

1953: Jennison and Das Gupta resolved Cygnus A into $2'$ double

1953-60: Period of consolidation. 3C catalogue published. Many more radio galaxies identified. Majority shown to be double. Some sources shown to be smaller than $1''$.

1960: 3C 48 was identified with a stellar like object. Thought to be a star but its spectrum was indecipherable

1963: 3C 273 also shown to be a radio star. Schmidt explains spectrum of 3C 273 assuming z=0.158 Greenstein and Matthews realise spectrum of 3C 48 is also highly redshifted. These quasi-stellar radio sources are christened *quasars*

1963: Quasars seem not to obey Hubble Law. Are their redshifts cosmological gravitational or local?

1964: Variability in CTA 102 claimed by Sholomitskii. A *super-civilisation* announced by TASS!

1965: Radio sources with flat spectra found. Several shown to be variable on time scale of years (Dent and Haddock). Radio-quiet quasars found by Sandage

1968-72: Flat spectrum radio cores found to occur generally in extended radio sources - implying that their nuclei are continually active

1969-72: Some compact radio sources deduced to be expanding faster than light ? "super-relativistic expansion"

1971: Rees proposes that extended sources are continuously supplied with energy from their nuclei

1976-79: Jets found in the lobes of some radio sources (umbilical energy cords)

1970s and 1980s: Multi-wavelength studies detect optical/radio correlations, which allow radio diagnostics to be combined with optical diagnostics to study interactions. Mounting evidence that extragalactic radio sources are powered by gravitational energy produced by accretion of matter onto rotating black hole

3. Specification of Radiation

3.1. *Flux Density and Brightness Temperature*

Consider a point source of radiation at distance r from the observer.

Flux F(A) = Energy which crosses area A per sec (W).

Flux density f(r) = Energy which crosses unit area per sec (W m^{-2}).

$[f(r_2)/f(r_1)] = A_1 / A_2 = 1/r^2$ (Inverse square law)

Monochromatic flux density f_ν or S_ν = Energy which crosses unit area/ sec/ Hz This is usually just called the flux density. 1 Jansky = 10^{-26} W m^{-2} Hz^{-1} (N.B. 1 Hertz (Hz) = 1 cycle per sec)

We can define similar parameters for an extended source of radiation:

Brightness B(x,y) = Flux density from source element which subtends solid angle Ω (W m^{-2} ster^{-1})

Monochromatic Brightness, $B_\nu(x,y)$ = flux density per unit solid angle.

Brightness Temperature, $T_{B,\nu}$, is defined as the temperature of a black body that would have the same monochromatic brightness at a given frequency ν.

According to Planck's Law $B_\nu = [(2 h\nu^3)/c^2] [e^{(h\nu/kT)} - 1]^{-1}$.

For radio waves $h\nu \ll kT$ and this reduces to the Rayleigh Jeans Law $B_\nu = [(2kT)/c^2]\nu^2$

Thus brightness temperature $T_{B,\nu} = (c^2/2k) [B_\nu/\nu^2]$

3.2. *Polarisation*

Linear polarisation was first detected in extragalactic radio sources in 1962 providing confirmatory evidence that the radio emission was caused by the synchrotron process (see later). It is now well known that most sources are significantly polarised and that the fractional polarisations are in general larger at higher frequencies. The observed integrated polarisations at frequencies $>\sim 1$ GHz vary from $\sim 0.5\%$ to $\sim 15\%$ with a typical value of $\sim 4\%$.

The polarisation of quasi-monochromatic radiation is conveniently specified by the four Stokes parameters I, Q, U, V. These may be defined in terms of the radiation intensities in pairs of orthogonal modes, measured in two sets of orthogonal Cartesian coordinate systems, the one (X, Y) oriented east- west/ north-south and the other (X=, Y=) at a relative angle of 45°. Then,

$$I = < E_X >^2 + < E_Y >^2 \tag{3.1}$$

$$Q = < E_X >^2 - < E_Y >^2 \tag{3.2}$$

$$U = < E_{X=} >^2 - < E_{Y=} >^2 \tag{3.3}$$

$$V = < E_R >^2 - < E_L >^2 \tag{3.4}$$

Here E^2 is the intensity in a particular mode of polarisation defined by the coordinates and E_L and E_R are the left and right hand circularly polarised components respectively. Because the Stokes parameters have the dimensions of power and obey the Fourier transformation relations of interferometry, they are a useful form of expressing radio polarisations.

Linearly polarised radiation is specified by Q and U. These vary under rotation as the real and imaginary parts of a complex quantity P= Q+iU.

This can be expressed as $\mathbf{P} = \sqrt{(Q^2 + U^2)}\, e^{2i\chi} = mIe^{2i\chi}$

where $m = \sqrt{(Q^2 + U^2)}/I$ is the fractional polarisation and $\chi = (0.5)\tan^{-1}(U/Q)$ is the angle defined by the direction in which the electric intensity $\mathbf{E}$ is polarised, measured in an anticlockwise direction from the north.

4. Single Dish Radio Telescopes

Simplified this consists of an antenna that receives radio waves from a location (x,y) on a source of monochromatic brightness $B_\nu(x,y)$.

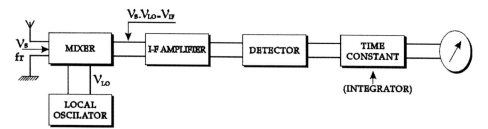

FIGURE 1. Schematic diagram of a typical receiver used in radio astronomy.

Amplification is often difficult at the high frequencies observed. The output of the antenna at frequency ν is usually fed to a *superheterodyne receiver* (see **Figure 1**) which *mixes* the signal with a local oscillator at frequency ν_L to give an output *intermediate frequency* (IF) $(\nu - nu_L)$. After IF amplification, the signal is *detected* and *integrated* for time t to reduce the noise by $\sqrt{t}$.

The limiting factor in most single dish observations is set by the stability of the receiver and ancillary electronics. Several tricks (modulation) are used to calibrate these, including *beam switching* between adjacent locations on the sky and *frequency switching* between adjacent frequencies.

An antenna of effective area A produces an output power in bandwidth $\Delta\nu$ of

$$P_\nu(x,y)\Delta\nu \propto AB_\nu(x,y)\Delta\nu \tag{4.5}$$

The antenna temperature of a source, T_A, is the temperature of a black screen that would produce the same power as the output power from the antenna due to the source

$$kT_A(x,y) = AB_\nu(x,y) \tag{4.6}$$

$$T_A(x,y) = [AB_\nu(x,y)]/k \tag{4.7}$$

Expressing B in terms of the source brightness temperature, T_B, we can write the antenna temperature as

$$T_A(x,y) = [2A/\lambda^2]T_B(x,y) \tag{4.8}$$

An antenna has an acceptance pattern (polar diagram), so that for a fixed antenna A is a function of the direction in the sky (x,y).

In the case of a parabaloidal antenna (*dish*) of diameter R the polar diagram corresponds to an angular resolution

$$\Delta\theta \sim \lambda/R \text{ radians} \tag{4.9}$$

e.g. For a large dish (e.g. MPI Bonn) R = 100m, implying $\Delta\theta \sim 1'$ at $\lambda = 3$ cm. This is only useful for largest sources. To study most extragalactic radio sources we need higher spatial resolution.

5. Techniques for Obtaining Higher Resolution

5.1. *Lunar occultations*

Variation of source flux density as it is occulted by moon gives one-dimensional brightness distribution convolved with diffraction pattern of moon. $\Delta\theta \sim 1''$ is readily achievable.

Advantages: Resolution not highly dependent on frequency. Disadvantages: Two-dimensional structure difficult to reconstruct. Can only be applied to sources that happen to lie on moon's path (within 5° of ecliptic, i.e. $-29° < \delta < +29°$). Most notable success was in pinpointing 3C 273 by Hazard and co-workers which led directly to discovery of quasars.

5.2. *Interplanetary scintillations*

Irregularities in electron density of solar corona (scale $\sim$100 km) cause small sources to flicker or "scintillate". Large sources are unaffected - frosted glass effect. Degree of scintillation and its variation with distance from Sun can give useful information on amount of small-scale ($<1''$) structure in a source.

Advantages: Easy. High resolution at low frequency. Disadvantages: Measures only power spectrum and does not result in *maps*. All but the crudest inferences about structure depend on assumed model for solar corona. Can only be applied to limited source list $\sim$50° from Sun.

It was in the course of developing this technique that Hewish and Bell discovered pulsars, one of the many important serendipitous discoveries of radio astronomy.

5.3. *Interferometry*

This is by far the most important tool used by radio astronomers. It was first developed by Ryle (in 1952) and independently by Christiansen (in 1953) and by Mills and Little (in 1953). In the next section we shall summarise some of the most important points stressing the jargon which is often used in articles.

6. Principle of the Radio Interferometer

6.1. *Response to point source*

A radio interferometer is analogous to Young's famous double slit experiment in optics.

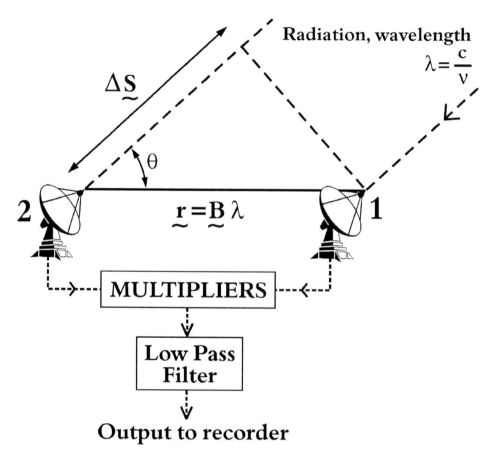

Radiation, wavelength

$$\lambda = \frac{c}{\nu}$$

$\Delta \underset{\sim}{S}$

θ

$\underset{\sim}{r} = \underset{\sim}{B}\lambda$

2   1

MULTIPLIERS

Low Pass Filter

Output to recorder

FIGURE 2. Schematic diagram of a interferometer.

Assumptions:
- 2 identical antennas separated by a distance $\mathbf{R} = \mathbf{B}\lambda$ (the baseline)
- Monochromatic and unpolarised radiation at frequency ν from point source of flux density S at large distance (infinity) and oriented at an angle θ to the baseline $\mathbf{R}$.

Output voltages from the antennas are sent through a multiplier equidistant from the antennas and subsequently through a low-pass filter to a recorder.

Radiation to antenna 2 travels extra distance $rm\Delta s = R\cos\theta$.

Voltages at the two multiplier inputs as a function of time t are

$$V_1 = V_{01}\cos(2\pi\nu t) \tag{6.10}$$

and

$$V_2 = V_{02}\cos(2\pi\nu t + \phi) \tag{6.11}$$

where

$$\phi = (2\pi \, \Delta s)/\lambda \qquad (6.12)$$

The output of the multiplier is

$$M = V_1 V_2 = V_{01} V_{02} \, \cos(2\pi\nu t) \, \cos(2\pi\nu t + \pi) \qquad (6.13)$$

$$M = 2V_{01} V_{02} \, \cos\phi + 2V_{01} V_{02}[2\pi(2\nu) + \phi] \qquad (6.14)$$

Where

$$2V_{01} V_{02} \, \cos\phi < 100 \text{ Hz}$$

and

$$2V_{01} V_{02}[2\pi(2\nu) + \phi] > 10^9 \text{ Hz}$$

This travels through a low-pass filter, which rejects the high-frequency term. The output is

$$L \sim V_{01} V_{02} \cos\phi \qquad (6.15)$$

But $V_{01} V_{02}$ = flux density = S and $\phi = (2\pi \, \Delta s)/\lambda = 2\pi \, B \, \cos\theta$

Hence

$$L \sim S \cos[2\pi B \, \cos\theta] \qquad (6.16)$$

This is the response of an interferometer whose physical baseline has length B to a point source of flux density S whose line of sight vector makes an angle θ with the physical baseline and is the fundamental equation of interferometry.

As the earth rotates the angle θ changes and the output of our interferometer is a set of "interference fringes" whose amplitude is proportional to the flux density of the source (**Figure 3**).

Depending on the position of the source in the sky, the output of an interferometer will be positive or negative. Thus the output of an interferometer can be visualised as a set of imaginary positive and negative lobes in the sky - the *interferometer polar diagram*. We can think of this as a comb-like structure fixed in the sky with a pattern determined entirely by the geometry of the physical baseline. The source can be visualised as *passing through* the lobes as the earth rotates. If the source is not a point source but has an angular size comparable to the separation between an adjacent maximum and minimum lobe, the fringes will be smeared and the amplitude of the output fringes will be reduced. Hence the interferometer can give information about the structure of a radio source.

6.2. Resolving power

A measure of the resolving power of an interferometer is the lobe separation or fringe pattern. From equation (6.16) it can be seen that a maximum will occur for every value of θ that satisfies the relation:

$$2\pi B \cos\theta = 2n\pi \qquad n = 0, 1, 2, ...\text{etc.}$$

If θ_1 and θ_2 are values of θ which correspond to two adjacent lobe maxima (fringe peaks), $\theta_1 \sim \theta_2$ and the resolution of the interferometer can be expressed as $\Delta\theta = \theta_2 - \theta_1$.

$$2\pi B(\cos\theta_2 - \cos\theta_1) = 2(n+1)\pi - 2n\pi = 2\pi \qquad (6.17)$$

But

$$\cos\theta_2 - \cos\theta_1 = \cos(\theta_1 + \Delta\theta) - \cos\theta_1 = \Delta\theta \, \sin\theta$$

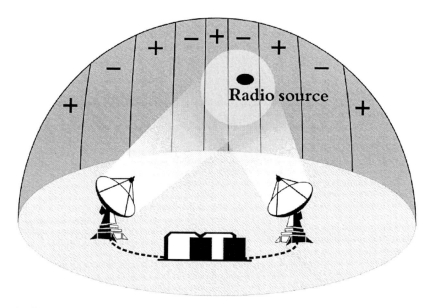

The interferometer "phase polar diagram"

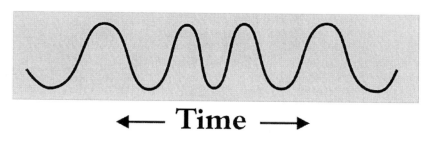

← Time →

The output fringes

FIGURE 3. Interference fringes.

Hence

$$\Delta\theta = 1/(\text{B } \sin\theta) = 1/\text{B}_\text{p} \qquad (6.18)$$

$\text{B}\sin\theta) = \text{B}_\text{p}$ is the component of the physical baseline projected in a plane perpendicular to the line of sight to the source, i.e. the *projected baseline*.

The angular resolution of an interferometer in radians is approximately equal to the inverse of the projected baseline expressed in number of wavelengths. This varies as the source moves through the sky and θ changes (e.g. for a 100 km baseline, $\lambda = 20$ cm and $\theta = 90°$, $\text{B}_\text{p} = 200\ 000\ \lambda$ and $\Delta\theta \sim (1/200\ 000)$ rad $\sim 1''$).

6.3. *Projected baseline - u,v ellipses*

Before discussing how an interferometer can measure the structure of an extended source, it is important to analyse in detail how the projected baseline (i.e. angular resolution) varies as the earth rotates causing the radio source to move through the sky.

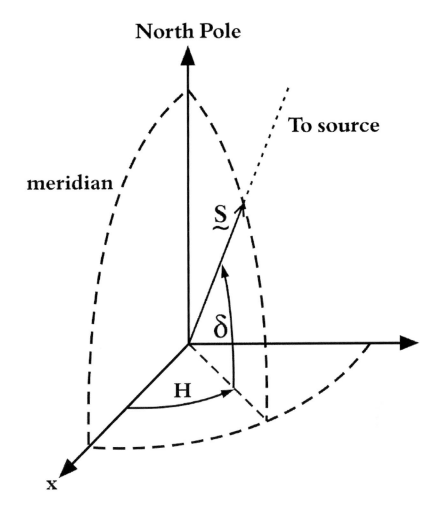

FIGURE 4. The co-ordinate system for a typical radio interferometer.

To do this we must express the physical baseline vector, **B**, and the source direction (unit vector **s**) in useful coordinates for a source with hour angle, H and declination δ (**Figure 4**).

We take a coordinate system with the z-axis pointing to the north celestial pole, the (x,y) axis in the equatorial plane with the x-axis coinciding with the meridian (H = 0).

We are interested in

$$\cos \theta = (\mathbf{B} \cdot \mathbf{s})/|\mathbf{B}|$$

Now

$$\mathbf{B} = (B_x, B_y, B_z)$$

and

$$\mathbf{s} = (s_x, s_y, s_z)$$

where

$$s_x = \cos \delta \cos H$$
$$s_y = \cos \delta \sin H$$
$$s_z = \sin \delta$$

Hence

$$\cos \theta = (1/B)(B_x \cos \theta \cos H + B_y \cos \delta \sin H + B_z \sin \delta) \qquad (6.19)$$

The projected baseline B_p is the baseline as seen from the source and is in the plane perpendicular to **s**, as shown in **Figure 5**. At any instant it can be resolved into a north-south component, v, and an east-west component, u. B_p is two-dimensional, with no component in the source direction.

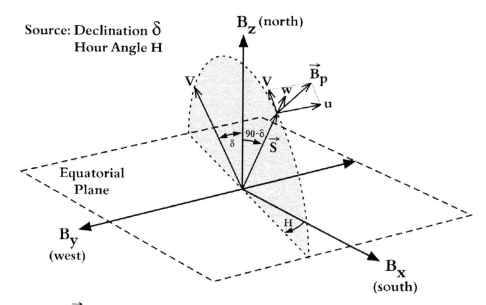

$\vec{B}_p$ describes an ellipse in the u-v plane **U-V Ellipse**

FIGURE 5. System of co-ordinates for the projected baseline in the interferometric observation of a radio source.

By inspecting the coordinate systems it follows that

$$u = B_y \cos H + B_x \sin H \qquad (6.20)$$

$$v = B_z \cos \delta - B_x \sin \delta \cos H - B_y \sin \delta \sin H \qquad (6.21)$$

It can easily be verified that
$$[(u^2/a^2) + (v^2 - v_0^2)/b^2] = 1$$
where
$$a = \sqrt{(B_x^2 + B_y^2)}$$
$$b = \sin \delta \sqrt{(B_x^2 + B_y^2)}$$
and
$$v_0 = B_z \cos \delta$$

This is the equation of an ellipse with semi-major axis a and semi-minor axis b. The ellipse is centred at the point $(0, v_0)$ and has eccentricity $\cos \delta$.

Thus as the hour angle of the source changes, the tip of the projected baseline vector describes an ellipse. An observer at the source looking at out interferometer would see one telescope appear to describe an ellipse about the other due to the rotation of the earth. (Of course the telescopes would be *hidden* during part of the trajectory.) Note that this ellipse depends on both the declination of the source and on the parameters of the baseline. For sources at the equator ($\delta = 0$), the ellipse degenerates to a straight line and for sources at the pole ($\delta = \pi/2$), the ellipse becomes a circle. The description of angular resolution in terms of this projected baseline u,v ellipse is important in deriving the structure of radio sources with such a tracking interferometer.

6.4. *Response to an Extended Source*

We are now fully equipped to examine the response of our interferometer to a spatially-extended source of radiation. We still assume that our source is monochromatic, unpolarised, and that it is much smaller than the primary beam pattern of the antenna. Such an extended source can be imagined to consist of a very large number of infinitesimal point sources, each having an hour angle H and a declination D.

Let the brightness centroid of the source have an hour angle H_0 and a declination δ_0. Since hour angle and declination are not orthogonal coordinates, it is convenient to define a Cartesian coordinate system, (l, m), whose origin is at the brightness centre of the source (H_0, δ_0). Then:

$$l = H \cos \delta - H_0 \cos \delta_0 \sim (H - H_0) \cos \delta_0 \qquad (6.22)$$

$$m = \delta - \delta_0 \qquad (6.23)$$

Suppose that an infinitesimal element of angular size (dl, dm) at the location (l,m) on the source contributes an intensity I(l,m) dl dm. We can treat this as a point source and write the response of our multiplying interferometer to it according to equation (6.16) as

$$dI = I(l, m) \cos(2\pi B \cos \theta) \, dl \, dm \qquad (6.24)$$

Expanding $\cos \theta$ about the source centroid in a Taylor series

$$\cos \theta = \cos \theta_0 + d/dl(\cos \theta)l + d/dm(\cos \theta) \, m$$
$$= \cos \theta_0 + (1/\cos \theta)[d/dH(\cos \theta)]l + [d/d\theta(\cos \theta)] \, m \qquad (6.25)$$

From Equations 6.19, 6.20 and 6.21, using the expressions for the projected baseline components (u,v), we can write

$$\cos \theta = \cos \Theta_0 + (u/B) \, l + (v/B) \, m \qquad (6.26)$$

Combining Eqs. (6.24) and (6.26):

$$dI(u, v) = I(l, m) \cos[2\pi B \cos \theta_0 + 2\pi(ul + vm)] \, dl \, dm \qquad (6.27)$$

Expanding this and putting $\phi_0 = 2\pi B \cos \theta_0$ (phase term due to point source at brightness centroid), we obtain

$$dI(u, v) = I(l, m) \cos [2\pi (ul + vm)] \cos \phi_0 - \sin [2\pi (ul + vm)] \sin \phi_0 \, dl \, dm \qquad (6.28)$$

Integrating gives the output response of an interferometer to a spatially-extended source

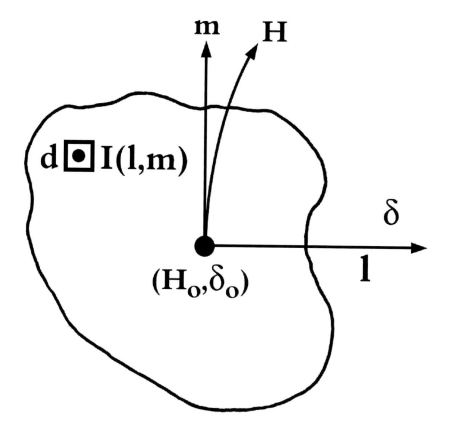

FIGURE 6. System of co-ordinates for the emission of an extended radio source.

$$I(u, v) = \cos \phi_0 \int \int I(l, m) \cos [2\pi (ul + vm)] \, dl \, dm$$
$$- \sin \phi_0 \int \int I(l, m) \sin [2\pi (ul + vm)] \, dl \, dm \qquad (6.29)$$

This can be written as

$$I(u, v) = C(u, v) \cos \phi_0 - S(u, v) \sin \phi_0 \qquad (6.30)$$

where

$$C(u, v) = \int \int I(l, m) \cos [2\pi (ul + vm)] \, dl \, dm$$

and

$$S(u, v) = \int \int I(l, m) \sin [2\pi (ul + vm)] \, dl \, dm$$

are the normal cosine and sine Fourier transforms of the radio source brightness distribution.

We can express the output of the interferometer in complex notation by

$$|I(u, v)| \, e^{i\phi(u,v)} = \int \int I(l, m) \, e^{2\pi i(ul + vm)} \, dl \, dm \qquad (6.31)$$

For an instantaneous value of the projected baseline (u,v) the output interference fringes have an amplitude $|I(u, v)|$ and phase $\phi(u,v)$ measured with respect to a point source at $l = 0$, $m = 0$.

This is the well-known relation that the amplitude and phase of the interferometer fringes are the amplitude and phase of the Fourier transform of the source brightness distribution on the sky.

The fringe visibility is defined to be the normalised fringe amplitude, i.e. $\gamma = |I(u, v)|/S$ where $S = \int \int |I(u, v)|$ dl dm is the total monochromatic flux density of the source.

Similar Fourier relations exist for the other 3 Stokes parameters, Q, U, V that represent the distributions of linearly and circularly polarised radiation in the source. By having two orthogonal linear or circularly polarised receivers on each telescope and by combining or correlating the 4 pairs, I -fringes, Q-fringes, U -fringes and V -fringes can be measured simultaneously. Interferometers make excellent polarimeters and therefore can be used to map the polarisation distribution in sources.

7. Aperture Synthesis

If the fringe amplitude $|I(u, v)|$ and phase $\phi(u, v)$ are measured at *all* projected baselines (measured in no. of λ) out to a given value (u_m, v_m), Fourier inversion gives a map of $I(1,m)$ with a resolution in radians $\sim 1/u_m$ (east-west) and $\sim 1/v_m$ (north-south). This application of interferometry is called *earth rotational aperture synthesis* because it uses the rotation if the earth to give a similar resolution to that of a filled aperture which covers the baseline (**Figure 7**).

Some important properties of aperture synthesis maps will now be mentioned.

7.1. *Sidelobes*

When aperture synthesis maps are made by Fourier transforming the fringe visibilities without further treatment, several artefacts and distortions are present in the maps. There are ripples or *sidelobes* due to incomplete baseline coverage (see **Figure 8**). Mathematically they can usually be expressed in terms of Bessel functions. Most important are:

(*a*) *near-in sidelobes* due to abrupt cut off at edge of array. Radii $= (1/u_m, 1/v_m)$. Most severe close to source. Can be minimised (at cost of resolution) by giving data from longer baselines relatively less weight, i.e. *tapering* the data with a *grading function* before Fourier transformation.

(*b*) *diffraction grating rings* produced by the discrete nature of the array. For a regularly spaced E-W array with adjacent elements spaced by Δu wavelengths, the grating rings are ellipses centred on the source with radii $\Delta l = 1/\Delta u$, $\Delta m = 1/(\Delta u \sin \delta)$ radians.

(*c*) effects due to unsampled short baselines. Sometimes (e.g. because it is physically impossible) the shortest spacings of an otherwise filled comb are unavailable.
Missing zero spacing $\rightarrow$ Constant negative offset in map.
Other missing short spacings $\rightarrow$ Variable negative offset or *bowl*.

7.2. *Field of view*

This is usually limited by the directional properties of the individual antennas in the array, i.e. attenuation of the *primary beam*. Another limitation is set by the size of visibility array that is transformed. To save computing time the Fourier transform is often carried out using the Fast Fourier Transform (FFT) algorithm. In this case the data must be convolved to a uniformly spaced rectangle grid of points before transformation. The effect of this *gridding*, or rectangular sampling of the data in the (u-v) plane, is that

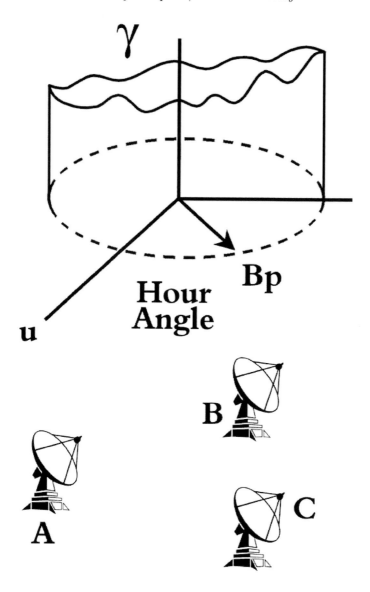

FIGURE 7. Fringe amplitude obtained for an aperture synthesis system as the Earth rotates.

responses of the instrument outside the required field of view are *aliased* or reflected into the picture.

7.3. *Synthesised beam*

This is the calculated response to a point source observed with exactly the same baseline coverage and transformed in exactly the same manner as the observed field. Includes effects of resolution and map distortions. Distinguish between *synthesised beam* and the *primary beam*.

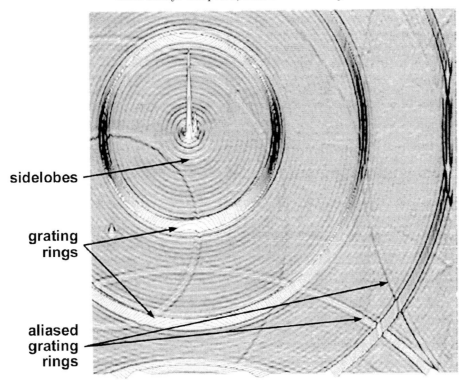

sidelobes

grating
rings

aliased
grating
rings

FIGURE 8. An aperture synthesis map showing both the true image and the spurious effects caused by sidelobes and grating rings.

7.4. *Cleaning the Map*

To remove the distortions due to sidelobes and non-uniform baseline coverage a technique called CLEAN was developed by Högbom at Leiden in the early seventies. This is an iterative technique which approximates an extended source by a number of point sources, subtracts the effects of each of these from the original *dirty map* using the known (imperfect) synthesised beam and subsequently restores the resultant delta-functions of intensity to the residual dirty map with an idealised clean restoring beam in which the imperfections have been removed. The restored clean map gives the source brightness distribution without distortions due to sidelobes, grating rings, etc. Cleaning not only removes distortions in the structures, but also often improves the dynamic range, allowing weaker sources to be detected. Although it is now a standard procedure carried out routinely on radio maps, one should treat cleaned maps with caution. The assumption that an extended source can be described by many points can introduce an artificial clumpiness in the restored structures.

7.5. *Calibration and Self-Calibration*

It is usual to calibrate the gains and phases of the interferometers by observing a point source of known fringe amplitude (flux density) and position (phase) before and after observing the field of interest. However, sometimes conditions change during observations, so that this *absolute* calibration is insufficient. Supplementary calibration procedure is to obtain relative calibration internal to the field of interest. Self-calibration is an iterative procedure whereby an initial map is used to derive a model brightness distribution. This is used to calculate a set of gain and phase corrections, which are implemented to make a new map. This is repeated until the calculated gain and phase corrections are small. To achieve the highest dynamic range, cleaning and self-calibration are usually combined.

7.6. *Model fitting*

For interferometers over long baselines, measurement of the phase of the Fourier transform is difficult. Moreover, these are usually measurements at only a few baselines (limited region of the u,v plane). Under these circumstances it is usual to propose a model for the source structure. The assumed I(l,m) is then Fourier transformed to derive the fringe amplitudes corresponding to the range of projected baselines observed. The predicted γs are then compared with the observed γs. Luckily, the basic features of extragalactic radio sources are simple. Sometimes an iterative model fitting procedure is used to produce convergence between the model and the fringe data. The resultant maps are usually not unique. So beware!

7.7. *Closure and Redundancy*

Where the baselines are too long to measure the absolute phase of fringes, and where there are more than one simultaneous baseline a parameter called the closure phase can be derived to additionally constrain the source structures. It can easily be shown that over any triangle of baselines, the sum of the differential phases must be zero. Thus for telescopes A, B, C forming the baselines AB, BC and CA, $\phi_{AB} + \phi_{BC} + \phi_{CA} = 0$. This requirement helps constrain source models. A similar closure constraint can be imposed on the fringe amplitudes. Closure constraints are present in self-calibration (Section 7.5).

8. Practical Interferometers

Radio interferometers can be classified into three types according to how the signal from the elements are transported over the baseline. There are three classes.

8.1. *Physical link interferometers. (e.g. Cable or wave-guide links)*

Fringe amplitudes and phases easily measured. Only practical for R < few km. (e.g. *VLA*, Westerbork, Cambridge, main AT) Angular resolution useful for studying overall structures of extragalactic radio sources at frequencies > 1GHz

8.2. *Radio link interferometers*

Fringe amplitudes and closure phases easily measured. Absolute phases sometimes. Practical for R < few hundred km. For larger R many repeater stations are needed. This means complicated and unreliable electronics. (Merlin/ Jodrell Bank, outer AT). Angular resolution useful for studying overall structures of extended extragalactic radio sources at frequencies < 1GHz.

8.3. *Non-linked Interferometers*

Signals from each antenna are recorded on tape and processed at a later date. Completely independent receivers at each antenna. Made possible due to development of accurate frequency standards and wide-band tape recorders. Used in *VLBI*: (Very Long Baseline Interferometry). Usually baseline R limited only by radius of earth. Recent Japanese satellite (*VSOP*) combined with ground-based telescopes is extending *VLBI* to longer baselines. Usually uses fringe amplitudes and closure phases combined with iterative model fitting techniques to derive structure information.

Angular resolution useful for studying compact extragalactic radio sources and the nuclear regions of extended extragalactic radio sources.

9. General Properties of Extragalactic Radio Sources

9.1. *Introduction*

Luminous extragalactic radio sources are the most significant constituents of the radio sky. They are interesting for several reasons. First, because they are amongst the most energetic (up to 10^{61} ergs) and largest (up to ~4 Mpc) objects in the known Universe, they are important laboratories in their own right. Secondly, they are unique probes of physical conditions in the early Universe. They emit luminous emission lines they can be seen to large distances and their distances can be measured. Thirdly, they can be used to derive information about the geometry of the Universe.

The starting point for radio source studies are surveys of the sky at fixed frequencies. From the surveys, catalogues of radio sources are compiled with positions and flux densities down to the relevant limiting flux densities, S_{min}. We shall see later that the finding frequency, ν_f at which the survey is made determines the sort of sources that are predominantly picked up.

To derive physically useful parameters for the sources it is important to determine their distances. To do this one attempts to identify them with objects seen at visible wavelengths (e.g. galaxies or QSOs). Using either the redshifts or the apparent magnitudes of the associated optical objects, the distances to the radio sources can be determined and their radio luminosities estimated.

Detailed studies of individual sources and statistical studies of radio source populations are complementary tools for studying extragalactic radio sources. There are several observable properties of radio sources from which we have learned about their nature. In the fifties and sixties much information was obtained from the integrated (whole source) properties of radio sources such as their flux densities and polarisations as a function of frequency. Subsequently the spatial distributions of these quantities determined with interferometric arrays were crucially important, particularly when the observed quantities were combined with redshift to yield intrinsic (physically relevant) parameters. The observed spectra and polarisation properties are as would be expected from synchrotron radiation produced by a relativistic plasma, i.e. relativistic electrons spiralling in a magnetic field.

9.2. *Spectra*

Radio spectra are usually shown as log-log plots of flux density, S, against frequency, ν. Note that in radio astronomy "S" rather than "F" (as in optical astronomy) is used to denote flux density. We shall define the radio spectral index, α by $S \propto \nu^{\alpha}$ i.e. the slope of the logarithmic spectral plot. [Caution α is sometimes defined by $S \propto \nu^{-\alpha}$].

Radio sources can be divided into three main classes according to their integrated spectra and sizes.

(*a*) Steep-spectrum sources (Extended sources): These have $-1.3 < \alpha < -0.6$ and are mostly associated with sources whose emission is dominated by components larger than the optical galaxy, i.e. scale of several hundred kiloparsec. The spectra are usually fairly straight between 100 MHz and 1 GHz (power- law). At lower frequencies they often *turn over* and cut off. Above 1 GHz they frequently steepen with $\alpha \to \alpha - 0.5$. They are identified both with radio galaxies and quasars.

(*b*) Flat-spectrum sources (Compact sources). These have $\alpha > -0.4$ and are mostly associated with sources dominated by compact cores (< 1pc) and variable. Spectra of these sources are frequently *lumpy*, each lump of which is believed to be due to separate components at different stages of evolution. They are predominantly associated with quasars.

(*c*) Peaked-spectrum sources (Kiloparsec-scale sources): About 10% of sources have a simple spectrum with a pronounced peak near 1 GHz (GPS sources) or at lower frequencies (CSS sources). These sources are dominated by components having sizes of a few hundred parsec.

Steep-spectrum sources are the dominant populations in surveys made at low frequencies ($<\sim 1$ GHz) but high frequency surveys (nu > 5 GHz) comprise mostly flat-spectrum sources. There have been various proposals as to how these various classes of radio source are related to each other (e.g. Section 11.2): These include

(*a*) orientation of the sources with respect to the line of sight,

(*b*) age

(*c*) environment in which they are produced.

9.3. *Morphology of Extended Sources*

Several aspects of the nature of extended radio sources can be deduced immediately from the radio maps.

(*a*) The structure of most radio sources is linear and double, implying that **the nucleus produces twin collimated beams of relativistic plasma** (see **Figure 9**).

(*b*) Most extended radio sources are found to have compact parsec-scale components (*cores*) at their nuclei implying that the nuclei are still active and have been active for the time it takes for electrons to travel from the nuclei to the extremities (up to $\sim 10^7$ - 10^8 **years**).

(*c*) In some cases, similar orientations are found (to within a few degrees) between nuclear and extended structures spanning size scales of factors of $> 10^5$ and reaching sizes of >4 Mpc. This implies that the **machine** that produces the collimated emission **has a memory for direction** in some cases **exceeding about** 10^8 **years**. Good arguments for a **rotating black hole**. (The clinching argument that active galaxies harbour black holes at their centres was provided recently by the beautiful *VLBI* study of the kinematics of the 23 GHz water maser emission from the galaxy NGC 4258 (Mlyoshi *et al.* (1995)). The motion is Keplerian to an amazing accuracy, implying $10^{7.5}M_\odot$ in a 0.13 pc radius.).

(*d*) **Radio jets** are frequently seen linking the nuclei to the extended radio lobes. These are believed to be channels along which **energy is supplied quasi-continuously** to the lobes and which were predicted by Rees in 1972 (Rees (1972)), before their discovery.

(*e*) C-shaped symmetry is sometimes seen the morphologies, with as extremes the so-called head-tail radio sources (see **Figure 10**). This implies that the radio-emitting **relativistic plasma is <u>distorted</u>** by the motion of the radio source through the ambient medium.

(*f*) S- and Z-shaped symmetry is seen in morphologies implying that there are **shear motions** in the ambient medium or that the **central machine can change its direction** (e.g. precess) during the lifetime of the source (see **Figure 11**).

(*g*) There is a luminosity dependence in the morphologies (see **Figure 11 and Figure 12**), first pointed out by Fanaroff and Riley.

- For monochromatic luminosities, $P_{178} <\sim 10^{25}$ W/Hz, the extended lobes are relatively low- brightness and <u>woofly</u>. Fanaroff - Riley Class 1 (e.g. Virgo A).
- For monochromatic luminosities, $P_{178} >\sim 10^{25}$ W/Hz, the extended lobes appear more collimated, with bright <u>hot spots</u> at the source extremities. Fanaroff - Riley Class II (e.g. 3C 390.3).

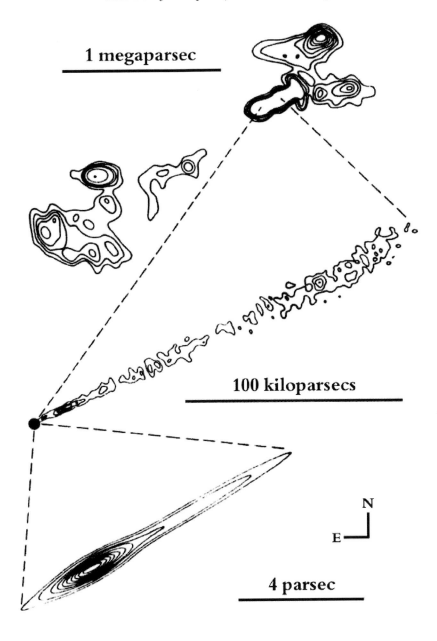

1 megaparsec

100 kiloparsecs

4 parsec

N

E

FIGURE 9. The morphology of NGC 6251 from radio mapping observations.

10. Physics of Radio Sources: Synchrotron Radiation

10.1. *Synchrotron Emission*

We have seen that the observed spectra and polarisations of extragalactic radio sources are consistent with synchrotron radiation. This is the most important radiation mech-

BENDING SEQUENCE

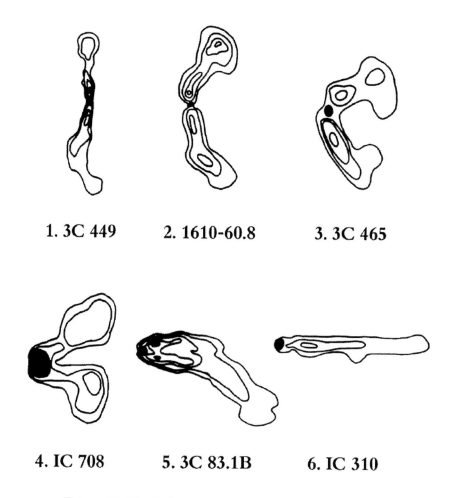

1. 3C 449   2. 1610-60.8   3. 3C 465

4. IC 708   5. 3C 83.1B   6. IC 310

FIGURE 10. The C-shape symmetry seen in some radio sources.

anism in extragalactic radio astronomy and is produced by relativistic electrons in a magnetic field.

A single relativistic electron of mass "m" gyrating in a magnetic field of induction **B** will emit a monochromatic power

$$P(E, \nu) = \text{constant } B_p \text{erp}(\nu/\nu_\perp) \int_{\nu/\nu(c)} K_{5/3}[E/(mc^2)]d[E/(mc^2)] \qquad (10.32)$$

where $K_{5/3}$ is a modified Bessel function.

The radiation power increases with frequency according to $\nu^{1/3}$ until a critical frequency $\nu_C = \text{constant } B_p \text{erp}E^2$, above which the spectrum cuts off exponentially.

(For a 3 GeV electron in a 10μg magnetic field, $\nu_C = 1500$ MHz, i.e. in the radio)

The radiation loses energy at a rate

Rotational Symmetry Sequence

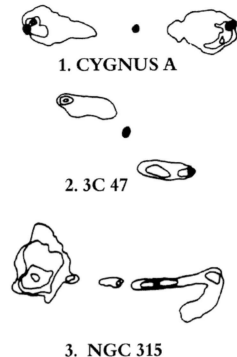

1. CYGNUS A

2. 3C 47

3. NGC 315

4. 3C 315

FIGURE 11. The rotational symmetry sequence seen in radio sources.

$$-\mathrm{dE/dt} = \int_0^\infty \mathrm{P(E}, \nu)\mathrm{d}\nu = \mathrm{constant}\mathrm{B}_\perp^2 \mathrm{E}^2 \qquad (10.33)$$

A 3 GeV electron in a 10μG magnetic field loses half its energy in a time $\mathrm{t}_{1/2} \sim 10^{7.5}$ yr. Assume:

(*a*) No. of electrons per unit volume with energies between E and E + dE is

Edge Brightening Sequence

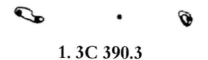

1. 3C 390.3

2. 3C 236

3. 3C 449

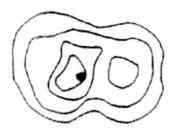

4. VIRGO A

FIGURE 12. The edge-brightening sequence seen in radio sources.

$$n(E)dE = n_0 E^{-\gamma} dE \qquad \text{for } E_1 < E < E_2 \qquad (10.34)$$

i.e. the power law observed for the distribution of cosmic ray energies

  (*b*) an isotropic distribution of pitch angles

  Power emitted per unit frequency per unit volume per unit solid angle is

$$\epsilon(\nu) = \int_0^\infty n(E) P(E, \nu) dE \qquad (10.35)$$

$$= B_\perp^{[(\gamma+1)/2]} \nu^{-[(\gamma-1)/2]} \tag{10.36}$$

$\epsilon(\nu) \propto \nu^\alpha$, where $\alpha = (1 - \gamma)/2$ is the *spectral index*.

Hence there is a direct relation between a power law cosmic ray energy spectrum and the power law radio synchrotron radiation produced by these electrons moving in a magnetic field.

The radiation is highly polarised, with a degree of polarisation dependant on the spectrum given by

$$\Pi = (\gamma + 1)/(\gamma + 7/3) = (1 - \alpha)/(5/3 - \alpha) \tag{10.37}$$

About 80% of extragalactic radio sources catalogued by surveys at radio frequencies <1 GHz typically have power spectra $-0.6 < \alpha < 1$ and are usually polarised, consistent with these predictions of synchrotron emission theory. The median observed spectral index is $\alpha \sim -0.8$ corresponding with an cosmic ray energy index of $\gamma \sim 2.6$.

10.2. *Modifications to Synchrotron Spectra*

The radio synchrotron spectrum will depart from a power law if

(*a*) the electron energy distribution of the cosmic rays is not a power law, (i.e. γ varies with E)

(*b*) the radiation undergoes absorption (produces cut-off of spectrum at low frequencies)

(*c*) there are differential energy losses and gains that are energy dependant (dE/dt is function of E)

10.2.1. *Absorption*

Possible absorption mechanisms include

(*a*) synchrotron self-absorption

(*b*) thermal absorption, and

(*c*) the Razin-Tsytovich effect

10.2.2. *Synchrotron Self-Absorption*

Synchrotron self-absorption occurs if the radiation is produced in such a small volume that it is reabsorbed by the source itself.

For synchrotron self-absorption to occur the source brightness temperature must exceed the kinetic temperature of the radiating electrons

$$T_B > T_{KIN}$$

Now

$$T_{B,\nu} = (c^2/2k)[B_\nu/\nu^2] = (c^2/2k)[(S/\theta^2)/\nu^2]$$

and

$$kT_{KIN} = E \propto \sqrt{\nu/B_\perp}$$

Synchrotron self absorption occurs at frequencies $\nu < \nu_s$, where

$$\nu_s^{5/2} \propto [S_m/\theta^2]/B_\perp^{1/2} \tag{10.38}$$

and S_m is the emitted flux density at the peak frequency ν_s

A <u>model</u> synchrotron self absorbed source has a spectrum of the form

$$S \propto \nu^\alpha \text{ for } \nu \gg \nu_s; \qquad S \propto \nu^{5/2} \text{ for } \nu \ll \nu_s \tag{10.39}$$

If the spectrum and θ are measured, synchrotron self absorption can be used to constrain the magnetic field strength.

Thermal Absorption. If a cold cloud of ionised gas is located in front of or within the source, then the observed flux density will fall off sharply below the frequency, ν_t, where the optical depth is unity. For an electron the temperature T_e,

$$\nu_t \sim 3.6 \times 10^5 T_e^{-1.5} \epsilon \tag{10.40}$$

where $\epsilon = \int n_e dl$ is the emission measure, and n_e the density of thermal electrons. If the material is in front of the source, the spectrum is

$$S \sim \nu^\alpha \exp - (\nu_t/\nu)^2 \tag{10.41}$$

If the material is mixed with the source, the spectrum is

$$S \sim \nu^\alpha \text{ for } \nu \gg \nu_t; \qquad S \sim \nu^{(\alpha+2)} \text{ for } \nu \ll \nu_t \tag{10.42}$$

The Razin-Tsytovich effect. If the density of thermal electrons is large enough, then at frequencies below a value ν_r, the index of refraction, η, is less than one, causing the velocity of a relativistic electron to be less than the phase velocity of light in the medium and attenuating the emitted energy. This Razin- Tsytovich effect causes the spectrum to cut off sharply for frequencies $\nu < \nu_r$, where

$$\nu_r \sim 20(n_e/B) \text{ MHz} \tag{10.43}$$

10.2.3. *Differential energy losses*

Energy losses include
(*a*) synchrotron and inverse Compton ($\propto E^2$)
(*b*) collision ($\propto E$) and
(*c*) ionisation (independent of E)
Each of these modify the spectral index α.

A synchrotron emitting electron that has been accelerated time t_S ago, undergoes synchrotron energy losses for frequencies $\nu > \nu_L$, where $\nu_L \propto B^{-3} t_S^{-2}$ MHz.

The age of the synchrotron emitting electron is $t_S \propto \nu^{-1/2} B^{3/2}$. For typical synchrotron sources $t_S \sim 10^7$ yr for radio frequencies and $t_S \sim 10^4$ yr in the optical.

10.3. *Modifications to Synchrotron Polarisation*

According to synchrotron theory, the radiation should be highly polarised ($\sim$70%) in a direction perpendicular to the magnetic field orientation. But radio sources are typically polarised by only a few percent. There are several effects, which can both modify the direction of polarisation and reduce its fraction.

10.3.1. *Faraday Rotation*

When radiation of wavelength λ passes through an ionised plasma of electron density n_e containing a magnetic field of strength H, there is a slight difference in refractive index for the left and right circularly polarised waves. The result is to rotate the direction of the plane of polarisation through an angle $\psi \propto \lambda^2 = R_M \lambda^2$.

The *rotation measure*, R_M, is usually measured in radians per square meter and is defined as

$$R_M = 8.1 \times 10^5 \int n_e H_L dL \tag{10.44}$$

where n_e is the electron density and H_L is the component of magnetic field parallel to the direction of propagation.

Differential Faraday rotation between different parts of a radio source can cause cancellation between polarisation vectors and *depolarise* the source radiation.

Measurement of the polarisation position angle and percentage at several frequencies gives information about the electron density in the ionised plasma and the magnetic field within it.

10.3.2. *Faraday Depolarisation*

Irregularities in the ionised plasma on the scale of the beam can result in significantly different Faraday rotations for polarised radiation. In the resultant addition of the vector polarisation the observed fraction is reduced.

10.3.3. *Bandwidth Depolarisation*

Because of its λ^2 dependence, a sufficiently large Faraday rotation can cause the direction of polarisation to be different at two edges of the observing band, resulting in vector cancellation.

10.3.4. *Beam Depolarisation*

If the projected magnetic field directions change on a scale smaller than the beam, cancellation occurs, reducing the observed fractional polarisation.

10.4. *Energy Required*

By analysing the energy budget of a synchrotron radio source we can constrain the minimum energy needed to produce it.

The total energy of a radio source is

$$U_{tot} = U_{magnetic} + U_{electrons} + U_{protons}$$

$$U_{magnetic} = V[B^2/8\pi] \text{ where V is the volume filled by the source}$$

$$U_{electrons} = \int_{E1}^{E2} N(E) \text{ E dE per unit volume} \tag{10.45}$$

Taking $N(E) = N_0 E^{-\gamma}$, $E_1 < E < E_2$, $U_{electrons} = N_0[E_2^{2-\gamma} - E_1^{2-\gamma}]/(2-\gamma)$
We can determine N_0 from the bolometric luminosity L

$$L = \int_{E1}^{E2} (-dE/dt)N(E)dE \tag{10.46}$$

Taking $-dE/dt \propto B_\perp^2 E^2 = C_3 B_\zeta^2 E^2$ and $N(E) = N_0 E^{-\gamma}$

$$U_{electrons} = [L/(CB_\perp^2)][(3-\gamma)/(2-\gamma)][E_2^{2-\gamma} - E_1^{2-\gamma}]/[E_2^{3-\gamma}E_1^{3-\gamma}] \tag{10.47}$$

Taking $\alpha = (1-\gamma)/2$ (i.e. approximating that each electron only radiates at the critical frequency $\nu_C = $ constant $B_\perp E^2 = C_1 B_\perp E^2$

$$U_{electrons} = [C_1^{1/2}/C_3][(2\alpha+2)/(2\alpha+1)][\nu_2^{\alpha+2} - \nu_1^{\alpha+2}]/[\nu_2^{\alpha+1}\nu_1^{\alpha+1}][LB^{-1.5}] \tag{10.48}$$

Take $U_{protons} \propto U_{electrons} = kU_{electrons}$

$$U_{tot} = U_{magnetic} + U_{electrons} + U_{protons} = U_{magnetic} + (1+k)U_{electrons}$$
$$= V[/8\pi]B^2 + \text{Constant} \times \text{Function } (\alpha, \nu_1, \nu_2) \text{ } LB^{-1.5} \tag{10.49}$$

Energy is minimum when $dU_{tot}/dt = 0$. The corresponding minimum energy value of $B = B_{me}$ where $B_{me} = $ Constant x Function $\alpha, \nu_1, \nu_2 (L/V)^{2/7}$

The corresponding minimum energy is $(U_{tot})_{me} = (7/3)[(B^2_{me})/(8\pi)]$

In terms of observable quantities, B_{me} can be written

$$B_{me} = 6 \times 10^{-5}[(1+k)/\xi(1+z)^{3-\alpha}(\nu_2^{\alpha+2} - \nu_1^{\alpha+2})/(\alpha+2)$$
$$S_0/\nu_0^\alpha 1/(\theta_x\theta_y \text{ s } \sin^{1.5}\phi)]^{2/7} \qquad (10.50)$$

where

- k is the ratio of energy in protons to that in electrons
- ξ is the volume filling factor of the radiation
- z is the redshift
- the source is assumed to radiate only between frequencies ν_1 and ν_2 with a power law of spectral index α
- the measured flux density at frequency ν_0 is S_0
- the source has angular size θ_x θ_y arcseconds and the line of sight thickness is s parsec
- ϕ is the angle between the uniform magnetic field and the line of sight.

In applying this equation to calculate the physical parameters, one should be aware of the multitude of assumptions that go into it. In particular, k, ξ, s, ϕ are all subject to large uncertainties.

The observed parameters of radio sources give B_{me} ~few μGauss for the extended (~100 kpc) emission and B_{me} ~few mGauss for compact parsec-scale emission. The corresponding values for minimum energies imply that at least ~10^{57} to 10^{61} ergs are needed to produce the most luminous extended radio sources. They are amongst the most energetic objects known in the Universe.

10.5. *Source Confinement*

In order to account for the non-spherical shapes of radio sources, the particles and field must be prevented from dispersing at the relativistic sound speed of c/%3. The pressure exerted by the relativistic gas in the radio source, u/3, must therefore be balanced by some pressure or by internal drag. Several mechanisms have been considered and in each case the pressure balance condition can be used to calculate limits for the various parameters that characterise the resisting forces. Some possible confining mechanisms and the relevant constraints are:

Internal Gravitational: Point mass M at centre of spherical source, diam. θ (arcsec), Distance D (Mpc) $M > 7 \times 10^7 (D\theta)^2 B_{me} M_\odot$

External Static Thermal: Temperature T $\rho_{ext} T > 2.5 \times 10^{-10} B^2_{me}$ g cm^{-3}

External Dynamic: Ram pressure, translational velocity, vt $\rho_{ext} T > 7 \times 10^{-23} B^2_{me} (v_t/c)^{-2}$ g cm^{-3}

External Magnetic: Field strength Bext $B_{ext} > B_{me}$

Hence the source parameters can be used to obtain information about the parameters of the medium in which the radio sources are located (**Figure 13, Figure 14**). In real life it is not always clear which confining mechanisms are applicable. Several may be occurring simultaneously.

10.6. *Ages - Localised Particle Acceleration*

A problem frequently considered in discussions of radio sources concerns the place where the observed synchrotron electrons are accelerated. Does this occur entirely in the nucleus, mainly in the hotspots, or is there an accelerating process at work throughout the entire source?

For the acceleration to have taken place at an angular distance θ (arcsec) from where the radiation is observed, the electron will have travelled to its present position in a time

$$t_d = 4.7 \times 10^6 \, D \, \theta/(v \sin \psi) \tag{10.51}$$

where D (Mpc) is the distance, v (km/s) is the velocity at which the electrons travel from the nucleus to the position under study and ψ is the angle between this direction and the line of sight. For an isotropic distribution of pitch angles the average age of a synchrotron electron (van der Laan & Perola (1969)) is

$$t_r = 0.82 B^{1/2} (B^2 + B_R^2)^{-1} (1 + z)^{-1/2} \nu *^{-1/2} \text{ yr} \tag{10.52}$$

where B(G) is the magnetic field strength, $B_R = 4 \times 10^{-6} (1 + z)^2 G$ is the equivalent magnetic field strength of the microwave background. ν^* is the frequency above which an exponential drop in flux occurs and usually exceeds ν, the observing frequency.

For the radiating electrons to have been accelerated at angular distance θ from the observed position, the following condition must hold: $t_d < t_r$, i.e.

$$5.7 \times 10^6 D_\theta \theta v^{-1} (\sin \psi)^{-1} B^{-1/2} (B^2 + B_R^2)(1 + z)^{1/2} \nu *^{1/2} < 1 \tag{10.53}$$

There are many cases in extragalactic radio sources where this condition does not hold, implying that localised particle acceleration occurs and is widespread.

11. Compact Radio Sources and Unification

11.1. *Super-relativistic Expansion*

Many flat spectrum radio sources are known to vary in flux- density on time scales of from months to years. The spectral variations are usually interpreted in terms of a series of expanding shells, each undergoing synchrotron self-absorption. As the shells expand, the cut-off frequency v_s, moves to lower frequencies and the flux decreases.

Structural variations are seen when these sources are mapped with *VLBI* usually in the form of blobs that move away from each other in a one-sided jet, often with an apparent velocity greater than that of light. This apparent superluminal expansion is usually explained in terms of the transverse Doppler effect.

Consider two plasma blobs separated by distance R moving away from each other with a velocity v along a direction making an angle θ with the line of sight (**Figure 15**).

In the emitted frame
the velocity is $\qquad v = \Delta R/(\Delta t)_{em}$
the transverse component is $\quad v_t = (\Delta R) \sin \theta/(\Delta t)_{em}$

Using special relativity:
In the observed frame $\qquad (\Delta t)_{obs} = (\Delta t)_{em} 1 - [(v/c) \cos \theta$

The apparent velocity is $\qquad (v_t)_{obs} = (v \sin \theta)/1 - (v \cos \theta)/c$

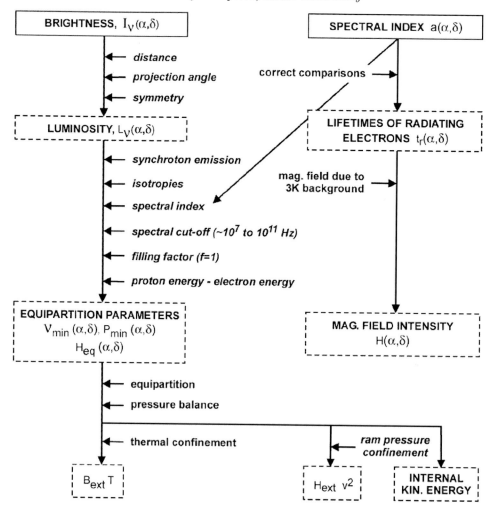

SCHEMATIC DIAGRAM OF THE MOST IMPORTANT PARAMETERS OF RADIO
SOURCES DERIVABLE FROM THE INTENSITY DISTRIBUTION MEASUREMENTS.
SOLID BOXES CONTAIN OBSERVED QUANTITIES AND DASHED BOXES
CONTAIN DERIVED QUANTITIES.

FIGURE 13.

If $v \sim c$ and θ is small, $(v_t)_{obs} > c$, producing apparent superluminal expansion. [e.g.
For $\gamma = 1 - (v/c)^{2^{-0.5}} \sim 5$, and $\theta \sim 10°$, $(v_t)_{obs} \sim 5c$.]

A consequence of such a model is that that the apparent flux densities of the approaching and receding components are altered by relativistic aberration

Consider approaching and receding components that when stationary both emit equal
flux density $S(\nu_{em})d\nu_{em}$

Remembering that the dimensions of flux density are (energy) (time)$^{-1}$ (distance)$^{-2}$,
the observed flux density of the approaching blob is

$$(S\Delta\nu)_{obs+} \sim S(\nu_{obs}/\delta)d\nu_{obs}(\delta)^3 = S(\nu_{obs})\delta^{3-\alpha} \tag{11.54}$$

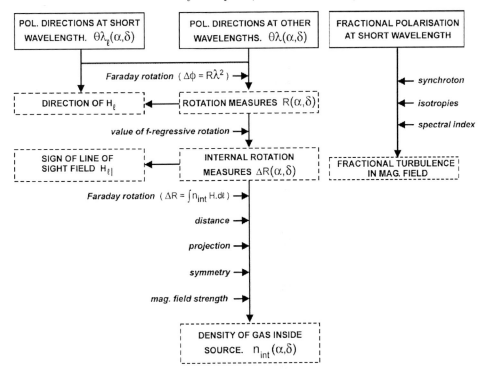

**SCHEMATIC DIAGRAM OF THE PARAMETERS DERIVABLE
FROM POLARISATION MEASUREMENTS.
SOLID BOXES (MEASURED QUANTITIES), DASHED
BOXES (DERIVED QUANTITIES).**

FIGURE 14.

where $\delta = (\nu_{em}/\nu_{obs}) = \gamma^{-1}[1 - (v/c)\cos\theta]^{-1}$ is the Doppler factor.

The increase in apparent flux density that is produced by the approaching relativistic speeds is called Doppler boosting and factors of 10 - 100 are relatively easily obtained.

(Note that the receding blob is attenuated by a similar factor.)

11.2. *Unification of compact and extended radio sources*

There have been several attempts to understand extended and compact radio sources as the same phenomenon. The following scenarios are often invoked to unify different classes of radio sources:

Orientation: Various models have been proposed which purport to explain the observed properties of two or more classes of active galaxies on the basis of a single population viewed at different orientations. In one class of orientation unification models, all radio sources are assumed to have quasars in their nuclei, but these quasars are obscured, except in a direction close to the radio source axis. Thus, if the radio source axis lies close to the plane of the sky, it is seen as a radio galaxy. As the angle between the axis and the line of sight decreases, the object is seen as a steep-spectrum extended quasar. When the radio axis is very close to the line of sight, the quasar component is enormously Doppler-boosted and the object is observed as a flat-spectrum compact quasar.

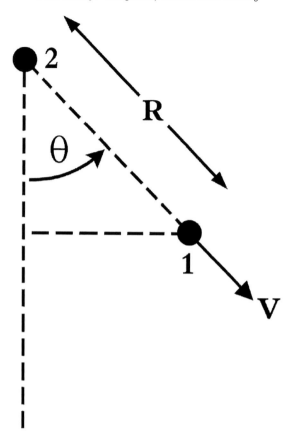

FIGURE 15. Super-relativistic expansion from two blobs of plasma at an angle to the line of sight of the observer.

Environment: It is known that spiral galaxies do not harbour powerful radio sources. Hence the host galaxies and the environment may well play an important role in determining the radio properties and the class of a radio source.

Variability: Compact radio sources and compact cores in extended radio sources can vary. There is also evidence that radio sources can undergo periods of recurrent activity. Hence extended radio galaxies could well go through several quasar phases during their lifetimes.

In practise, it is likely that orientation, environment and variability all play a role in defining the observed properties of radio sources.

12. Radio Source Statistics and Cosmology

Because of their enormous luminosities, radio sources have been important cosmological probes since the dawn of radio astronomy. The statistics of radio sources have provided several diagnostics for studying cosmical evolution. Here I shall describe some widely used statistical tools.

12.1. *Source Counts and Evolution*

Consider a static Euclidean Universe populated by a uniform distribution of radio sources. Suppose that the number of sources per unit volume with luminosity P is given by $\phi(P)$, the luminosity function.

The observed flux density from a source of luminosity P at distance D is

$$S = P/(4\pi D^2)$$

The radius out to which a source of luminosity P has a flux density exceeding S is

$$D(P, > S) = P/(4\pi S)^{1/2}$$

The number of objects of power P in the shell $D \to D + dD$ is

$$N(P) = \phi(P)dV(D) = \phi(P) \,.\, 4\pi D^2 \, dD \tag{12.55}$$

The number of sources with luminosity P and flux density $>S$ is

$$N(P, > S) = \phi(P) \int_0^{D(S)} 4\pi D^2 \, dD$$

$$= (4\pi/3)\phi(P)P^{3/2}(4\pi S^{-3/2}) \, dP \tag{12.56}$$

The total number of sources (all P, with flux density $>S$) in a simple Euclidean Universe is

$$N(> S) = (16\pi^2/3) \int_0^4 \phi(P)P^{3/2} \, dP.S^{-3/2}$$

$$N(> S) \propto S^{-3/2} \qquad \text{...a power law} \tag{12.57}$$

In the case of a real relativistic Friedmann-Walker Universe, D is replaced by the luminosity distance, D_L expressed in terms of redshift z, deceleration parameter, q_0. One must also take account of the K- correction in the fluxes introduced by the redshift.

$$D \to D_L(1 + z)^{(1-\alpha)} \tag{12.58}$$

where α is the spectral index $(S \propto \nu^\alpha)$

The expression for the total number of sources observed with flux densities exceeding S then becomes:

$$N(> S) = \int_0^4 dP \int_0^{z(s)} \phi(P)4\pi D_L^2 dr \tag{12.59}$$

where $dr = (c \, dz)/H_0 \, (1+z)(\Omega z+1)^{1/2}$ is the element of co-moving coordinate distance

For $z \ll 1$, this reduces to $N(>S) \propto S^{-3/2}$ and for $z \gg 1$ to $N(>S) \propto S^{-\beta}$, where $\beta < 3/2$.

So we might expect to find $N(>S)$ with a slope of -3/2 for bright sources gradually flattening for fainter sources.

The observed $N(>S)$ differs substantially from this expected behaviour assuming a uniform distribution of sources., e.g. For bright S the counts are steeper than -3/2, indicating a <u>deficit</u> of sources. The observed $N(>S)$ has long been important evidence that the Universe is evolving.

There have been many detailed studies of evolution, using a combination of redshift data and source counts. The luminosity function is then assumed to be a function of redshift $\phi(P,z)$ and for a given geometry of the Universe the source counts and luminosity distributions are predicted and compared with observation.

12.2. *Radio luminosity function and Evolution*

The radio luminosity function (*RLF*) of galaxies (defined in the same way as the optical luminosity function) is a measure of the relative space density of galaxies as a function of radio luminosity. Because it is difficult to obtain *bolometric* radio luminosities, usually the radio luminosity functions are monochromatic, referenced to a specified frequency.

The probability that a galaxy has a monochromatic radio luminosity between P and (P + dP) within a volume element between V and (V+dV) is $\phi(P)$ dP dV where $\phi(P)$ is the *differential* luminosity function.

Expressed in integral or cumulative form N(>P), the probability that a galaxy has a monochromatic radio luminosity >P within a volume element between V and (V+dV) is

$$N(> P)dV = \mu_P \int_0^4 \phi(P)dP \qquad (12.60)$$

The differential form of the luminosity function has the advantage that each point is independent and the uncertainties easier to quantify. Information can be obtained about the *RLF* both from flux-limited radio samples and from flux-limited optical samples.

Let us now consider how the *RLF* can be determined in practise. One can approach this either using a complete radio sample or a complete optical sample.

12.2.1. *The radio sample method*

A radio survey produces a sample of radio sources complete to a given radio flux density level, S_{min}. If the distance to such source is determined (either from the redshift or apparent magnitude of the associated optical object) the luminosity function can be calculated

The calculation must correct for the fact that we can see more luminous sources over a much larger volume than the less luminous ones. The result of plotting the raw luminosity distribution of sources from the flux-limited sample is therefore not representative of the true luminosity function of sources in a given volume. Hence a flux-limited radio sample is biased to more distant and higher luminosity objects than a volume-limited sample (**Figure 16**).

To take this distance effect into account, one calculates the maximum volume, V_i at which a each source "i" with luminosity P_i would still be included in the flux-limited sample.

For luminosities between P and P+ΔP this i-th source contributes $1/V_i$ to the luminosity function

$$\phi(P)\Delta P = \sum_{i=1}^{n}(1/V_i)\Delta P \qquad (12.61)$$

where "n" is the number of sources with luminosities between P and P+ΔP.

In practise not every source in a complete sample can be optically identified and its distance determined. It is usually assumed that the unidentified sources are at distances beyond the deepest plates of the area, i.e. they affect the high-luminosity end of the *RLF*.

12.2.2. *The optical sample method*

A second method is to start with an optically selected sample of galaxies at known distance and to measure the radio flux densities of each member. This removes the volume vs. radio power bias and gives good information on the weak end of the radio

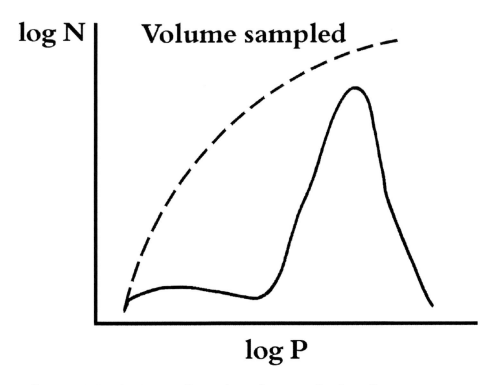

FIGURE 16. Luminosity sampling against volume sampling for radio source counts.

luminosity function, but because there are relatively few large-distance sources in such a sample, this is not useful for studying the high end of the *RLF*.

12.2.3. *The Observed RLF*

The *RLF* is normally plotted in log - log form and expressed in the form $\phi(P) \propto P^{-\alpha}$, where α is the slope. The luminosity distribution obtained in a given survey is critically dependent on the slope of the luminosity function. The number of sources n(P) having luminosity P observed in a survey with flux limit S_{min} is

$$n(P) = (4/3)\pi(P/S_{min})^{1.5}\phi(P) \tag{12.62}$$

Suppose we write the measured n(P) % $P^{(1.5-\alpha)}$

$\alpha < 1.5$, implies that the number of sources per unit volume increases with luminosity If $\alpha = 1.5$, the number of sources per unit volume is equally likely at every luminosity If $\alpha > 1.5$, the number of sources per unit volume decreases with luminosity

In practise determination of the radio luminosity function is difficult. At the high luminosity end the result is intimately bound up with cosmology since the volume seen depends on the cosmology used to calculate it and there is evidence that it is somewhat redshift-dependent.

Radio luminosities of extragalactic radio sources are found to span 8 - 9 orders of magnitude. The *RLF* has a form $\phi(P)\%P^{-0.5}$ for $P_{1.4GHz} < 10^{25}$ W/Hz (normal elliptical galaxies), steepening to $\phi(P)\%P^{-1.2}$ for $P_{1.4GHz} > 10^{25}$ W/Hz (radio-loud galaxies/quasars)

The radio luminosity function of elliptical galaxies is found to depend on the optical luminosity of the galaxies. The *bivariate* radio/optical luminosity function $\phi(\text{P}_{\text{Rad}}, \text{L}_{\text{Opt}})$ shows a dependence on L_{Opt} of the form $\phi\%L_{\text{opt}}^{-1.5}$, implying that more massive ellipticals have a larger chance of undergoing radio activity.

13. Distant Radio Galaxies

I shall conclude these lectures with mention of a field of research with which I have been closely involved in, namely distant radio galaxies.

The study of the first few billion years poses one of the most fascinating and fundamental problems in modern astrophysics. How were galaxies and clusters of galaxies formed? The extremely smooth observed microwave background revealed that the Big Bang was remarkably uniform. Nonetheless we see (clumpy) galaxies and quasars out to redshifts of 5, corresponding to 90% of the distance to the Big Bang and showing that they must have formed in less than a billion years. Did small clumps form first and larger structures subsequently or vice versa?

A related problem is presented by the dramatic increase and decrease in the population of quasars and radio galaxies that occurred during the first 20% of the lifetime of the Universe, indicated by statistical studies. During this quasar era it is likely that galaxies formed their first stars. Nuclear activity may well be a process that every galaxy experienced in the early Universe and represent a fundamental phase in the evolution of galaxies. The sudden end of this *quasar era* after 3 billion years is not fully understood.

Radio galaxies are amongst the most massive of all known galaxies and are particularly important probes of the early Universe. They have several distinct emitting components, which provide diagnostics about different physical constituents of galaxies. A non-exhaustive list of these is given in the table. All the components are spatially resolvable and therefore provide geometrical information about the internal processes within the objects. In this case the fact that all these diagnostics are simultaneously present and that the interrelationships and interactions between them can be studied make distant radio galaxies unique laboratories for probing the early Universe.

13.1. *Aligned and Clumpy Optical Continua*

One of the most puzzling aspects of high-redshift radio galaxies is the nature of their UV/optical/IR continuum emission. Deep continuum images with the HST show that several distant radio galaxies are composed of many (> 20) distinct sub-kiloparsec clumps distributed within the giant Lyα halos and resemble model simulations of galaxy formation. Unlike nearby radio galaxies, the radio emission of $z> 0.6$ radio galaxies is roughly aligned with the optical/IR continuum Several models have been proposed or considered to account for this alignment effect (e.g. see McCarthy (1993)). The most promising models are scattering of light from a hidden quasar by electrons or dust, star formation stimulated by the radio jet as it propagates outward from the nucleus and nebular continuum emission associated with strong emission line regions. Other scenarios involve

(*a*) inverse Compton scattering of *CMB* photons,

(*b*) enhancement of radio luminosity by interaction of the jet with an anisotropic parent galaxy, and

(*c*) alignment of the angular momentum of the nuclear black hole with an anisotropic protogalactic distribution.

Although polarisation results suggest that dust scattering is occurring, as yet no single model for the radio/optical alignment is satisfactory. Disentangling the various components (e.g. old stellar, young stellar, dust scattering, electron scattering, thermal bremstrahlung), although difficult, is crucial in establishing the ages of the objects and in understanding how they have been produced so soon after the Big Bang.

OBSERVABLE COMPONENT	CONSTITUENT	PROPERTIES
UV/Optical continuum	Stars	Stellar populations and their evolution
	Dust	Scattering properties
Optical emission lines	Ionised Gas $10^4 - 10^5$K	Kinematics, Mass Densities, Temperatures Ionisation, Ages
Optical absorption lines	Neutral hydrogen (Lyman α)	Kinematics, Mass, Column Density
IR-millimetre continuum	Dust	Temperature, Mass
IR-millimetre lines	Molecular Gas	Temperatures, Densities Masses, Stellar Evolution
Radio coninuum	Relativistic Plasma	Magnetic Fields Energetics, Pressures Particle acceleration, Jet collimation and propagation
	Thermal Plasma	Densities, magnetic fields, (Faraday Depolarisation)
Radio Absorption Lines	Neutral Hydrogen	Kinematics, Spin temperatures, Column Densities

13.2. *Giant Gas Halos*

One of the most remarkable features of distant radio galaxies is that they usually possess enormous luminous halos of ionised gas. The halos can extend to > 150 kpc, with velocity dispersions of typically $\sim$1000 km/s, and are clumpy. The dominant emission line in these spectra is Lyα. A gas halo has a characteristic mass of 10^9 M$_\odot$ and is typically composed of 10^{12} clouds, each having a size of about 40 light days, i.e. comparable with that of the solar system. These clouds may well delineate an early formation stage of individual stars. Other lines that are often present, but with fainter intensities ($< 10\%$ of Lyα) are C IV, HeII, C III]. Not only can the spatial distribution of the physical conditions within these halos be readily studied, but the halos also provide a unique tool for measuring the **kinematic** properties of material in the early Universe. The inner regions of the gas halos are often morphologically related to the radio jets, showing that the jets and gas often undergo vigorous interactions. On the larger scale the gas halos are rounder and these gigantic structures are likely to play an important role in the formation of massive galaxies.

What are the kinematic and physical properties of the gas and what part does the gas play in the galaxy formation and evolution? It is still unclear whether the gas halo is the product of accretion or cooling flow processes and to what extent and how the gas is eventually converted to stars. Large scale organised motion in the gas has been observed with velocities of a few hundred km/s

13.3. *Probes of Forming Proto-clusters*

Not only are radio galaxies prime targets for investigating the formation of massive galaxies, but they also provide one of the best potential method for studying the formation and evolution of clusters of galaxies in the early Universe. Distant radio galaxies are extremely clumpy. There is convincing evidence that distant radio galaxies are the massive central elliptical or cD galaxies at the centres of protoclusters of galaxies and the ultimate product of a hierarchical evolutionary process in which small clumps in the universe conglomerate to form bigger and bigger units.

14. The Future

During these lectures I have tried to touch on some of the highlights of extragalactic radio astronomy, emphasising the tools that are commonly used by radio astronomers. However, in 5 hours it is impossible to do justice to the enormous contribution that radio astronomy has made to our knowledge of the Universe and I have had to omit many important topics. Spectral line radio astronomy is the most glaring example.

The study of distant radio galaxies touched on in the preceding section illustrates the change in character of radio astronomy over the last 20 years. Previously extragalactic radio sources were a class of objects studied in the radio regime by radio astronomers, with some complementary observations generously supplied by optical astronomers. Since then this study has evolved from its technique-oriented beginnings into a problem-oriented field in which comparison of diagnostics from different spectral regimes are essential in unravelling the nature of the objects. Advances in CCD and infrared array technology, the launching of the HST and the availability of the Keck Telescope has had at least as large an impact on our understanding of the radio source phenomenon and the use of radio sources as cosmological probes as the *VLA*.

We have learned a great deal about extragalactic radio sources during the half century since they were discovered. Still there are several intriguing questions about extragalactic radio sources and their host galaxies that still remain to be tackled. These include the following:

(*a*) Why do luminous radio sources never occur within spiral galaxies, although spirals often contain active nuclei?

(*b*) What caused the dramatic evolution of the space density of radio galaxies and quasars during the first 2 billion years after the Big Bang? Why is the evolution of radio-loud and radio-quiet objects roughly similar?

(*c*) What is the process by which radio jets are produced and collimated by the central rotating black hole in the nucleus of their parent galaxies?

(*d*) Is there a connection between nuclear activity and the generation of stars in galaxy formation and evolution?

During the next 20 years I expect the multi-spectral trend in astronomy to be reinforced. Large gains in sensitivity can be expected in the millimetre (using large interferometric arrays) and in the infrared (exploiting larger arrays and adaptive optics). The hosts of extragalactic radio sources will undoubtedly continue to be prime targets of study for such arrays. Only a small number of galaxies seen optically in the Hubble Deep Field were detected by the deepest radio maps produced by the *VLA*. There are already plans for a next generation radio array, having an effective area of as much as 1 sq. km. and incorporating new technology such as multiplexing beam-switching which would be at least an order of magnitude more sensitive and likely detect almost all *normal* galaxies visible on the HST fields. One of the many exciting task for such an instrument would be to make a three-dimensional map of the structure of the Universe in neutral hydrogen out to high redshifts using the 21 cm line. Another relatively unexplored region of the spectrum which could well produce unexpected dividends during the next two decades is the lowest radio frequencies between 50 MHz and 20 MHz, below which the ionospheric window closes.

I thank Rudolf Le Poole and Philip Best for critically reading this manuscript.

REFERENCES

MCCARTHY, P.J., 1993 *ARA&A* **31**, 639

MLYOSHI, M. *et al.* 1995 *Nature* **373**, 127

REES, M.J. 1972 *Mon. Not. Roy. Astron. Soc.* **159**, 11

VAN DER LAAN, H., & PEROLA, G.C., 1969 *A&A* **3**, 468

GENERAL REFERENCES:

BRACEWELL, R., 1986 *The Fourier Transform and Its Applications, 2nd Edition, McGraw-Hill*

BURKE, B. F., GRAHAM-SMITH, F., 1997 *An Introduction to Radio Astronomy, Cambridge Univ Press (Cambridge)*

KRAUS J. D., 1966 *Radio Astronomy (McGraw Hill Text), 2nd Edition*

ROHLFS K., WILSON, T. L., 1996 *Tools of Radio Astronomy 2nd Edition, Springer Verlag*

THOMPSON, A., R., MORAN, J., M., AND SWENSON, G., W., 1994 *Interferometry and Synthesis in Radio Astronomy, Krieger Publishing Company*

VERSCHUUR, G., KELLERMANN, K. I., 1991 *Galactic and extra-galactic radio astronomy, 2nd Edition, Springer-Verlag*

Internet Resources for Radio Astronomy

By HEINZ ANDERNACH[1]

[1]Depto. de Astronomía, IFUG, Universidad de Guanajuato, Guanajuato, C.P. 36000, Mexico
Email: heinz@astro.ugto.mx

A subjective overview of Internet resources for radio-astronomical information is presented. Basic observing techniques and their implications for the interpretation of publicly available radio data are described, followed by a discussion of existing radio surveys, their level of optical identification, and nomenclature of radio sources. Various collections of source catalogues and databases for integrated radio source parameters are reviewed and compared, as well as the WWW interfaces to interrogate the current and ongoing large-area surveys. Links to radio observatories with archives of raw (uv-) data are presented, as well as services providing images, both of individual objects or extracts ("cut-outs") from large-scale surveys. While the emphasis is on radio continuum data, a brief list of sites providing spectral line data, and atomic or molecular information is included. The major radio telescopes and surveys under construction or planning are outlined. A summary is given of a search for previously unknown optically bright radio sources, as performed by the students as an exercise, using Internet resources only. Over 200 different links are mentioned and were verified, but despite the attempt to make this report up-to-date, it can only provide a snapshot of the situation as of mid-1998.

1. Introduction

Radio astronomy is now about 65 years old, but is far from retiring. Karl Jansky made the first detection of cosmic static in 1932, which he correctly identified with emission from our own Milky Way. A few years later Grote Reber made the first rough map of the northern sky at metre wavelengths, demonstrating the concentration of emission towards the Galactic Plane. During World War II the Sun was discovered as the second cosmic radio source. It was not until the late 1940s that the angular resolution was improved sufficiently to allow the first extragalactic sources be identified: Centaurus A (NGC 5128) and Virgo A (M 87). Interestingly, the term *radio astronomy* was first used only in 1948 (Haynes *et al.* (1996), p. 453, item 2). During the 1950s it became obvious that not only were relativistic electrons responsible for the emission, but also that radio galaxies were reservoirs of unprecedented amounts of energy. Even more impressive radio luminosities were derived once the quasars at ever-higher redshifts were found to be the counterparts of many radio sources. In the 1950s radio astronomers also began to map the distribution of neutral hydrogen in our Galaxy and find further evidence for its spiral structure.

Radio astronomy provided crucial observational data for cosmology from early on, initially based on counts of sources and on their (extremely isotropic) distribution on the sky, and since 1965 with the discovery and precise measurement of the cosmic microwave background (*CMB*). Only now are the deepest large-area surveys of discrete radio sources beginning to provide evidence for anisotropies in the source distribution, and such surveys continue to be vital for finding the most distant objects in the Universe and studying their physical environment as it was billions of years ago. If this were not enough, today's radio astronomy not only provides the highest angular resolution achieved in astronomy (fractions of a milliarcecond, or mas), but it also rivals the astrometric precision of optical astronomy (∼2 mas; Sovers *et al.* (1998)). The *relative* positions of neighbouring sources can even be measured to a precision of a few micro-arcsec (μas), which allows detection of relative motions of ∼20 μas per year. This is comparable to the angular "velocity" of the growth of human fingernails as seen from the distance of the Moon.

The "radio window" of the electromagnetic spectrum for observations from the ground is limited at lower frequencies mainly by the ionosphere, making observations below ∼30 MHz difficult near maxima of solar activity. While Reber was able to measure the emission from the Galactic Centre at 0.9 MHz from southern Tasmania during solar minimum in 1995, observations below about 1 MHz are generally only possible from space. The most complete knowledge of the radio sky has been achieved in the frequency range between 300 (λ=1 m) and 5000 MHz (λ=6 cm). At higher frequencies both meteorological conditions as well as receiver sensitivity become problems, and we have good data in this range only for the strongest sources in the sky. Beyond about 1000 GHz ($\lambda \leq 0.3$ mm) we reach the far infrared. Like the optical astronomers, who named their wavebands with certain letters (e.g. U, B, V, R, I, ...), radio astronomers took over the system introduced by radio engineers. Jargon like P-, L-, S-, C-, X-, U-, K- or Q-band can still be found in modern literature and stands for radio bands near 0.33, 1.4, 2.3, 4.9, 8.4, 15, 23 and 40 GHz (see Reference Data for Radio Engineers, 1975). The *CRAF* Handbook for Radio Astronomy gives a detailed description of the allocation and use of the various frequency bands allocated to astronomers (excluding the letter codes).

Unlike optical astronomers with their photographic plates, radio astronomers have used electronic equipment from the outset. Given that they had nothing like the "finding charts" used in optical astronomy to orient themselves in the radio sky, they were used to working with maps showing co-ordinates, which were rarely seen in optical research papers. Nevertheless, the display and description of radio maps in older literature shows some rare features. Probably due to the recording devices like analogue charts used up to the early 1980s, the terms "following" and "preceding", were frequently used rather than "east" and "west". Thus, e.g. "Nf" stands for "NE", or "Sp" for "SW". Sometimes the aspect ratio of radio maps was deliberately changed from being equi-angular, just to make the telescope beam appear round (Graham (1970)). Neither were radio astronomers at the forefront of archiving their results and offering publicly available databases. Happily all this has changed dramatically during the past decade, and the present report hopes to give a convincing taste of this.

As these lectures are aimed at professional astronomers, I do not discuss services explicitly dedicated to amateurs. I leave it here with a mention of the well-organised WWW site of the "Society of Amateur Radio Astronomers" (*SARA*; irsociety.com/sara.html). Note that in all addresses on the World-Wide-Web (WWW) mentioned here (the so-called " *URL*"s) I shall omit the leading characters "http://" unless other strings like "ftp://" need to be specified. The *URL*s listed were verified to be correct in May 1998.

2. Observing Techniques and Map Interpretation

Some theoretical background of radio radiation, interferometry and receiver technology has been given in G. Miley's contribution to these proceedings. In this section I shall briefly compare the advantages and limitations of both single dishes and radio interferometers, and mention some tools to overcome or alleviate some of their limitations. For a discussion of various types of radio telescopes see Christiansen & Högbom (1985). Here I limit myself to those items which appear most important to take into account when trying to make use of, and to interpret, radio maps drawn from public archives.

2.1. *Single Dishes versus Interferometers*

The basic relation between the angular resolution θ and the aperture (or diameter) D of a telescope is $\theta \approx \lambda/D$ radians, where λ is the wavelength of observation. For the radio domain λ is ∼10^6 times larger than in the optical, which would imply that one has to build a radio telescope a million times larger than an optical one to obtain the same angular resolution. In the early days of radio astronomy, when the observing equipment was based on radar dishes no longer required by the military after World War II, typical angular resolutions achieved were of the order of degrees. Consequently interferometry developed into an important and successful technique by the early 1950s (although arrays of dipoles, or Yagi antennas were used, rather than parabolic dishes, because the former were more suited to the metre-wave band used in the early experiments). Improved economic conditions and technological advance also permitted a significant increase in the size of single dishes. However, the sheer weight of the reflector and its support structure has set a practical limit of about 100 metres for fully steerable parabolic single dishes. Examples are the Effelsberg 100-m dish (`www.mpifr-bonn.mpg.de/effberg.html`) near Bad Münstereifel in Germany, completed in 1972, and the Green Bank Telescope (*GBT*; §8) in West Virginia, USA, to be completed in early 2000. The spherical 305-m antenna near Arecibo (Puerto Rico; `www.naic.edu/`) is the largest single dish available at present. However, it is not steerable; it is built in a natural and close-to-spherical depression in the ground, and has a limiting angular resolution of $\sim 1'$ at the highest operating frequency (8 GHz). Apart from increasing the dish size, one may also increase the observing frequency to improve the angular resolution. However, the D in the above formula is the aperture within which the antenna surface is accurate to better than $\sim 0.1\lambda$, and the technical limitations imply that the bigger the antenna, the less accurate the surface. In practice this means that a single dish never achieves a resolution of better than $\sim 10''$–$20''$, even at sub-mm wavelengths (cf. Fig. 6.8 in Rohlfs & Wilson (1996)).

Single dishes do not offer the possibility of instantaneous imaging as with interferometers by Fourier transform of the visibilities. Instead, several other methods of observation can be used with single dishes. If one is interested merely in integrated parameters (flux, polarisation, variability) of a (known) point source, one can use "cross-scans" centred on the source. If one is very sure about the size and location of the source (and its neighbourhood) one can even use "on–off" scans, i.e. point on the source for a while, then point to a neighbouring patch of "empty sky" for comparison. This is usually done using a pair of feeds and measuring their difference signal. However, to take a real image with a single dish it is necessary to raster the field of interest, by moving the telescope e.g. along right ascension (*RA*), back and forth, each scan shifted in declination (*DEC*) with respect to the other by an amount of no more than $\sim 40\%$ of the half-power beam width (*HPBW*) if the map is to be fully sampled. At decimetre wavelengths this has the advantage of being able to cover a much larger area than with a single "pointing" of an interferometer (unless the interferometer elements are very small, thus requiring large amounts of integration time). The biggest advantage of this raster method is that it allows the map size to be adjusted to the size of the source of interest, which can be several degrees in the case of large radio galaxies or supernova remnants (*SNRs*). Using this technique a single dish is capable of tracing (in principle) all large-scale features of very extended radio sources. One may say that it "samples" spatial frequencies in a range from the the map size down to the beam width. This depends critically on the way in which a baseline is fitted to the individual scans. The simplest way is to assume the absence of sources at the map edges, set the intensity level to zero there, and interpolate linearly between the two opposite edges of the map. A higher-order baseline is able to remove the variable atmospheric effects more efficiently, but it may also remove real underlying source structure. For example, the radio extent of a galaxy may be significantly underestimated if the map was made too small. Rastering the galaxy in two opposite directions may help finding emission close to the map edges using the so-called "basket-weaving" technique (Sieber *et al.* (1979)). Different methods in baseline subtraction and cut-offs in source size have led to two different versions of source catalogues (Becker *et al.* (1991) and Gregory & Condon (1991)), both drawn from the 4.85-GHz Green Bank survey. The fact that the surface density of these sources does not change towards the Galactic plane, while in the very similar southern *PMN* survey (Tasker *et al.* (1993)) it *does*, is entirely due to differences in the data reduction method (§3.3).

In contrast to single dishes, interferometers often have excellent angular resolution (again $\theta \approx \lambda/D$, but now D is the maximum distance between any pair of antennas in the array). However, the field of view is FOV$\approx \lambda/d$, where d is the size of an individual antenna. Thus, the smaller the individual antennas, the larger the field of view, but also the worse the sensitivity. Very large numbers of antennas increase the design cost for the array and the on-line correlator to process the signals from a large number of interferometer pairs. An additional aspect of interferometers is their reduced sensitivity to extended source components, which depends essentially on the smallest distance, say D_{min}, between two antennas in the interferometer array. This is often called the *minimum spacing* or *shortest baseline*. Roughly speaking, source components larger than $\sim \lambda/D_{\mathrm{min}}$ radians will be attenuated by more than 50% of their flux, and thus practically be lost. Figure 1 gives an extreme example of this, showing two images of the radio galaxy with the largest apparent size in the sky (10°). It is instructive to compare this with a high-frequency single-dish map in Junkes *et al.* (1993).

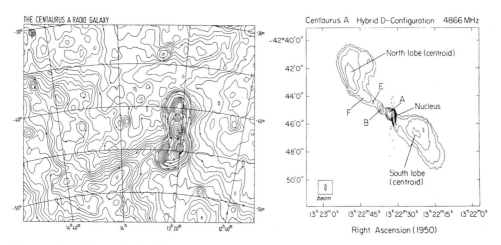

FIGURE 1.   Map of the Centaurus A region from the 408 MHz all-sky survey (Haslam *et al.* (1981), showing the full north-south extent of ∼10° of the radio structure and an emission feature due south east, apparently "connecting" Cen A with the plane of our Galaxy (see Combi *et al.* 1998). Right: A 1.4 GHz map obtained with the *VLA* (from Burns *et al.* 1983) showing the inner 10′ of Cen A. Without a single-dish map the full size of Cen A would not have been recognised.

The limitation in sensitivity for extended structure is even more severe for *Very Long Baseline Interferometry* (*VLBI*) which uses intercontinental baselines providing $\sim 10^{-3}$ arcsec (1 mas) resolution. The minimum baseline is often several hundred km, making the largest detectable component much smaller than an arcsec.

McKay & McKay (1998) created a WWW tool that simulates how radio interferometers work. This *Virtual Radio Interferometer* (VRI; `www.jb.man.ac.uk/~dm/vri/`) comes with the "VRI Guide" describing the basic concepts of radio interferometry. The applet simulates how the placement of the antennas affects the uv-coverage of a given array and illustrates the Fourier transform relationship between the accumulated radio visibilities and the resultant image.

The comparatively low angular resolution of single dish radio telescopes naturally suggests their use at relatively high frequencies. However, at centimetre wavelengths atmospheric effects (e.g. passing clouds) will introduce additional emission or absorption while scanning, leaving a stripy pattern along the scanning direction (so-called "scanning effects"). Rastering the same field along (DEC rather than RA, would lead to a pattern perpendicular to the first one. A comparison and subsequent combination of the two maps, either in the real or the Fourier plane, can efficiently suppress these patterns and lead to a sensitive map of the region (Emerson & Gräve (1988)).

A further efficient method to reduce atmospheric effects in single-dish radio maps is the so-called "multi-feed technique". The trick is to use pairs of feeds in the focal plane of a single dish. At any instant each feed receives the emission from a different part of the sky (their angular separation, or "beam throw", is usually 5–10 beam sizes). Since they largely overlap within the atmosphere, they are affected by virtually the same atmospheric effects, which then cancel out in the difference signal between the two feeds. The resulting map shows a positive and negative image of the same source, but displaced by the beam throw. This can then be converted to a single positive image as described in detail by Emerson *et al.* (1979). One limitation of the method is that source components larger than a few times the largest beam throw involved will be lost. The method has become so widely used that an entire symposium has been dedicated to it (Emerson & Payne (1995)).

From the above it should be clear that single dishes and interferometers actually complement each other well, and in order to map both the small- and large-scale structures of a source it may be necessary to use both. Various methods for combining single-dish and interferometer data have been devised, and examples of results can be found in Brinks & Shane (1984), Landecker *et al.* (1990), Joncas *et al.* (1992), Landecker *et al.* (1992), Normandeau *et al.* (1992) or Langer *et al.* (1995). The *Astronomical Image Processing System* (*AIPS*; `www.cv.nrao.edu/aips`), a widely used reduction package in radio astronomy, provides the task IMERG (cf. `www.cv.nrao.edu/aips/cook.html`) for this purpose. The package *Miriad* (`www.atnf.csiro.au/computing/software/miriad`) for reduction of radio interferometry data offers two programs (*immerge* and *mosmem*) to realise this combination of single dish and interferometer data (§2.3). The first one works in the Fourier plane and uses the single dish and mosaic data for the short and long spacings, respectively. The second one compares the single dish and mosaic images and finds the "Maximum Entropy" image consistent with both.

2.2. *Special Techniques in Radio Interferometry*

A multitude of "cosmetic treatments" of interferometer data have been developed, both for the "uv-" or visibility data and for the maps (i.e. before and after the Fourier transform), mostly resulting from 20 years of experience with the most versatile and sensitive radio interferometers currently available, the *Very Large Array* (*VLA*) and its more recent *VLBI* counterparts the *European VLBI Network* (*EVN*), and the *Very Large Baseline Array* (*VLBA*); see their WWW pages at `www.nrao.edu/vla/html/VLAhome.shtml`, `www.nfra.nl/jive/evn/evn.html`, and `www.nrao.edu/vlba/html/VLBA.html`. The volumes edited by Perley *et al.* (1989), Cornwell & Perley (1991), and Zensus *et al.* (1995) give an excellent introduction to these effects, the procedures for treating them, as well as their limitations. The more prominent topics are bandwidth and time-average smearing, aliasing, tapering, uv-filtering, (*CLEAN*ing, self-calibration, spectral-line imaging, wide-field imaging, multi-frequency synthesis, etc.

2.3. *Mosaicing*

One way to extend the field of view of interferometers is to take "snapshots" of several individual fields with adjacent pointing centres (or *phase centres*) spaced by no further than about one (and preferably half a) "primary beam", i.e. the *HPBW* of the individual array element. For sources larger than the primary beam of the single interferometer elements the method recovers interferometer spacings down to about half a dish diameter shorter than those directly measured, while for sources that fit into the primary beam mosaicing (also spelled "mosaicking") will recover spacings down to half the dish diameter (Cornwell (1988), or Cornwell (1989)). The data corresponding to shorter spacings can be taken either from other single-dish observations, or from the array itself, using it in a single-dish mode. The "Berkeley Illinois Maryland Association" (*BIMA*; `bima.astro.umd.edu/bima/`) has developed a *homogeneous array* capability, which is the central design issue for the planned *NRAO* Millimeter Array (*MMA*; `www.mma.nrao.edu/`). The strategy involves mosaic observations with the *BIMA* compact array during a normal 6–8 hour track, coupled with single-antenna observations with all array antennas mapping the same extended field (see Pound *et al.* (1997) or `bima.astro.umd.edu:80/bima/memo/memo57.ps`).

Approximately 15% of the observing time on the *Australia Telescope Compact Array* (*ATCA*; `www.narrabri.atnf.csiro.au/`) is spent on observing mosaics. A new pointing centre may be observed every 25 seconds, with only a few seconds of this time consumed by slewing and other overheads. The largest mosaic produced on the *ATCA* by 1997 is a 1344 pointing-centre spectral-line observation of the Large Magellanic Cloud. Joint imaging and deconvolution of these data produced a $1997 \times 2230 \times 120$ pixel cube (see `www.atnf.csiro.au/research/lmc_h1/`). Mosaicing is heavily used in the current large-scale radio surveys like *NVSS*, *FIRST*, and *WENSS* (§3.7).

2.4. *Map Interpretation*

The *dynamic range* of a map is usually defined as the ratio of the peak brightness to that of the "lowest reliable brightness level", or alternatively to that of the rms noise of a source-free region of the image. For both interferometers and single dishes the dynamic range is often limited by sidelobes occurring near strong sources, either due to limited uv-coverage, and/or as part of the diffraction pattern of the antenna. Sometimes the dynamic range, but more often the ratio between the peak brightness of the sidelobe and the peak brightness of the source, is given in dB, this being ten times the decimal logarithm of the ratio. In interferometer maps these sidelobes can usually be reduced using the (*CLEAN* method, although more sophisticated methods are required for the strongest sources (cf. Noordam & de Bruyn (1982), Perley (1989)), for which dynamic ranges of up to 5×10^5 can be achieved (de Bruyn & Sijbring (1993)). For an Alt-Az single dish the sidelobe pattern rotates with time on the sky, so a simple average of maps rastered at different times can reduce the sidelobe level. But again, to achieve dynamic ranges of better than a few thousand the individual scans have to be corrected independently before they can be averaged (Klein & Mack (1995)).

Confusion occurs when there is more than one source in the telescope beam. For a beam area Ω_b, the *confusion limit* S_c is the flux density at which this happens as one considers fainter and fainter sources. For an integral source count N(S), i.e. the number of sources per sterad brighter than flux density S, the number of sources in a telescope beam Ω_b is Ω_b N(S). S_c is then given by Ω_b N(S_c) ≈ 1. A radio survey is said to be *confusion-limited* if the expected minimum detectable flux density S_{min} is lower than S_c. Clearly, the confusion limit decreases with increasing observing frequency and with smaller telescope beamwidth. Apart from estimating the confusion limit theoretically from source counts obtained with a telescope of much lower confusion level (see Condon (1974)), one can also derive the confusion limit *empirically* by subsequent weighted averaging of N maps with (comparable) noise level σ_i, and with each of them *not* confusion-limited. The weight of each map should be proportional to σ_i^{-2}. In the absence of confusion, the *expected* noise, $\sigma_{N,exp}$, of the average map should then be

$$\sigma_{N,exp} = \left(\sum_{i=1}^{N} \sigma_i^{-2} \right)^{-1/2}$$

If this is confirmed by experiment, we can say that the "confusion noise", σ_c, is negligible, or at least that $\sigma_c \ll \sigma_N$. However, if σ_N approaches a saturation limit with increasing N, then σ_c, can be estimated according to $\sigma_c^2 = \sigma_{obs}^2 - \sigma_{N,exp}^2$. As an example, the confusion limit of a 30-m dish at 1.5 GHz (λ=20 cm) and a beam width of *HPBW*=34' is ~400 mJy. For a 100-m telescope at 2.7, 5 and 10.7 GHz (λ=11 cm, 6 cm and 2.8 cm; *HPBW*=4.4', 2.5' and 1.2'), the confusion limits are ~2, 0.5, and $\lesssim$0.1 mJy. For the *VLA* D-array at 1.4 GHz (*HPBW*=50'') it is ~0.1 mJy. For radio interferometers the confusion noise is generally negligible owing to their high angular resolution, except for deep maps at low frequencies where confusion due to sidelobes becomes significant (e.g. for *WENSS* and *SUMSS*, see §3.7). Note the semantic difference between "confusion noise" and "confusion limit". They can be related by saying that in a confusion-limited survey, point sources can be reliably detected only above the confusion limit, or 2–3 times the confusion noise, while coherent extended structures can be reliably detected down to lower limits, e.g. by convolution of the map to lower angular resolution. There is virtually no confusion limit for polarised intensity, as the polarisation position angles of randomly distributed, faint background sources tend to cancel out any net polarisation (see Rohlfs & Wilson (1996), p. 216 for more details). Examples of confusion-limited surveys are the large-scale low frequency surveys e.g. at 408 MHz (Haslam *et al.* (1982)), at 34.5 MHz (Dwarakanath & Udaya Shankar (1990)), and at 1.4 GHz (Condon & Broderick (1986a)). Of course, confusion becomes even more severe in crowded areas like the Galactic plane (Kassim (1988)).

When estimating the error in flux density of sources (or their significance) several factors have to be taken into account. The error in absolute calibration, Δ_{cal}, depends on the accuracy of the adopted flux density scale and is usually of the order of a few per cent. Suitable absolute calibration sources for single-dish observations are listed in Baars *et al.* (1977) and Ott *et al.* (1994) for intermediate frequencies, and in Rees (1990a) for low frequencies. Note that for the southern hemisphere older flux scales are still in use, e.g. Wills (1975). Lists of calibrator sources for intermediate-resolution interferometric observations (such as the *VLA*) can be found at the *URL* www.nrao.edu/~gtaylor/calib.html, and those for very-high resolution observations (such as the *VLBA*) at magnolia.nrao.edu/vlba_calib/vlbaCalib.txt. When comparing different source lists it is important to note that, especially at frequencies below ~400 MHz, there are still different "flux scales" being used which may differ by $\gtrsim$10%, and even more below ~100 MHz. The "zero-level error" is important mainly for single-dish maps and is given by $\Delta_o = m\sigma/\sqrt{n}$, where m is the number of beam areas contained in the source integration area, n is the number of beam areas in the area of noise determination, and σ is the noise level determined in regions "free of emission" (and includes contributions from the receiver, the atmosphere, and confusion). The error due to noise in the integration area is $\Delta_\sigma = \sigma\sqrt{m}$. The three errors combine to give a total flux density error of (see Klein & Emerson (1981)) $\Delta S = \Delta_{cal} + \sqrt{\Delta_o^2 + \Delta_\sigma^2}$. Clearly, the relative error grows with the extent of a source. This also implies that the upper limit to the flux density of a non-detected source depends on the size assumed for it: while a point source of ten times the noise level will clearly be detected, a source of the same flux, but extending over many antenna beams may well remain undetected. In interferometer observations the non-zero size of the shortest baseline limits the sensitivity to extended sources. At frequencies $\gtrsim 10\,\mathrm{GHz}$ the atmospheric absorption starts to become important, and the measured flux S will depend on elevation ϵ approximately according to $S=S_o \exp(-\tau \csc \epsilon)$, where S_o is the extra-atmospheric flux density, and τ the optical depth of the atmosphere. E.g., at 10.7 GHz and at sea level, typical values of τ are 0.05–0.10, i.e. 5–10% of the flux is absorbed even when pointing at the zenith. The values of τ increase with frequency, but decrease with altitude of the observatory. Uncertainties in the zenith-distance dependence may well dominate other sources of error above $\sim 50\,\mathrm{GHz}$.

When estimating flux densities from interferometer maps, the maps should have been corrected for the polar diagram (or "primary beam") of the individual antennas, which implies a decreasing sensitivity with increasing distance from the pointing direction. This so-called "primary-beam correction" divides the map by the attenuation factor at each map point and thus raises both the intensity of sources, and the map noise, with increasing distance from the phase centre. Some older source catalogues, mainly obtained with the *Westerbork Synthesis Radio Telescope* (*WSRT*; e.g. Oort & van Langevelde (1987), or Righetti *et al.* (1988)) give both the (uncorrected) "map flux" and the (primary-beam corrected) "sky flux". The increasing uncertainty of the exact primary beam shape with distance from the phase centre may dominate the flux density error on the periphery of the field of view.

Care should be taken in the interpretation of structural source parameters in catalogues. Some catalogues list the "map-fitted" source size, θ_m, as drawn directly from a Gaussian fit of the map. Others quote the "deconvolved" or "intrinsic" source size, θ_s. All of these are model-dependent and usually assume both the source and the telescope beam to be Gaussian (with full-width at half maximum, FWHM=θ_b), in which case we have $\theta_b^2 + \theta_s^2 = \theta_m^2$. Values of "0.0" in the size column of catalogues are often found for "unresolved" sources. Rather than zero, the intrinsic size is smaller than a certain fraction of the telescope beam width. The fraction decreases with increasing signal-to-noise (S/N) ratio of the source. The estimation of errors in the structure parameters derived from 2-dimensional radio maps is discussed in Condon (1997). Sometimes flux densities are quoted which are smaller than the error, or even negative (e.g. Dressel & Condon (1978), and Klein *et al.* (1996)). These should actually be converted to, and interpreted as *upper limits* to the flux density.

2.5. *Intercomparison of Different Observations and Pitfalls*

Two main emission mechanisms are at work in radio sources (e.g. Pacholczyk (1970)). The *non-thermal* synchrotron emission of relativistic electrons gyrating in a magnetic field is responsible for supernova remnants, the jets and lobes of radio galaxies and much of the diffuse emission in spiral galaxies (including ours) and their haloes. The *thermal* free-free or *bremsstrahlung* of an ionised gas cloud dominates e.g. in H II regions, planetary nebulae, and in spiral galaxies at high radio frequencies. In addition, individual stars may show "magneto-bremsstrahlung", which is synchrotron emission from either mildly relativistic electrons ("gyrosynchrotron" emission) or from less relativistic electrons ("cyclotron" or "gyroresonance" emission). The historical confirmation of synchrotron radiation came from the detection of its polarisation. In contrast, thermal radiation is unpolarised, and characterised by a very different spectral shape than that of synchrotron radiation. Thus, in order to distinguish between these mechanisms, multi-frequency comparisons are needed. This is trivial for unresolved sources, but for extended sources care has to be taken to include the entire emission, i.e. *integrated* over the source area. Peak fluxes or fluxes from high-resolution interferometric observations will usually underestimate their total flux. Very-low frequency observations may overestimate the flux by picking up radiation from neighbouring (or "blending") sources within their wide telescope beams. Compilations of integrated spectra of large numbers of extragalactic sources have been prepared e.g. by Kühr *et al.* (1979), Herbig & Readhead (1992), and Bursov *et al.* (1997) (see `cats.sao.ru/cats_spectra.html`).

An important diagnostic of the energy transfer within radio sources is a two-dimensional comparison of maps observed at different frequencies. Ideally, with many such frequencies, a spectral fit can be made at each resolution element across the source and parameters like the relativistic electron density and radiation lifetime, magnetic field strength, separation of thermal and non-thermal contribution, etc. can be estimated (cf. Klein *et al.* (1989) or Katz-Stone & Rudnick (1994)). However, care must be taken that the observing instruments at the different frequencies were sensitive to the same range of "spatial frequencies" present in the source. Thus interferometer data which are to be compared with single-dish data should be sensitive to components comparable to the entire size of the source. The *VLA* has a set of antenna configurations with different baseline lengths that can be matched to a subset of observing frequencies in order to record a similar set of spatial frequencies at widely different wavelengths – these are called "scaled arrays". For example, the B-configuration at 1.4 GHz and the C-configuration at 4.8 GHz form one such pair of arrays. Recent examples of such comparisons for very extended radio galaxies can be found in Mack *et al.* (1997) or Sijbring & de Bruyn (1998). Maps of the spectral indices of Galactic radio emission between 408 and 1420 MHz have even been prepared for the entire northern sky (Reich & Reich (1988)). Here the major limitation is the uncertainty in the absolute flux calibration.

2.6. *Linear Polarisation of Radio Emission*

As explained in G. Miley's lectures for this winter school, the linear polarisation characteristics of radio emission give us information about the magneto-ionic medium, both within the emitting source *and* along the line of sight between the source and the telescope. The plane of polarisation (the "polarisation position angle") will rotate while passing through such media, and the fraction of polarisation (or "polarisation percentage") will be reduced. This "depolarisation" may occur due to cancellation of different polarisation vectors within the antenna beam, or due to destructive addition of waves having passed through different amounts of this "Faraday" rotation of the plane of polarisation, or also due to significant rotation of polarisation vectors across the bandwidth for sources of high rotation measure (RM). More detailed discussions of the various effects affecting polarised radio radiation can be found in Pacholczyk (1970, 1977), Gardner *et al.* (1966), Burn (1966), and Cioffi & Jones (1980).

During the reduction of polarisation maps, it is important to estimate the ionospheric contribution to the Faraday rotation, which increases in importance at lower frequencies, and may show large variations at sunrise or sunset. Methods to correct for the ionospheric rotation depend on model assumptions and are not straightforward. E.g., within the *AIPS* package the "Sunspot" model may be used in the task FARAD. It relies on the mean monthly sunspot number as input, available from the US National Geophysical Data Centre at `www.ngdc.noaa.gov/stp/stp.html`. The actual numbers are in files available from `ftp://ftp.ngdc.noaa.gov/STP/SOLAR_DATA/SUNSPOT_NUMBERS/` (one per year: filenames are year numbers). Ionospheric data have been collected at Boulder, Colorado, up to 1990 and are distributed with the *AIPS* software, mainly to be used with *VLA* observations. Starting from 1990, a dual-frequency *GPS* receiver at the *VLA* site has been used to estimate ionospheric conditions, but the data are not yet available (contact `cflatter@nrao.edu`). Raw *GPS* data are available from `ftp://bodhi.jpl.nasa.gov/pub/pro/y1998/` and from `ftp://cors.ngs.noaa.gov/rinex/`. The *AIPS* task `GPSDL` for conversion to total electron content (*TEC*) and rotation measure (*RM*) is being adapted to work with these data.

A comparison of polarisation maps at different frequencies allows one to derive two-dimensional maps of *RM* and depolarisation (*DP*, the ratio of polarisation percentages between two frequencies). This requires the maps to be sensitive to the same range of spatial frequencies. Generally such comparisons will be meaningful only if the polarisation angle varies linearly with λ^2, as it indeed does when using sufficiently high resolution (e.g. Dreher *et al.* (1987)). The λ^2 law may be used to extrapolate the electric field vector of the radiation to $\lambda = 0$. This direction is called the "intrinsic" or "zero-wavelength" polarisation angle (χ_o), and the direction of the homogeneous component of the magnetic field at this position is then perpendicular to χ_o (for optically thin relativistic plasmas). Even then a careful analysis has to be made as to which part of *RM* and *DP* is intrinsic to the source, which is due to a "cocoon" or intracluster medium surrounding the source, and which is due to our own Galaxy. The usual method to estimate the latter contribution is to average the integrated *RM* of the five or ten extragalactic radio sources nearest in position to the source being studied. Surprisingly, the most complete compilations of *RM* values of extragalactic radio sources date back many years (Tabara & Inoue (1980), Simard-Normandin *et al.* (1981), or Broten *et al.* (1988)).

An example of an overinterpretation of these older low-resolution polarisation data is the recent claim (Nodland & Ralston (1997)) that the Universe shows a birefringence for polarised radiation, i.e. a rotation of the polarisation angle not due to any known physical law, and proportional to the cosmological distance of the objects emitting linearly polarised radiation (i.e. radio galaxies and quasars). The analysis was based on 20-year old low-resolution data for integrated linear polarisation (Clarke *et al.* (1980)), and the finding was that the difference angle between the intrinsic ($\lambda=0$) polarisation angle and the major axis of the radio structure of the chosen radio galaxies was increasing with redshift. However, it is now known that the distribution of polarisation angles at the smallest angular scales is very complex, so that the integrated polarisation angle may have little or no relation with the exact orientation of the radio source axis. Although the claim of birefringence has been contested by radio astronomers (Wardle *et al.* (1997)), and more than a handful of contributions about the issue have appeared on the *LANL/SISSA* preprint server (`astro-ph/9704197, 9704263, 9704285, 9705142, 9705243, 9706126, 9707326, 9708114`) the original authors continue to defend and refine their statistical methods (`astro-ph/9803164`). Surprisingly, these articles neither explicitly list the data actually used, nor do they discuss their quality or their appropriateness for the problem (cf. the comments in sect. 7.2 of Trimble & McFadden (1998)).

2.7. *Cross-Identification Strategies*

While the nature of the radio emission can be inferred from the spectral and polarisation characteristics, physical parameters can be derived only if the distance to the source is known. This requires identification of the source with an optical object (or an IR source for very high redshift objects) so that an optical spectrum may be taken and the redshift determined. By adopting a cosmological model, the distance of extragalactic objects can then be inferred. For sources in our own Galaxy kinematical models of spiral structure can be used to estimate the distance from the radial velocity, even without optical information, e.g. using the H I line (§6.4). More indirect estimates can also be used, e.g. emission measures for pulsars, apparent sizes for H I clouds, etc.

The strategies for optical identification of extragalactic radio sources are very varied. The easiest case is when the radio position falls within the optical extent of a galaxy. Also, a detailed radio map of an extended radio galaxy usually suggests the position of the most likely optical counterpart from the symmetry of the radio source. Most often two extended radio lobes straddle a point-like radio core which coincides with the optical object. However, various types of asymmetries may complicate the relation between radio morphology and location of the parent galaxy (see e.g. Figs. 6 and 7 of Miley (1980)). These may be wiggles due to precession of the radio jet axis, or bends due to the movement of the radio galaxy through an intracluster medium (see www.jb.man.ac.uk/atlas/icon.html for a fine collection of real maps). For fainter and less extended sources the literature contains many different methods to determine the likelihood of a radio-optical association (Notni & Fröhlich (1975), Richter (1975), Padrielli & Conway (1977), de Ruiter *et al.* (1977)). The last of these papers proposes the dimensionless variable $r = \sqrt{(\Delta\alpha/\sigma_\alpha)^2 + (\Delta\delta/\sigma_\delta)^2}$ where $\Delta\alpha$ and $\Delta\delta$ are the positional differences between radio and optical position, and σ_α and σ_δ are the combined radio and optical positional errors in RA (α) and DEC (δ), respectively. The likelihood ratio, LR, between the probability for a real association and that of a chance coincidence is then $LR(r) = (1/ 2\lambda) \, exp\left(r^2 (2\lambda - 1)/2\right)$, where $\lambda = \pi \, \sigma_\alpha \, \sigma_\delta \, \rho_{opt}$, with ρ_{opt} being the density of optical objects. The value of ρ_{opt} will depend on the Galactic latitude and the magnitude limit of the optical image. Usually, for small sources, $LR \gtrsim 2$ is regarded as sufficient to accept the identification, although the exact threshold is a matter of "taste". A method that also takes into account also the extent of the radio sources, and those of the sources to be compared with (be it at optical or other wavelengths), has been described in Hacking *et al.* (1989)). A further generalisation to elliptical error boxes, inclined at any position angle (like those of the IRAS satellite), is discussed in Condon *et al.* (1995).

A very crude assessment of the number of chance coincidences from two random sets of N_1 and N_2 sources distributed all over the sky is $N_{cc} = N_1 \, N_2 \, \theta^2 / 4$ chance pairs within an angular separation of less than θ (in radians). In practice the decision on the maximum θ acceptable for a true association can be drawn from a histogram of the number of pairs within θ, as a function of θ. If there is any correlation between the two sets of objects, the histogram should have a more or less pronounced and narrow peak of true coincidences at small θ, then fall off with increasing θ up to a minimum at θ_{crit}, before rising again proportional to θ^2 due to pure chance coincidences. The maximum acceptable θ is then usually chosen near θ_{crit} (cf. Bischof & Becker (1997) or Boller *et al.* (1998)). At very faint (sub-mJy) flux levels, radio sources tend to be small ($\ll 10''$), so that there is virtually no doubt about the optical counterpart, although very deep optical images, preferably from the Hubble Space Telescope (HST), are needed to detect them (Fomalont *et al.* (1997)).

However, the radio morphology of extended radio galaxies may be such that only the two outer "hot spots" are detected without any trace of a connection between them. In such a case only a more sensitive radio map will reveal the position of the true optical counterpart, by detecting either the radio core between these hot spots, or some "radio trails" stretching from the lobes towards the parent galaxy. The paradigm is that radio galaxies are generally ellipticals, while spirals only show weak radio emission dominated by the disk, but with occasional contributions from low-power active nuclei (AGN).

Recently an unusual exception has been discovered: a disk galaxy hosting a large double-lobed radio source (Figure 2), almost perpendicular to its disk, and several times the optical galaxy size (Ledlow *et al.* (1998)).

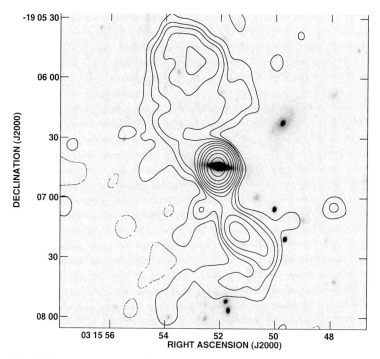

FIGURE 2. *VLA* contours at 1.5 GHz of B1313−192 in the galaxy cluster A 428, overlaid on an R-band image. The radio source extends $\approx 100\,h_{75}^{-1}$kpc north and south of the host galaxy, which is disk-like rather than elliptical (from Ledlow *et al.* 1998, courtesy M. Ledlow).

An approach to semi-automated optical identification of radio sources using the Digitized Sky Survey is described in Haigh *et al.* (1997). However, Figure 3 shows one of the more complicated examples from this paper. Note also that the concentric contours near the centre of the radio source encircle a *local minimum*, and not a maximum. To avoid such ambiguities some software packages (e.g. "NOD2", Haslam (1974)) produce arrowed contours indicating the direction of the local gradient in the map.

Morphological considerations can sometimes lead to interesting misinterpretations. A linear feature detected in a Galactic plane survey with the Effelsberg 100-m dish had been interpreted as probably being an optically obscured radio galaxy behind our Galaxy (Seiradakis *et al.* (1985)). It was not until five years later (Landecker *et al.* (1990)) that interferometer maps taken with the Dominion Radio Astrophysical Observatory (*DRAO*; www.drao.nrc.ca) revealed that the linear feature was merely the straighter part of the shell of a weak and extended supernova remnant (G 65.1+0.6).

One of the most difficult classes of source to identify optically are the so-called "relic" radio sources, typically occurring in clusters of galaxies, with a very steep radio continuum spectrum, and without clear traces of association with any optical galaxy in their host cluster. Examples can be found in Giovannini *et al.* (1991), Feretti *et al.* (1997), or Röttgering *et al.* (1997). See astro-ph/9805367 and 9902105 for ideas on their origin.

Generally source catalogues are produced only for detections above the 3–$5\,\sigma$ level. However, Lewis (1995) and Moran *et al.* (1996) have shown that a cross-identification between catalogues at different wavelengths allows the "detection" of real sources even down to the 2σ level.

3. Radio Continuum Surveys

3.1. *Historical Evolution*

Our own Galaxy and the Sun were the first cosmic radio sources to be detected due the work of K. Jansky, G. Reber, G. Southworth, and J. Hey in the 1930s and 1940s. Several other regions in the sky had been found to emit strong discrete radio emission, but in these early days the angular resolution of radio telescopes was far too poor to uniquely identify the sources with something "known", i.e. with an optical object, as there were simply too many of the latter within the error box of the radio position. It was not until 1949 that Bolton *et al.* (1949) identified three further sources with optical objects. They associated Tau A with the "Crab Nebula", a supernova remnant in our Galaxy, Vir A with M 87, the central galaxy in the Virgo cluster, and Cen A with NGC 5128, a bright nearby elliptical galaxy with a prominent dust lane. By 1955, with the publication of the "2C" survey (Shakeshaft *et al.* (1955)) the majority of radio sources were still thought to be Galactic stars, albeit faint ones, since no correlation with bright stars was observed. However, in the previous year, the bright radio source Cyg A had been identified with a very faint ($\sim16^m$) and distant (z=0.057) optical galaxy (Baade & Minkowski (1954)).

Excellent accounts of early radio astronomy can be found in the volumes by Hey (1971, 1973), Graham-Smith (1974), Edge & Mulkay (1976), Sullivan III (1982), Sullivan III (1984), Kellermann & Sheets (1984), Robertson (1992), and in Haynes *et al.* (1996), the latter two describing the Australian point of view. The growth in the number of discrete source lists from 1946 to the late 1960s is given in Appendix 4 of Pacholczyk (1970).

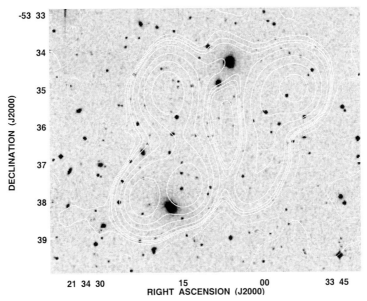

FIGURE 3. 408 MHz contours from the Molonglo Observatory Synthesis Telescope (*MOST*) of a complex radio source in the galaxy cluster A 3785, overlaid on the Digitized Sky Survey. The source is a superposition of two wide-angle tailed *WAT*) sources associated with the two brightest galaxies in the image, as confirmed by higher-resolution *ATCA* maps (from Haigh *et al.* 1997, ctsy. A. Haigh)

A comprehensive list of radio source catalogues published from 1950 to 1975 has been compiled by Collins 1977. Many of the major source surveys carried out during the late 1970s and early 1980s (6C, UTR, TXS, B2, MRC, WSRT, GB, PKS, S1–S5) are described in Jauncey (1977). The proceedings volume by Condon & Lockman (1990) includes descriptions of several large-scale surveys in the continuum, H I, recombination lines, and searches for pulsars and variable sources.

3.2. *Radio Source Nomenclature: The Good, the Bad and the Ugly*

As an aside, Appendix 4 of Pacholczyk (1970) explains the difficulty (and liberty!) with which radio sources were designated originally. In the early 1950s, with only a few dozen radio sources known, one could still afford to name them after the constellation in which they were located followed by an upper case letter in alphabetic sequence, to distinguish between sources in the same constellation. This method was abandoned before even a couple of sources received the letter B. Curiously, even in 1991, the source PKS B1343−601 was suggested *a posteriori* to be named "Cen B" as it is the second strongest source in Centaurus (McAdam (1991)). Apparently the name has been adopted (see Tashiro *et al.* (1998)). Sequential numbers like 3C NNN were used in the late 1950s and early 1960s, sorting the sources in *RA* (of a given equinox, like B1950 at that time and until rather recently). But when the numbers exceeded a few thousand, with the 4C survey (Pilkington & Scott (1965) and Gower *et al.* (1967)) a naming like 4C DD.NN was introduced, where DD indicates the declination strip in which the source was detected and NN is a sequence number increasing with *RA* of the source, thus giving a rough idea of the source location (although the total number of sources in one strip obviously depends on the declination). A real breakthrough in naming was made with the Parkes (PKS) catalogue (Bolton *et al.* (1964)) where the "*IAU* convention" of co-ordinate-based names was introduced. Thus e.g. a name PKS 1234−239 would imply that the source lies in the range $RA= 12^h34^m...12^h35^m$ and $DEC=-23° 54'...-24° 0'$. Note that to construct the source name the exact position of the source is truncated, not rounded. An even number of digits for *RA* or *DEC* would indicate integer hours, minutes or seconds (respectively of time and arc), while odd numbers of digits would indicate the truncated (i.e. downward-rounded) tenth of the unit of the preceding pair of digits. Since the co-ordinates are equinox-dependent and virtually all previous co-ordinate-based names were based on B1950, it has become obligatory to precede the co-ordinate-based name with the letter J if they are based on the J2000 equinox. Thus e.g. PKS B0000−506 is the same as PKS J0002−5024, and the additional digit in *DEC* merely reflects the need for more precision nowadays. Vice versa, the *lack* of a fourth digit in the B1950 name reflects the recommendation to never change a *name* of a source even if its position becomes better known later. The current sensitivity of surveys and the resulting surface density of sources implies much longer names to be unique. Examples are *NVSS* B102023+252903 or *FIRST* J102310.0+251352 (which are actually the same source!). Authors should follow *IAU* recommendations for object names (§8.8). The origin of existing names, their acronyms and recommended formats can be traced with the on-line "Dictionary of Nomenclature of Celestial Objects" (vizier.u-strasbg.fr/cgi-bin/Dic; Lortet *et al.* (1994)). A query for the word "radio" (option "Related to words") will display the whole variety of naming systems used in radio astronomy, and will yield what is perhaps the most complete list of radio source literature available from a single WWW site. Authors of future radio source lists, and project leaders of large-scale surveys, are encouraged to consult the latter *URL* and register a suitable acronym for their survey well in advance of publication, so as to guarantee its uniqueness, which is important for its future recognition in public databases.

3.3. *Major Radio Surveys*

Radio surveys may be categorised into *imaging* and *discrete source* surveys. *Imaging surveys* were mostly done with single dishes and dedicated to mapping the extended emission of our Galaxy (e.g. Haslam *et al.* (1982), Dwarakanath & Udaya Shankar (1990)) or just the Galactic plane (Reich *et al.* (1984), Jonas *et al.* (1985)). Only some of them are useful for extracting lists of discrete sources (e.g. Reich *et al.* (1997)). The semi-automatic procedure of source extraction implies that the derived catalogues are usually limited to sources with a size of at most a few beamwidths of the survey. The highest-resolution radio imaging survey covering the full sky, and containing Galactic foreground emission on all scales, is still the 408 MHz survey (Haslam *et al.* (1982)) with $HPBW \sim 50'$. Four telescopes were used and it has taken 15 years from the first observations to its publication. The 22 MHz survey by Roger et al. (1999) took 30 years to publish, and gives a comprehensive list of low-frequency surveys of Galactic emission. A 1.4 GHz all-sky survey (see Reich (1982), Reich & Reich (1986) for the northern hemisphere) is being completed in the south with data from the 30-m dish at Instituto Argentino de Radioastronomía, Argentina.

The *discrete source surveys* may be done either with interferometers or with single dishes. Except for the most recent surveys (*FIRST*, *NVSS* and *WENSS*, see §3.7) the interferometer surveys tend to cover only small parts of the sky, typically a single field of view of the array, but often with very high sensitivities reaching a few μJy in the deepest surveys. The source catalogues extracted from discrete source surveys with single dishes depend somewhat on the detection algorithm used to find sources from two-dimensional maps. There are examples where two different source catalogues were published, based on the same original maps. Both the "87GB" (Gregory & Condon (1991)) and "*BWE*" (Becker *et al.* (1991)) catalogues were drawn from the same 4.85 GHz maps (Condon *et al.* (1989)) obtained with the Green Bank 300-ft telescope. The authors of the two catalogues (published on 510 pages of the same volume of ApJS), arrived at 54 579 and 53 522 sources, respectively. While the 87GB gives the peak flux, size and orientation of the source, the *BWE* gives the integrated flux only, plus a spectral index between 1.4 and 4.85 GHz from a comparison with another catalogue. Thus, while being slightly different, both catalogues complement each other. The same happened in the southern hemisphere, using the same 4.85 GHz receiver on the Parkes 64-m antenna: the "*PMN*" (Griffith *et al.* (1994)) and "*PMNM*" catalogues (Gregory *et al.* (1994)) were constructed from the same underlying raw scan data, but using different source extraction algorithms, as well as imposing different limits in both signal-to-noise for catalogue source detection, and in the maximum source size. The larger size limit for sources listed in the *PMN* catalogue, as compared to the northern 87GB, becomes obvious in an all-sky plot of sources from both catalogues: the Galactic plane is visible only in the southern hemisphere (Tasker *et al.* (1993)), simply due to the large number of extended sources near the plane which have been discarded in the northern catalogues (Becker *et al.* (1991)). Baleisis *et al.* (1998) have also found a 2%–8% mismatch between 87GB and *PMN*. Eventually, a further coverage of the northern sky made in 1986 (not available as a separate paper) has been averaged with the 1987 maps (which were the basis for 87GB) to yield the more sensitive GB6 catalogue (Gregory *et al.* (1996)). Thus, a significant difference in source peak flux density between 87GB and GB6 may indicate variability, and Gregory *et al.* (1998) have indeed confirmed over 1400 variables.

If single-dish survey maps (or raster scans) are sufficiently large, they may be used to reveal the structure of Galactic foreground emission and discrete features like e.g. the "loops" or "spurs" embedded in this emission. These are thought to be nearby supernova remnants, an idea supported by additional evidence from X-rays (Egger & Aschenbach 1995) and older polarisation surveys (Salter (1983)). Surveys of the linear polarisation of Galactic emission will not be dealt with here. As pointed out by Salter & Brown (1988), an all-sky survey of linear polarisation, at a consistent resolution and frequency, is still badly needed. No major polarisation surveys have been published since the compendium of Brouw & Spoelstra (1976), except for small parts of the Galactic plane (Junkes *et al.* (1987)). This is analogous to a lack of recent surveys for discrete source polarisation (§2.6). Apart from helping to discern thermal from non-thermal features, polarisation maps have led to the discovery of surprising features which are not present in the total intensity maps (Wieringa *et al.* (1993b), Gray *et al.* (1998)). Although the *NVSS* (§3.7) is not suitable to map the Galactic foreground emission and its polarisation, it offers linear polarisation data for ∼2 million radio sources. Many thousands of them will have sufficient polarisation fractions to be followed up at other frequencies, and to study their Faraday rotation and depolarisation behaviour.

3.4. *Surveys from Low to High Frequencies: Coverage and Content*

There is no concise list of all radio surveys ever made. Purton & Durrell (1991) used 233 different articles on radio source surveys, published 1954–1991, to prepare a list of 386 distinct regions of sky covered by these surveys (`cats.sao.ru/doc/SURSEARCH.html`). While the source lists themselves were not available to these authors, the list was the basis for a software allowing queries to determine which surveys cover a given region of sky. A method to retrieve references to radio surveys by acronym has been mentioned in §3.2. In §4.1 a quantitative summary is given of what is available electronically.

In this section I shall present the "tip of the iceberg": in **Table 1**, I have listed the largest surveys of discrete radio sources which have led to source catalogues available in electronic form. The list is sorted by frequency band (col. 1), and the emphasis is on finder surveys with more than ∼800 sources *and* more than ∼0.3 sources/deg². However, some other surveys were included if they constitute a significant contribution to our knowledge of the source population at a given frequency, like e.g. re-observations of sources originally found at other frequencies. It is supposedly complete for source catalogues with ≳ 2000 entries, whereas below that limit a few source lists may be missing for not fulfilling the above criteria. Further columns give the acronym of the survey or observing instrument, the year(s) of publication, the approximate range of *RA* and *DEC* covered (or Galactic longitude l and latitude b for Galactic plane surveys), the angular resolution in arcmin, the approximate limiting flux density in mJy, the total number of sources listed in the catalogue, the average surface density of sources per square degree, and a reference number which is resolved into its "bibcode" in the Notes to the table. Three famous series of surveys are excluded from **Table 1**, as they are not contiguous large-area surveys, but are dedicated to many individual fields, either for Galactic or for cosmological studies (e.g. source counts at faint flux levels). These are the source lists from various individual pointings of the interferometers at *DRAO* Penticton (P), Westerbork (W) and the Cambridge One-Mile telescope (5C).

Both single-dish and interferometer surveys are included in **Table 1**. Interferometers usually provide much higher absolute positional accuracy, but there is one major interferometer survey (*TXS* at 365 MHz; Douglas *et al.* (1996), `utrao.as.utexas.edu/txs.html`), for which one fifth of its ∼67 000 catalogued source positions suffer from possible "lobe-shifts". These sources have a certain likelihood to be located at an alternative, but precisely determined position, about 1′ from the listed position. It is not clear *a priori* which of the two positions is the true one, but the ambiguity can usually be solved by comparison with other sufficiently high resolution maps (see Fig. B1 of Vessey & Green (1998)

Table 1. Major Surveys of Discrete Radio Sources †

Freq (MHz)	Name	Year of publ	RA(h) or l(d)	Decl(deg) or b(d)	HPBW (')	S_min (mJy)	N of objects	n/ sq.deg	Ref	Electr Status
10-25	UTR-2	78-95	0-24	> -13	25-60	10000	1754	0.2	54	A C
31	NEK	88	350<l<250	|b|<~2.5	13x 11	4000	703	0.7	51	A C
38	8C	90/95	0-24	> +60	4.5	1000	5859	1.7	1	A C n
80#	CUL1	73/95	0-24	-48,+35	3.7	2000	999	0.04	41	A C
80#	CUL2	75/95	0-24	-48,+35	3.7	2000	1748	0.06	42	A C
82	IPS	87	0-24	-10,+83	27x350	500	1789	0.08	52	C
151	6CI	85	0-24	> +80	4.5	200	1761	5.7	2	A C
151	6CII	88	8.5-17.5	+30,+51	4.5	200	8278	4.1	3	A C
151	6CIII	90	5.5-18.3	+48,+68	4.5	200	8749	4.5	4	A C
151	6CIV	91	0-24	+67,+82	4.5	200	5421	3.8	28	A C
151	6CVa	93	1.6- 6.2	+48,+68	4.5	~300	2229	3.0	39	A C
151	6CVb	93	17.3-20.4	+48,+68	4.5	~300	1229	2.6	39	A C
151	6CVI	93	22.6- 9.1	+30,+51	4.5	~300	6752	2.7	40	A C
151*	7CI	90	(10.5+41)	(6.5+45)	1.2	80	4723	9.7	21	C
151	7CII	95	15-19	+54,+76	1.2	~100	2702	6.5	49	A C n
151	7CIII	96	9-16	+20,+35	1.2	~150	5526	4.0	56	A C N
151	7C(G)	98	80<l<180	|b|<5.5	1.2	~100	6262	4.8	55	C n
160#	CUL3	77/95	0-24	-48,+35	1.9	1200	2045	0.08	43	A C
178	4C	65/67	0-24	-7,+80	~23.	2000	4844	0.2	57	A C N M
232	MIYUN	96	0-24	+30,+90	3.8	~100	34426	3.3	24	A C
325	WENSS	97/98	0-24	+30,+90	0.9	18	229420	~22.	58	C
327*	WSRT	91/93	5 fields	(+40,+72)	~1.0	3	4157	~50.	32	A C
327	WSRTGP	96	43<l<91	|b|<1.6	~1.0	~10	3984	~25.	30	A C n
365	UTRAO36	92	0-24	+31,+41	~0.1	250	3196	~2.	38	C
365	TXS	96	0-24	-35.5,+71.5	~0.1	250	66841	~2.	22	A C n
408	MRC	81/91	0-24	-85,+18.5	~3.	700	12141	0.5	6	A C N
408	B2	70-74	0-24	+24,+40	3 x10	250	9929	3.1	7	A C M
408	B3	85	0-24	+37,+47	3 x 5	100	13354	5.2	8	A C N
408	MC1	73	1-17	-22,-19	2.7	100	1545	2.3	9	A C M
408	MC4	76	0-18	-74,-62	2.7	130	1257	1.0	10	A C M
408	MDS2	84	5-23	-21,-20	2.8	60	799	2.7	11	C
611	NAIC	75	22-13	-3,+19	12.6	350	3122	0.6	12	C
608*	WSRT	91/93	sev.fields	(~40,~72)	0.5	3	1693	~50.	32	A C
1400	GB	72	7-16	+46,+52	10 x11	90	1086	2.0	13	C
1400	GB2	78	7-17	+32,+40	10 x11	90	2022	2.2	14	C
1400	WB92	92	0-24	-5,+82	10 x11	~150	31524	0.7	27	A C N
1400	GPSR	90	20<l<120	|b|<0.8	0.08	25	1992	8.9	33	A C
1400	NVSS34	98	0-24	-40,+90	0.9	2.0	1807317	~55.	60	C
1400	FIRST5	98	7.3,17.4	+22.2,57.6	0.1	1.0	382892	~90.	59	A C
1400	FIRST5	98	21.3,3.3	-11.5,+1.6	0.1	1.0	54537	~90.	59	A C
1408	RRF	90	357<l<95.5	|b|<4.0	9.4	98	884	1.1	29	A C
1420	RRF	98	95.5<l<240	-4<|b|<+5	9.4	80	1830	1.5	44	A C
1420*	PDF	98	B0112-46	r=1deg	0.1	0.1	1079	~340.	62	C
1400	ELAISR	98	3 fields	+32,+55	0.25	0.14	867	205.	61	C
1400	GPSR	92	350<l<40	|b|<1.8	0.08	25	1457	8.1	37	C
1500	VLANEP	94	17.4,18.5	63.6,70.4	0.25	0.5	2436	83.	47	A C n
2700	PKS	(90)	0-24	-90,+27	~8.	~50	8264	0.3	15	A C N M
2700	F3R	90	357<l<240	|b|< 5	4.3	40	6483	2.7	34	A C
3900	Z	89	0-24	0,+14	1.2x52	50	8503	1.7	16	A C
3900	Z2	91	0-24	0,+14	1.2x52	40	2944	0.6	5	A C
3900	RC	91-93	0-24	4.5,5.5	1.2x52	4	1189	3.2	26	C n
4775#	NAIC-GB	83	22.3-13	-3,+19	2.8	~20	2661	0.6	17	C
4760	GBdeep	86		~33	2.8	15	882	6.6	18	C
4850	MG1-4	86-91	var.	-0.5,+51	2.8	40	24180	1.5	20	C n
4850	87GB	91	0-24	0,+75	~3.5	25	54579	2.7	19	A C N
4850	BWE	91	0-24	0,+75	~3.5	25	53522	2.7	23	A C N
4850	GB6	96	0-24	0,+75	~3.5	18	75162	3.7	53	A C
4850	PMNM	94	0-24	-88,-37	4.9	25	15045	1.8	45	A C N
4850	PMN-S	94	0-24	-87.5,-37	4.2	20	23277	2.8	31a	A C N
4850	PMN-T	94	0-24	-29,-9.5	4.2	42	13363	2.0	31b	A C N
4850	PMN-E	95	0-24	-9.5,+10	4.2	40	11774	1.9	48	A C N
4850	PMN-Z	96	0-24	-37,-29	4.2	72	2400	1.1	50	A C N
4875	ADP79	79	357<l< 60	|b|<1	2.6	~120	1186	9.4	25	C
5000	HCS79	79	190<l< 40	|b|<2	4.1	260	915	1.1	46	A C
5000	GT	86	40<l<220	|b|<2	2.8	70	1274	1.8	35	C
5000	GPSR	94	350<l< 40	|b|<0.4	~0.07	3	1272	26.	36	A C

† A total of 66 surveys are listed with 3 058 035 entries altogether.
See the explanations and references in the Notes to this Table.

References and Notes to Table 1

1a	1995MNRAS.274..447Hales+	29	1990A&AS...83..539Reich W.+
1b	1990MNRAS.244..233Rees	30	1996ApJS..107..239Taylor+
2	1985MNRAS.217..717Baldwin+	31a	1994ApJS...91..111Wright+
3	1988MNRAS.234..919Hales+	31b	1994ApJS...90..179Griffith+
4	1990MNRAS.246..256Hales+	32	1993BICDS..43...17Wieringa +PhD Leiden
5	1991SoSAO..68...14Larionov+	33	1990ApJS...74..181Zoonematkermani+
6a	1991Obs...111...72Large+	34	1990A&AS...85..805Fuerst+
6b	1981MNRAS.194..693Large+	35	1986AJ.....92..371Gregory & Taylor
7a	1970A&AS....1..281Colla+	36	1994ApJS...91..347Becker+
7b	1972A&AS....7....1Colla+	37	1992ApJS...80..211Helfand+
7c	1973A&AS...11..291Colla+	38	1992ApJS...82....1Bozyan+
7d	1974A&AS...18..147Fanti+	39	1993MNRAS.262.1057Hales+
8	1985A&AS...59..255Ficarra+	40	1993MNRAS.263...25Hales+
9	1973AuJPA..28....1Davies+	41a	1973AuJPA..27....1Slee & Higgins
10	1976AuJPA..40....1Clarke+	41b	1995AuJPh..48..143Slee
11	1984PASAu...5..290White	42a	1975AuJPA..36....1Slee & Higgins
12	1975NAICR..45.....Durdin+	42b	1995AuJPh..48..143Slee
	NAIC Internal Report	43a	1977AuJPA..43....1Slee
13	1972AcA....22..227Maslowski	43b	1995AuJPh..48..143Slee
14	1978AcA....28..367Machalski	44	1997A&AS..126..413Reich, P.+
15	1991PASAu...9..170Otrupcek+Wright	45a	1994ApJS...90..173Gregory+
16	1989MIRpubl.......Amirkhanyan+	45b	1993AJ....106.1095Condon+
	MIR Publ., Moscow	46	1979AuJPA..48....1Haynes+
17	1983ApJS...51...67Lawrence+	47	1994ApJS...93..145Kollgaard+
18	1986A&AS...65..267Altschuler	48	1995ApJS...97..347Griffith+
19	1991ApJS...75.1011Gregory+Condon	49	1995A&AS..110..419Visser+
20a	1986ApJS...61....1Bennett+	50	1996ApJS..103..145Wright+
20b	1990ApJS...72..621Langston+	51	1988ApJS...68..715Kassim
20c	1990ApJS...74..129Griffith+	52	1987MNRAS.229..589Purvis+
20d	1991ApJS...75..801Griffith+	53	1996ApJS..103..427Gregory+
21	1990MNRAS.246..110McGilchrist+	54	1995Ap&SS.226..245Braude+ +older refs
22	1996AJ....111.1945Douglas+	55	1998MNRAS.294..607Vessey & Green D.A.
23	1991ApJS...75....1Becker+	56	1996MNRAS.282..779Waldram+
24	1997A&AS..121...59Zhang+	57a	1965MmRAS..69..183Pilkington & Scott
25	1979A&AS...35...23Altenhoff+	57b	1967MmRAS..71...49Gower+
26a	1991A&AS...87....1Parijskij+	58	1997A&AS..124..259Rengelink+ and WWW
26b	1992A&AS...96..583Parijskij+	59	1997ApJ..475..479White+ and WWW
26c	1993A&AS...98..391Parijskij+	60	1998AJ....115.1693Condon+ and WWW
27	1992ApJS...79..331White & Becker	61	1998MNRAS.302..222Ciliegi+
28	1991MNRAS.251...46Hales+	62	1998MNRAS.296..839Hopkins+ +PhD Sydney

Notes to Table 1. #: not a finder survey, but re-observations of previously catalogued sources. *: circular field, central co-ordinates and radius are given. The catalogue electronic status is coded as follows: A: available from *ADC/CDS* (§4.1); C: (all of them!) searchable simultaneously via *CATS* (§4.2); N: fluxes are in *NED*; n: source positions are in *NED* (cf. §4.3); M: included in *MSL* (§4.1). An update of this table is kept at `cats.sao.ru/doc/MAJOR_CATS.html`.

for an example). For a reliable cross-identification with other catalogues these alternative positions obviously have to be taken into account.

The angular resolution of the surveys tends to increase with observing frequency, while the lowest flux density detected tends to decrease (but increase again above ~ 8 GHz). In fact, until recently the relation between observing frequency, ν, and limiting flux density, S_{lim}, of large-scale surveys between 10 MHz and 5 GHz followed rather closely the power-law spectrum of an average extragalactic radio source, $S \sim \nu^{-0.7}$. This implied a certain bias against the detection of sources with rare spectra, like e.g. the "compact steep spectrum" (CSS) or the "GHz-peaked spectrum" (*GPS*) sources (O'Dea (1998)). With the new, deep, large-scale radio surveys like *WENSS*, *NVSS* and *FIRST* (§3.7), with a sensitivity of 10–50 times better than previous ones, one should be able to construct much larger samples of these cosmologically important type of sources (cf. Snellen *et al.* (1996)). A taste of some cosmological applications possible with these new radio surveys has been given in the proceedings volume by Bremer *et al.* (1998).

Table 1 also shows that there are no appreciable source surveys at frequencies higher than 5 GHz, mainly for technical reasons: it takes large amounts of telescope time to cover a large area of sky to a reasonably low flux limit with a comparatively small beam. New receiver technology as well as new scanning techniques will be needed. For example, by continuously (and slowly) slewing with all elements of an array like the *VLA*, an adequately dense grid of phase centres for mosaicing could be simulated using an appropriate integration time. More probably, the largest gain in knowledge about the mm-wave radio sky will come from the imminent space missions for microwave background studies, *MAP* and *PLANCK* (see §8.3). Currently there is no pressing evidence for "new" source populations dominating at mm waves (cf. sect. 3.3 of Condon *et al.* (1995)), although some examples among weaker sources were found recently (Crawford *et al.* (1996), Cooray *et al.* (1998)). Surveys at frequencies well above 5 GHz are thus important to quantify how such sources would affect the interpretation of the fluctuations of the microwave background. Until now, these estimates rely on mere extrapolations of source spectra at lower frequencies, and certainly the information content of the surveys in **Table 1** has not at all been fully exploited for this purpose.

Table 1 is an updated version of an earlier one (Andernach (1992)) which listed 38 surveys with ~450 000 entries. In 1992 I speculated that by 2000 the number of measured flux densities would have quadrupled. The current number (in 1998!) is already seven times the number for 1992.

3.5. *Optical Identification Content*

The current information on sources within our Galaxy is summarized in §3.6. The vast majority of radio sources more than a few degrees away from the Galactic plane are extragalactic. The latest compilation of optical identifications of extragalactic radio sources dates back to 1983 (Véron-Cetty & Véron (1983), hereafter *VV83*) and lists 14 585 entries for 10 173 different sources, based on 917 publications. About 25% of these are listed as "empty", "blank" or "obscured" fields (EF, BF, or OF), i.e. no optical counterpart has been found to the limits of detection. The *VV83* compilation has not been updated since 1983, and is not to be confused with the "Catalogue of Quasars and Active Galactic Nuclei" by the same authors. Both compilations are sometimes referred to as the ("well-known") "Véron catalogue", but usually the latter is meant, and only the latter is being updated (Véron-Cetty & Véron (1998) or "*VV98*"). The only other (partial) effort of a compilation similar to *VV83* was *PKSCAT90* (Otrupcek & Wright (1991)), which was restricted to the 8263 fairly strong PKS radio sources and, contrary to initial plans, has not been updated since 1990. It also lacks quite a few references published before 1990.

For how many radio sources do we know an optical counterpart? From **Table 1** we may very crudely estimate that currently well over 2 million radio sources are known (~3.3 million individual measurements are available electronically). A compilation of references (not included in *VV83*) on optical identifications of radio sources maintained by the present author currently holds ~560 references dealing with a total of ~56 000 objects. This leads the author to estimate that an optical identification (or absence thereof) has been reported for ~20 000–40 000 sources. Note that probably quite a few of these will either occur in more than one reference or be empty fields. Most of the information contained in *VV83* is absent from pertinent object databases (§4.3), given that these started including extragalactic data only since 1983 (*SIMBAD*) and 1988 (*NED*). However, most of the optical identifications published since 1988 can be found in *NED*.

Moreover, numerous optical identifications of radio sources have been made quietly (i.e. outside any explicit publication) by the *NED* team. Currently (May 98) *NED* contains ∼9800 extragalactic objects which are also radio sources. Only 57% of these have a redshift in *NED*. Even if we add to this some 2000–3000 optically identified Galactic sources (§3.6) we can state fairly safely that of all known radio sources, we currently know the optical counterpart for *at most half a percent*, and the distance for no more than *a quarter percent*. The number of counterparts is likely to increase by thousands once the new large radio survey catalogues (*WENSS, NVSS, FIRST*), as well as new optical galaxy catalogues, e.g. from *APM* (www.ast.cam.ac.uk/~apmcat), *SuperCOSMOS* (www.roe.ac.uk/scosmos.html) or *SDSS* (§3.7.3), become available. Clearly, more automated identification methods and multifibre spectroscopy (like e.g. *2dF*, *FLAIR*, and *6dF*, all available from www.aao.gov.au/) will be the only way to reduce the growing gap between the number of catalogued sources and the knowledge about their counterparts.

3.6. *Galactic Plane Surveys and Galactic Sources*

Some of the major discrete source surveys of the Galactic plane are included in **Table 1** (those for which a range in l and b are listed in columns 4 and 5, and several others covering the plane). Lists of "high"-resolution surveys of the Galactic radio continuum up to 1987 have been given in Kassim (1988) and Reich (1991). Due the high density of sources, many of them with complex structure, the Galactic plane is the most difficult region for the preparation of discrete source catalogues from maps. The often unusual shapes of radio continuum sources have led to designations like the "snake", the "bedspring" or "tornado", the "mouse" (cf. Gray (1994a)) or a "chimney" (Normandeau *et al.* (1996)). For extractions of images from some of these surveys see §6.3.

3.6.1. *What kind of discrete radio sources can be found in our Galaxy?*

Of the 100 000 brightest radio sources in the sky, fewer than 20 are stars. A compilation of radio observations of ∼3000 **Galactic stars** has been maintained until recently by Wendker (1995). The electronic version is available from *ADC/CDS* (catalogue #2199, §4.1) and includes flux densities for about 800 detected stars and upper limits for the rest. This compilation is not being updated any more. The most recent push for the detection of new radio stars has just come from a cross-identification of the *FIRST* and *NVSS* catalogues with star catalogues. In the *FIRST* survey region the number of known radio stars has tripled with a few dozen *FIRST* detections (S≳1 mJy at 1.4 GHz, Helfand *et al.* (1997)), and 50 (mostly new) radio stars were found in the *NVSS* (Condon *et al.* (1997)), many of them radio variable.

A very complete WWW page on **Supernovae** (*SNe*), including *SNRs*, is offered by Marcos J. Montes at cssa.stanford.edu/~marcos/sne.html. It provides links to other supernova-related pages, to catalogues of *SNe* and *SNR*, to individual researchers, as well as preprints, meetings and proceedings on the subject. D.A. Green maintains his "Catalogue of Galactic Supernova Remnants" at www.mrao.cam.ac.uk/surveys/snrs/. The catalogue contains details of confirmed Galactic *SNRs* (almost all are radio *SNRs*), and includes bibliographic references, together with lists of other possible and probable Galactic *SNRs*. From a Galactic plane survey with the RATAN-600 telescope (Trushkin (1996)) S. Trushkin derived radio profiles along *RA* at 3.9, 7.7, and 11.1 GHz for 70 *SNRs* at cats.sao.ru/doc/Atlas_snr.html (cf. Trushkin (1996)). Radio continuum spectra for 192 of the 215 *SNRs* in Green's catalogue (Trushkin (1998)) may be displayed at cats.sao.ru/cats_spectra.html.

Planetary nebulae (*PNe*), the expanding shells of stars in a late stage of evolution, all emit free-free radio radiation. The deepest large-scale radio search of *PNe* has been performed by Condon & Kaplan (1998), who cross-identified the "Strasbourg-*ESO* Catalogue of Galactic Planetary Nebulae" (*SESO*, available as *ADC/CDS* #5084) with the *NVSS* catalogue. To do this, some of the poorer optical positions in *SESO* for the 885 *PNe* north of $\delta=-40°$ had to be re-measured on the Digitized Sky Survey (*DSS*; `archive.stsci.edu/dss/dss_form.html`). The authors detect 680 (77%) *PNe* brighter than about S(1.4 GHz) = 2.5 mJy/beam. A database of Galactic Planetary Nebulae is maintained at Innsbruck (`ast2.uibk.ac.at/`). However, the classification of *PNe* is a tricky subject, as shown by several publications over the past two decades (e.g. Kohoutek (1997), Acker *et al.* (1991), or Acker & Stenholm (1990)). Thus the presence of a Planetary Nebula in a catalogue should not be taken as ultimate proof of its classification.

H II regions are clouds of almost fully ionised hydrogen found throughout most late-type galaxies. Major compilations of H II regions in our Galaxy were published by Sharpless (1959) (N=313) and Marsalkova (1974) (N=698). A graphical tool to create charts with objects from 17 catalogues covering the Galactic Plane, the *Milky Way Concordance* (`cfa-www.harvard.edu/~peterb/concord`), has already been mentioned in my tutorial in this volume. Methods to find candidate H II regions based on IR colours of IRAS Point Sources have been given in Hughes & MacLeod (1989) and Wood & Churchwell (1989), and were further exploited to confirm ultracompact H II regions (UC H II) via radio continuum observations (Kurtz *et al.* (1994)) or 6.7 GHz methanol maser searches (Walsh *et al.* (1997)). Kuchar & Clark (1997) merged six previous compilations to construct an all-sky list of 1048 Galactic H II regions, in order to look for radio counterparts in the 87GB and *PMN* maps at 4.85 GHz. They detect about 760 H II regions above the survey threshold of $\sim$30 mJy (87GB) and $\sim$60 mJy (*PMN*). These authors also point out the very different characteristics of these surveys, the 87GB being much poorer in extended Galactic plane sources than the *PMN*, for the reasons mentioned above (§3.3).

The "Princeton Pulsar Group" (`pulsar.princeton.edu/`) offers basic explanations of the pulsar phenomenon, a calculator to convert between dispersion measure and distance for user-specified Galactic co-ordinates, software for analysis of pulsar timing data, links to pulsar researchers, and even audio-versions of the pulses of a few pulsars. The largest catalogue of known **pulsars**, originally published with 558 records by Taylor *et al.* (1993) is also maintained and searchable there (with currently 706 entries). Pulsars have very steep radio spectra (e.g. Malofeev (1996), Shrauner et al. (1998), Toscano et al. (1998)) are point-like and polarised, so that pulsar candidates can be found from these criteria in large source surveys (Kouwenhoven *et al.* (1996)). Data on pulsars, up to pulse profiles of individual pulsars, from dozens of different papers can be found at the "European Pulsar Network" (`www.mpifr-bonn.mpg.de/pulsar/data/`). They have developed a flexible data format for exchange of pulsar data (Lorimer *et al.* (1998)), which is now used in an on-line database of pulse profiles as well as an interface for their simultaneous observations of single pulses. The database can be searched by various criteria like equatorial and/or Galactic co-ordinates, observing frequency, pulsar period and dispersion measure (DM). Further links on radio pulsar resources have been compiled at `pulsar.princeton.edu/rpr.shtml`, including many recent papers on pulsar research. Kaplan *et al.* (1998) have used the *NVSS* to search for phase-averaged radio emission from the pulsars north of $\delta_{2000}=-40°$ in the Taylor *et al.* (1993) pulsar catalogue. They identify 79 of these pulsars with a flux of S(1.4 GHz) $\gtrsim$ 2.5 mJy, and 15 of them are also in the *WENSS* source catalogue.

An excellent description of the various types of Galactic radio sources, including masers, is given in several of the chapters of Verschuur & Kellermann (1988).

Last, but not least, Galactic plane radio sources can point us to galaxies and clusters in the "**Zone of Avoidance**" (ZOA). In fact, in a large number of surveys for discrete radio sources, the Galactic plane does not show any excess number density of (usually compact) sources, e.g. in *TXS* (Douglas *et al.* (1996)) or *BWE* (Becker *et al.* (1991)). In a 2.7 GHz survey of the region $-3° < \ell < 240°$, $|b| < 5°$, with the Effelsberg 100-m telescope, the density of unresolved sources ($\lesssim 1'$ intrinsic size) was *not* found to vary with Galactic latitude (Fürst *et al.* (1990)). At much higher resolution (5″), using the *VLA* to cover the area $-10° < \ell < 40°$, $|b| < 1.8°$, a concentration of compact sources (size $\lesssim 20''$) towards the Galactic plane becomes noticeable, but only for $|b| < 0.4°$ (Helfand *et al.* (1992)). Becker *et al.* (1994) have shown that at 5 GHz this distribution has a width of only $10'$–$15'$. Most of the extragalactic sources will be far too optically faint to be ever identified. Two notable counter-examples are the prototype "head-tail" radio source 3C 129 (z=0.021; Miley *et al.* (1972)), now known to be a member of the Perseus supercluster of galaxies (Hauschildt (1987)), and "Centaurus B" (see §3.2). Other examples are two tailed radio sources, PKS B1610−608 and PKS B1610−605 in Abell cluster A 3627, which is thought to be the central clump of the "Great Attractor" (Kraan-Korteweg *et al.* (1997)). A typical wide-angle tailed *WAT*) radio source (G 357.30+01.24) has been found very close to the Galactic centre, indicating the presence of a cluster of galaxies in that direction (Gray (1994b)). Due to the extreme optical obscuration, there is little hope of optically identifying this cluster and determining its distance.

3.7. *Modern Large-Scale Discrete Source Surveys: NVSS, FIRST, WENSS and SUMSS*

Some of the first large-scale contiguous surveys with interferometers had become available in the 1980s. These were made with arcmin resolution at low frequencies where the large fields of view required only a few pointings (e.g. 6C or 8C; `www.mrao.cam.ac.uk/surveys/`). Only in the 1990s, however, has the increase in computing power allowed such surveys to be made at even sub-arcmin resolution with the most powerful interferometers like the *VLA* and the *WSRT*, requiring up to a quarter of a million pointings. Four ongoing or recently finished surveys in this category are described below.

3.7.1. *The "WENSS" Survey at 325–350 MHz*

The "Westerbork Northern Sky Survey" (*WENSS*; `www.strw.leidenuniv.nl/wenss`) is a radio survey made with the *WSRT* (`www.nfra.nl/wsrt/wsrtpage.htm`) from late 1990 to 1996, at frequencies 325 and 610 MHz (λ 92 and 49 cm). The entire sky north of declination +30° has been covered with ~6000 pointings, using a central frequency of 325 MHz below $\delta = 74°$, and 350 MHz for the polar region. Only 2000 square degrees (~20% of the sky north of +30°) were mapped at 610 MHz, and for the time being only the 325 MHz data have been made available to the public. At 325 MHz the resolution is $54'' \times 54'' \csc(\delta)$, and the positional accuracy for strong sources is 1.5″. The limiting flux density is ~18 mJy ($5\,\sigma$) at both frequencies. The final products of *WENSS*, a 325-MHz atlas of $6° \times 6°$ maps centred on the new *POSS* plate positions (5° grid) as well as a source catalogue are now available. *FITS* maps can be drawn from the anonymous `ftp` server at `ftp://vliet.strw.leidenuniv.nl/pub/wenss/HIGHRES/`. A postage stamp server for extraction of smaller images is planned for the near future. Also total intensity maps at a lower resolution of $4.2'$ will soon be made available. Ionospheric instabilities make the generation of polarisation maps (Stokes Q, U and V) formidably difficult and their production is not currently foreseen.

Presently there are two source catalogues at 325 MHz. The main catalogue contains 211,235 sources for $28° < \delta < 76°$. The polar catalogue contains 18,341 sources above 74°. These can be browsed at the *URL* `www.strw.leidenuniv.nl/wenss/search.html`.

A detailed description of the survey and the contents of the source lists are given in Rengelink *et al.* (1997).

Due to its low frequency and sensitivity to extended structure the *WENSS* survey is well-suited to detect very extended or low surface brightness objects like giant radio galaxies (cf. Schoenmakers *et al.* (1998)), cluster radio haloes, and nearby galaxies. Comparing *WENSS* with other surveys at higher frequencies allows one to isolate candidates for high-redshift radio galaxies, GHz-peaked spectrum (*GPS*) sources, flat-spectrum sources (e.g. high-redshift quasars), and pulsars.

3.7.2. *The "NRAO VLA Sky Survey" at 1.4 GHz (NVSS)*

The *VLA* has been used from 1993 to 1997 to map the entire sky north of $\delta=-40°$ (82% of the sky) in its most compact (D) configuration, giving an angular resolution of 45″ at 1.4 GHz. About 220 000 individual snapshots (phase centres) have been observed. They were of a mere 23.5 sec duration each, except at low elevation when they were increased to up to 60 sec to make up for the loss of sensitivity due to ground radiation and air mass. A detailed description is given in Condon *et al.* (1998) (`ftp://www.cv.nrao.edu/pub/nvss/paper.ps`). The principal data products are:

- A set of 2326 continuum map "cubes," $4°\times4°$ with images of Stokes parameters I, Q, and U. The noise level is $\sim$0.45 mJy/beam in I, and 0.29 mJy/beam in Q and U. Positional accuracy varies from $<1″$ for strong (S>15 mJy) point sources to 7″ for the faintest ($\sim$2.3 mJy) detectable sources.
- A catalogue of $\sim$2 000 000 discrete sources detected in the entire survey
- Processed uv-data (visibilities) for each map cube constructed from over 100 individual pointings, for users wishing to investigate the data underlying the images.

The *NVSS* is accessible from `www.cv.nrao.edu/~jcondon/nvss.html`, and is virtually complete at the time of writing. The latest version of the *NVSS* catalogue (#34, May 98) is a single 152 Mb *FITS* file with 1.8×10^6 sources. It can be downloaded via anonymous `ftp`, but users interested in exploiting the entire catalogue may consider requesting a tape copy from *NRAO*. The publicly available program `NVSSlist` can extract selected portions of the catalogue very rapidly and is easily installed on the user's local disk for extensive cross-identification projects. The catalogue can also be browsed at `www.cv.nrao.edu/NVSS/NVSS.html` and a "postage stamp server" to extract *NVSS* images is available at `www.cv.nrao.edu/NVSS/postage.html`. Images are also available from Skyview (`skyview.gsfc.nasa.gov/`), but they neither are as up-to-date as those at *NRAO*, nor do they have the same *FITS* header (§6.2). As always, care must be taken in the interpretation of these images. Short integration times and poor uv-coverage can cause grating residuals and limited sensitivity to extended structure (see Fig. 4).

3.7.3. *The "FIRST" Survey at 1.4 GHz*

The *VLA* has been used at 1.4 GHz (λ=21.4 cm) in its B-configuration for another large-scale survey at 5″ resolution. It is called *FIRST* ("Faint Images of the Radio Sky at Twenty-centimeters") and is designed to produce the radio equivalent of the Palomar Observatory Sky Survey over 10 000 square degrees of the North Galactic Cap. An automated mapping pipeline produces images with 1.8″ pixels and a typical rms noise of 0.15 mJy. At the 1 mJy source detection threshold, there are $\sim$90 sources per square degree, about a third of which have resolved structure on scales from 2″–30″.

Individual sources have 90% confidence error circles of radius <0.5″ at the 3 mJy level and 1″ at the survey threshold. Approximately 15% of the sources have optical counterparts at the limit of the *POSS-I* plates (E$\sim$20.0), and unambiguous optical identifications are achievable to $m_v \sim$24. The survey area has been chosen to coincide with that of the

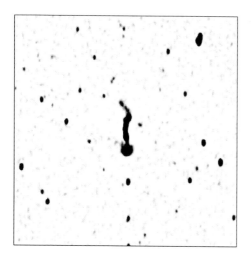 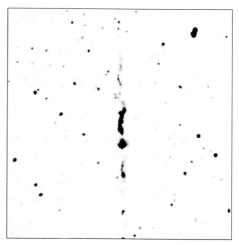

FIGURE 4. Reality and "ghosts" in radio maps of 1.5°×1.5° centred on the radio galaxy 3C 449. Left: a 325 MHz *WENSS* map shows the true extent (~23′) of 3C 449. Right: The 1.4 GHz *NVSS* map shows additional weak ghost images extending up to ~40′, both north and south of 3C 449. This occurs for very short exposures (here 23 sec) when there is extended emission along the projection of one of the *VLA* "arms" (here the north arm). Note that neither map shows any indication of the extended foreground emission found at 1.4 GHz coincident with an optical emission nebula stretching from NE to SW over the entire area shown (Fig. 4c of Andernach *et al.* 1992). However, with longer integrations, the uv-coverage of interferometers is sufficient to show the feature (Leahy *et al.* 1998).

Sloan Digital Sky Survey (*SDSS*; `www-sdss.fnal.gov:8000/`). This area consists mainly of the north Galactic cap ($|b| > +30°$) and a smaller region in the south Galactic hemisphere. At the m_v ~24 limit of *SDSS*, about half of the optical counterparts to *FIRST* sources will be detected.

The homepage of *FIRST* is `sundog.stsci.edu/`. By late 1997 the survey had covered about 5000 square degrees. The catalogue of the entire region (with presently ~437 000 sources) can be searched interactively at `sundog.stsci.edu/cgi-bin/searchfirst`. A postage stamp server for *FIRST* images (presently for 3000 square degrees) is available at `third.llnl.gov/cgi-bin/firstcutout`. For 1998 and 1999, the *FIRST* survey was granted enough time to cover an additional 3000 square degrees.

The availability of the full *NVSS* data products has reduced the enthusiasm of parts of the community to support the finishing of *FIRST*'s goals. However, only the *FIRST* survey (and less so the *NVSS*) provides positions accurate enough for reliable optical identifications, particularly for the cosmologically interesting faint and compact sources. On the other hand, in Figure 5 I have shown an extreme example of the advantage of *NVSS* for studies of extended sources. In fact, the Figure shows the complementary properties of *NVSS* and *FIRST*. A very extended source, perhaps just recognisable with *NVSS*, will be broken up by *FIRST* into apparently unrelated components. Thus, it would be worthwhile to look into the feasibility of merging the uv data of *NVSS* and *FIRST* to create maps at 10″–15″ resolution in the region covered by both surveys.

3.7.4. The "SUMSS" 843 MHz Survey with "MOST"

Since 1994 the "Molonglo Observatory Synthesis Telescope" (*MOST*) has been upgraded from the previous 70′ field of view to a 2.7° diameter field of view. As *MOST*'s aperture is almost filled, the image contains Fourier components with a wide range of

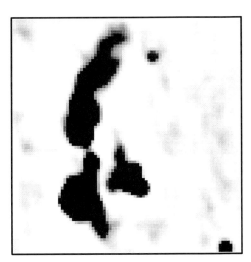 

FIGURE 5. 1.4 GHz maps of the radio galaxy 3C 40 (PKS B0123−016). Left: *NVSS* map of 20′×20′. The point-like component at the gravity centre of the radio complex coincides with NGC 547, a dominant dumb-bell galaxy in the core of the A 194 galaxy cluster (cf. Fig. 2a in my tutorial); the head-tail like source ∼5′ due SW is NGC 541. Right: *FIRST* map of 12′×12′: only the strongest parts of each component are detected and show fine structure, but appear unrelated. The radio core of NGC 547 is unresolved. Some fainter components appear to be artefacts.

angular scales, and has low sidelobes. In mid-1997, the *MOST* started the "Sydney University Molonglo Sky Survey" (*SUMSS*; Hunstead *et al.* (1998)). The entire sky south of $DEC=-30°$ and $|b| >10°$ will be mapped at 843 MHz, a total of 8000 square degrees covered by 2713 different 12-h synthesis field centres. It complements northern surveys like *WENSS* and *NVSS*, and it overlaps with *NVSS* in a 10° strip in declination (−30° to −40°), so as to allow spectral comparisons. *SUMSS* is effectively a continuation of *NVSS* to the southern hemisphere (see Table 2). However, with its much better uv coverage it surpasses both *WENSS* and *NVSS* in sensitivity to low surface brightness features, and it can fill in some of the "holes" in the uv plane where it overlaps with *NVSS*. The *MOST* is also being used to perform a Galactic plane survey (§8).

SUMSS positions are uncertain by no more than 1″ for sources brighter than 20 mJy, increasing to ∼2″ at 10 mJy and 3–5″ at 5 mJy, so that reliable optical identifications of sources close to the survey limit may be made, at least at high Galactic latitude. In fact, the identification rate on the *DSS* (§3.6) is ∼30% down to b_J ∼22 (Sadler (1998)). Observations are made only at night, so the survey rate is ∼1000 deg^2 per year, implying a total period of 8 years for the data collection. The south Galactic cap ($b < -30°$) should be completed by mid-2000. The release of the first mosaic images (4°×4°) is expected for late 1998. The *SUMSS* team at Univ. Sydney plans to use the *NVSS* WWW software, so that access to *SUMSS* will look similar to that for *NVSS*. For basic information about *MOST* and *SUMSS* see `www.physics.usyd.edu.au/astrop/SUMSS/`.

4. Integrated Source Parameters on the Web

In this section I shall describe the resources of information on "integrated" source parameters like position, flux density at one or more frequencies, size, polarisation, spectral

TABLE 2. Comparison of the new large-area radio surveys

	WENSS	SUMSS	NVSS	FIRST
Frequency	325 MHz	843 MHz	1400 MHz	1400 MHz
Area (deg^2)	10 100	8 000	33 700	10 000
Resolution ($''$)	$54 \times 54 \csc(\delta)$	$43 \times 43 \csc(|\delta|)$	45	5
Detection limit	15 mJy	<5 mJy	2.5 mJy	1.0 mJy
Coverage	$\delta > +30°$	$\delta < -30°, |b| > 10°$	$\delta > -40°$	$|b| > 30°, \delta > -12°$
Sources / deg^2	21	>40	60	90
No. of sources	230 000	320 000	2 000 000	900 000

index, etc. This information can be found in two distinct ways, either from individual source catalogues, each of which have different formats and types of parameters, or from "object databases" like *NED*, *SIMBAD* or *LEDA* (§4.3). The latter have the advantage of providing a "value-added" service, as they attempt to cross-identify radio sources with known objects in the optical or other wavebands. The disadvantage is that this is a laborious process, implying that radio source catalogues are being integrated at a slow pace, often several years after their publication. In fact, many valuable catalogues and compilations never made it into these databases, and the only way for the user to complement this partial information is to search the available catalogues separately on other servers. Due to my own involvement in providing the latter facilities, I shall briefly review their history.

4.1. *The Evolution of Electronic Source Catalogues*

Radio astronomers have used electronic equipment from the outset and already needed powerful computers in the 1960s to make radio maps of the sky by Fourier transformation of interferometer visibilities. Surprisingly radio astronomers have *not* been at the forefront of archiving their results, not even the initially rather small-sized catalogues of radio sources. It is hard to believe that the *WSRT* maintained one of the earliest electronic and publicly searchable archives of raw interferometer data (see www.nfra.nl/scissor/), but at the same time the source lists of 65 *WSRT* single-pointing surveys, published from 1973 to 1987 with altogether 8200 sources, had not been kept in electronic form. Instead, 36 of them with a total of 5250 sources were recovered in 1995–97 by the present author, using page-scanners and "Optical Character Recognition" (*OCR*) techniques.

During the 1970s, R. Dixon at Ohio State Univ. maintained what he called the "Master Source List" (*MSL*). The first version appeared in print almost 30 years ago (Dixon (1970)), and contained ~25 000 entries for ~12 000 distinct sources. Each entry contained the *RA*, *DEC* and flux density of a source at a given observing frequency; any further information published in the original tables was not included. The last version (# 43, Nov. 1981) contained 84,559 entries drawn from 179 references published 1953–1978. The list gives ~75 000 distinct source names, but the number of distinct sources is much smaller, though difficult to estimate. It was typed entirely by hand, for which reason it is affected by numerous typing errors (Andernach (1989)). Also, it was meant to collect positions and fluxes only from new finder surveys, not to update information on already known sources.

Although the 1980s saw a "renaissance" of radio surveys (e.g. MRC, B3, 6C, MIT-GB, GT, NEK, IPS in **Table 1**) that decade was a truly "dark age" for radio source databases (Andernach (1992)). The *MSL*, apart from being distributed on tape at cost, was not being updated any more, and by the end of the 1980s there was not a single radio source catalogue among the then over 600 catalogues available from the archives of the two established astronomical data centres, the "Astronomical Data Center" (*ADC*; `adc.gsfc.nasa.gov/adc.html`) at *NASA-GSFC*, and the "Centre de Données astronomiques de Strasbourg" (*CDS*; `cdsweb.u-strasbg.fr/CDS.html`). This may explain why even in 1990 the *MSL* was used to search for high-redshift quasars of low radio luminosity, simply by cross-correlating it with quasar catalogues (Hutchings *et al.* (1991), *HDP91* in what follows). These authors (using a version of *MSL* including data published up to 1975!) noted that the *MSL* had 23 coincidences within 60″ from QSOs in the HB 89 compilation (Hewitt & Burbidge (1989)) which were not listed as "radio quasars" in HB 89. However, *HDP91* failed to note that 13 of these 23 objects were already listed with an optical identification in *VV83*, published seven years before! From the absence of *weak* ($\lesssim 100$ mJy) radio sources associated with $z \gtrsim 2.5$ quasars, *HDP91* concluded that there were no high-z quasars of low radio luminosity. However, had the authors used the 1989 edition of *VV98* (Véron-Cetty & Véron (1989), *ADC/CDS* #7126) they would have found about ten quasars weaker than ~ 50 mJy at 5-GHz, from references published *before* 1989. This would have proven the existence of the objects searched for (but not found) by *HDP91* from compilations readily available at that time. The most recent studies by Bischof & Becker (1997) and Hooper *et al.* (1996)), however, indicate that these objects are indeed quite rare.

Alerted by this deficiency of publicly available radio source catalogues, I initiated, in late 1989, an email campaign among radio astronomers world-wide. The response from several dozen individuals (Andernach (1990)) was generally favourable, and I started to actively collect electronic source catalogues from the authors. By the time of the *IAU* General Assembly in 1991, I had collected the tabular data from about 40 publications totalling several times the number of records in the *MSL*. However, it turned out that none of the major radio astronomical institutes was willing to support the idea of a public radio source database with manpower, e.g. to continue the collection effort and prepare the software tools. As a result, the *EINSTEIN On-line Service EINLINE* or *EOLS*), designed to manage X-ray data from the *EINSTEIN* satellite, offered to serve as a test-bed for querying radio source catalogues. Until mid-1993 some 67 source tables with ~523 000 entries had been integrated in collaboration with the present author (Harris *et al.* (1995)). These are still searchable simultaneously via a simple telnet session (`telnet://einline@einline.harvard.edu`). However, in 1994 *NASA*'s funding of *EOLS* ceased, and no further catalogues have been integrated since then. A similar service is available from *DIRA2* (`www.ira.bo.cnr.it/dira/gb/`), providing 54 radio catalogues with 2.3 million records, including older versions of the *NVSS* and *FIRST* catalogues, as well as many items from the present author's collection. However, due to lack of manpower, *DIRA*'s catalogue collection is now outdated, and many items from **Table 1** are missing. In late 1993, Alan Wright (*ATNF*) and the present author produced a stand-alone package (called "*COMRAD*") of 12 major radio source catalogues with some 303,600 entries. It comes with `dBaseIV` search software for PCs and can still be downloaded from *URL* `wwwpks.atnf.csiro.au/databases/surveys/comrad/comrad.html`. Several other sites offer more or less "random" and outdated sets of catalogues (a few radio items included) and are less suitable when seeking up-to-date and complete information. Among these are *ESO*'s *STARCAT* (`arch-http.hq.eso.org/starcat.html`), *ASTROCAT* at *CADC* (`cadcwww.dao.nrc.ca/astrocat`), and *CURSA* within the Starlink project (`www.roe.ac.uk/acdwww/cursa/home.html`). *CURSA* is actually designed to be copied to the user's machine, and to work with local catalogues in a *CURSA*-compatible format.

From late 1989 until the present, I have continued my activities of collecting source catalogues, and since 1995, I have also employed *OCR* methods to convert printed source lists into electronic form, among them well-known compilations such as that of Kühr *et al.* (1979) of 250 pages, for which the electronic version had not survived the transition through various storage media. Recovery by *OCR* requires careful proof-reading, especially for those published in tiny or poorly printed fonts (e.g. Harris & Miley (1978), Walterbos *et al.* (1985), and Bystedt *et al.* (1984)). In many cases the original publications were impossible to recover with *OCR*. For some of these I had kept preprints (e.g. for Tabara & Inoue (1980)) whose larger fonts facilitated the *OCR*. Numerous other tables (e.g. Braude *et al.* (1979), Quiniento *et al.* (1988) or Altenhoff *et al.* (1979)) were patiently retyped by members of the *CATS* team (see below). Since about 1996, older source tables are also actively recovered with *OCR* methods at *CDS*. Unfortunately, due to poor proof-reading methods, errors are found quite frequently in tables prepared via *OCR* and released by *CDS*. Occasionally, tables were prepared independently by two groups, allowing the error rate to be further reduced by inter-comparison of the results. Up to now, radio source tables from 177 articles with a total of 75 000 data records and many thousand lines of text (used as documentation for the tables) were prepared via *OCR*, mostly by the present author. Surprisingly, about half of the tables (including those received directly from the authors) show some kind of problem (e.g. in nomenclature, internal consistency, or formatting, etc.) that requires attention, before they are able to be integrated into a database or searchable catalogue collection. This shows, unfortunately, that not enough attention is paid to the data section by the referees of papers. While the data section may appear uninteresting to them, one should keep in mind that in future re-analyses based on old published data, it is exactly the data section which remains as a heritage to future researchers, and not the interpretations given in the original papers. In the early 1990s, most of the tables were received from the original authors upon request, but currently about half of the tables can be collected from the *LANL/SISSA* electronic preprint server (`xxx.lanl.gov`). However, this has the danger that they may not be identical to the actual publication. The vast majority of tabular data sets are received in TEX format, and their conversion to (*ASCII* requires substantial efforts.

Currently my collection of radio source lists (`cats.sao.ru/~cats/doc/Andernach.html`) contains source lists from 500 articles, but only ~22% of the tables are also available from *ADC* or *CDS* (§4.2). While the collection started in 1989, half the 500 data sets were collected or prepared since 1996, and the current growth rate is ~80 data sets per year. About three dozen further source lists exist in the *CDS* archive (§4.2), most of which are either from the series of nine *AAS* CDROMs, issued twice a year from 1994 to 1997 with tables from the ApJ and AJ journals, or from recent volumes of the A&AS journal, thanks to a 1993 agreement between the Editors of A&AS and *CDS* to archive all major tables of A&AS at the *CDS*. Unfortunately such an agreement does not exist with other journals, for which reason my collection efforts will probably continue until virtually all astronomical journals provide electronic editions. Presently some tabular data (e.g. in the electronic A&A) are offered only as images, while other journals offer only hypertext versions of their tables, which frequently need further treatment to be converted to plain (*ASCII* format, required for their ingestion into databases.

The size distribution of electronic radio source catalogues (including my collection and that of *CDS*) is plotted in the left panel of Figure 6. For catalogues with more than ~200 entries the curve follows a power law with index near −0.6, a manifestation of *Zipf's law* in bibliometrics (Nicholls (1987)). The decline for smaller catalogues is due to the fact that many of them simply do not exist in electronic form. In the right panel of Fig. 6 the growth over time of the cumulative number of records of these catalogues is plotted.

The three major increases are due to the *MSL* in 1981, to the 87GB/GB6/*PMN* surveys in 1991, and, more recently, to the release of *NVSS*, *FIRST* and *WENSS* in 1996/97.

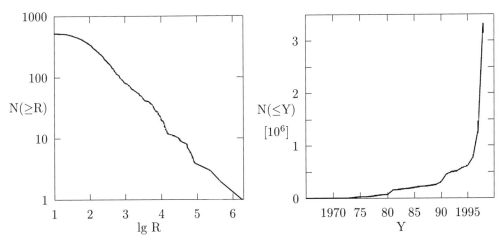

FIGURE 6. Left: Size distribution of radio source catalogues available in electronic form. R is the number of records in a source catalogue, and $N(\geq R)$ is the number of radio source catalogues with R records or more. The bottom right corner corresponds to the *NVSS* catalogue. Right: The growth in time of the number of continuum radio source measurements. Y is the year of publication of a radio source catalogue, and $N(\leq Y)$ is the cumulative number of records (in millions) contained in catalogues published up to and including year Y.

Already, since the early 1990s, the author's collection of radio catalogues has been the most comprehensive one stored at a single site. However, the problem of making this heterogeneous set of tables searchable with a common user interface was only solved in 1996, when the author started collaborating with a group of radio astronomers at the Special Astrophysical Observatory (*SAO*, Russia), who had built such an interface for their "Astrophysical **CAT**alogs support **S**ystem" (*CATS*; §4.2). Their common interests in radio astronomy stimulated the ingestion of a large number of items from the collection. By late 1996, *CATS* had surpassed *EOLS* in size and scope, and in mid-1997 an email service was opened by *CATS*, allowing one to query about 200 different source lists simultaneously for any number of user-specified sky positions, with just a single and simple email request.

4.2. *Searching in Radio Catalogues: VizieR and CATS*

The largest collections of astronomical catalogues, and published tabular data in general, are maintained at the *CDS* and *ADC*. The "Astronomer's Bazaar" at *CDS* (cdsweb.u-strasbg.fr/Cats.html) has over 2200 catalogues and tables for downloading via anonymous ftp. The full list of items (ftp://cdsarc.u-strasbg.fr/pub/cats/cats.all) may be queried for specific catalogues by author name, keyword, wavelength range, or by name of (space) mission. At *NASA*'s *ADC* (adc.gsfc.nasa.gov/adc.html) a similar service exists. Despite the claims that "mirror copies" exist in Japan, India and Russia, *CDS* and *ADC* are the only ones keeping their archives *current*. Both have their own catalogue browsers: VizieR at *CDS* (vizier.u-strasbg.fr/cgi-bin/VizieR), and Catseye at *ADC* (tarantella.gsfc.nasa.gov/catseye/ViewerTopPage.html), but currently none of them allows one to query large numbers of catalogues at the same time, although such a system is in preparation within VizieR at the *CDS*. Presently ∼ 200 catalogues appear when VizieR is queried for the waveband "radio". This includes many lists of H II regions, masers, etc., but excludes many of the major radio continuum surveys listed in **Table 1**. For radio source catalogues, the *CATS* system currently has the largest collection, and *CATS* is definitely preferable when radio continuum data are needed.

The *CATS* system (cats.sao.ru) currently permits searches through about 200 radio source catalogues from about 150 different references, with altogether over 3 million entries, including current versions of the *NVSS*, *FIRST* and *WENSS* catalogues. Many further radio source lists are available via anonymous FTP, as they have not yet been integrated into the search facility (e.g. when only source names, and not positions, are given in the available electronic version of the catalogue). Documentation is available for most of the source lists, and in many cases even large parts of the original paper text were prepared from page scans.

Catalogues in *CATS* may be selected individually from cats.sao.ru/cats_search.html, or globally by wavelength range. One may even select *all* searchable catalogues in *CATS* (including optical, IR, X-ray), making up over 4 million entries. They may be searched interactively on the WWW, or by sending a batch job via email. To receive the instructions about the exact format for such a batch request, send an empty email to cats@sao.ru (no subject required). The output can be delivered as a homogeneous table of sources from the different catalogues, or each catalogue in its native format. The latter assures that all columns as originally published may be retrieved, except when the "homogenised table format" is requested. However, currently the user has to check the individual catalogue documentation to find out what each column means. With the select option one may retrieve sources from a single sky region, either a rectangle or a circle in different co-ordinate systems (equatorial B1950 or J2000, or Galactic), while the match option allows a whole list of regions to be searched in order to find all the objects in each region. It is then the responsibility of the user to find out which of these data represent the object (or parts of an object, depending on the telescope characteristics) and may be used for inter-comparisons.

CATS offers a few other useful features. For several multi-frequency radio catalogues (or rather compilations of radio sources) *CATS* allows radio spectra to be plotted on-the-fly, e.g. for Kühr *et al.* (1979), Kühr *et al.* (1981), (Otrupcek & Wright (1991), *PKSCAT90*, Trushkin (1996), Kallas & Reich (1980), and Bursov *et al.* (1997). Various options for fitting these spectra and weighting the individual flux errors are provided. Examples for two sources are shown in Fig. 7. Note that *PKSCAT90* includes data obtained at only one epoch per frequency, while the Kühr compilations include several epochs at a given frequency. Therefore the variability of QSO 2216−03 (=PKS B2216−038) becomes obvious only in the lower right panel of Fig. 7.

Note that *CATS* (at least at present) is a searchable collection of catalogues, and not a relational database, i.e. no cross-identifications have yet been made between catalogues (except for a few, which resulted in yet other catalogues). However, given its vastly larger collection of radio source data, it is an indispensable tool that complements the information on radio sources found e.g. in *NED* or *SIMBAD*.

In future it is planned that the user may display both the sky distribution and a spectral energy distribution (radio spectrum) of all entries found for a (sufficiently narrow) positional search. The sky plot will indicate the angular resolution, the positional error box, and (if available) the shape of each catalogued source, so that the user may interactively discard possibly unrelated sources, and arrive at the radio spectrum of the object of interest, as mentioned by Verkhodanov *et al.* (1997).

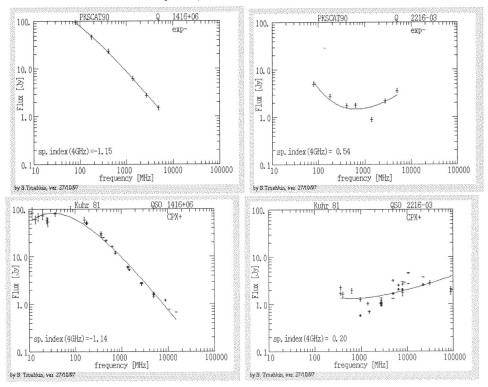

FIGURE 7. Radio Spectra of the two PKS sources PKS 1416+06 (left) and PKS 2216-03 (right) plotted with *CATS*. Upper row: data from *PKSCAT90* (one epoch per frequency); lower row: data from the multi-epoch compilation by Kühr *et al.* (1981).

4.3. *Object Databases: NED, SIMBAD, and LEDA*

These databases have already been described in my tutorial for this winter school, so I shall concentrate here on their relevance for radio astronomy. All three databases were originally built around catalogues of optical objects (galaxies in the case of *NED* and *LEDA*, and stars in the case of *SIMBAD*). It is quite natural that information on otherwise unidentified radio sources is not their priority. Also, being an extragalactic database, *NED* tends to provide more information on radio sources than *SIMBAD*, which was originally dedicated to stars, which constitute only a negligible population of radio continuum sources in the sky. The fact that before being included into *NED* or *SIMBAD*, every new (radio or other) source has to be checked for its possible identification with another object already in these databases, implies that the integration of large catalogues may take years from their publication. The rightmost column of **Table 1** gives an idea of this problem. A further obstacle for database managers is that they have to actively collect the published data from the authors or other resources. If you wish to see your data in databases soonest, the best thing is to send them (preferably in plain (*ASCII* format) to the database managers directly after publication.

    SIMBAD is accessible via password from `simbad.u-strasbg.fr/Simbad`, and has its priority in maintaining a good bibliography for astronomical objects (not necessarily those detected as a radio source only). *NED* can be accessed freely through the *URL* `nedwww.ipac.caltech.edu` and tends to make an effort to also populate its various "data frames" (like optical magnitude, fluxes at various frequencies, etc.) with recently published measurements.

Searches by object name rely on rather strict rules. In databases these may not always conform to *IAU* recommendations (cdsweb.u-strasbg.fr/iau-spec.html), mainly due to deviations from these recommendations by individual authors. In case of doubt about the exact name of a source, it is wise to start searching the databases by position.

The "Lyon-Meudon Extragalactic Database" (*LEDA*; www-obs.univ-lyon1.fr/leda) is primarily intended for studying the kinematics of the local Universe, and as such has little interest in radio continuum data on galaxies. However, *LEDA* is the ideal place to look for integrated neutral hydrogen (H I) data of nearby ($z \lesssim 0.2$) galaxies. These H I measurements play an important role for distance estimates of galaxies, independent of their radial velocities. This allows their "peculiar motions" to be calculated, i.e. the deviations of their radial velocities from the Hubble law.

Finally, one should keep in mind that *SIMBAD* and *NED* started to include references on extragalactic objects only since 1983 and 1988 respectively, although a few major references before these dates have now been included. In the following I give just one example in which the consequences of this have *not* been considered by users of *NED* or *SIMBAD*. The X-ray source RX J15237+6339 was identified (from *NED*) with the radio source 4C+63.22 by Brinkmann & Siebert (1994), and these authors comment that "One object (4C+63.22) is classified as 'Radio Source' only in the *NED* data base, so, strictly speaking, it belongs to the class of unidentified objects." However, according to *VV83* (their ref. 603=Porcas *et al.* (1980)) the source 4C+63.22 had actually been identified with an 18.5 m galaxy. This object is within 5″ of the brightest source in the *NVSS* catalogue within a radius of 90″. However, the *NVSS* map shows a large triple radio galaxy with a North-South extent of ∼ 4′. Later re-observation at 5 GHz with the *VLA* at 1.3″ resolution (Laurent-Muehleisen *et al.* (1997)) detected only a radio core coincident with a 16.6-mag object which the 1980 authors had already mentioned in their notes as 1′ offset from their prime candidate identification for the source 1522+638. I should add that the data table of the 1980 paper was published on microfiche.

5. Miscellaneous Databases and Surveys of Radio Sources

Several WWW sites offering topical databases of special types of radio source research will be mentioned briefly in this section.

5.1. *Clusters of Galaxies*

A collection of well-chosen radio source catalogues has been used, together with optical sky surveys like the *DSS*, to develop a database of radio-optical information on Abell/ACO (Abell *et al.* (1989)) clusters of galaxies. It is managed by A. Gubanov at the St.-Petersburg State University in collaboration with the present author (Gubanov & Andernach 1997). At the *URL* future.astro.spbu.ru/Clusters.html, the user may interactively prepare schematic radio-optical overlays, charts from the *FIRST*, *NVSS*, *APM* or *DSS* survey data, or retrieve references for cluster data. Radio continuum spectra of cluster radio galaxies may be displayed and fitted with user-specified functions. Source luminosities may be derived assuming cluster membership.

5.2. *VLBI and Astrometric Surveys*

The *VLBA* Calibrator Survey (magnolia.nrao.edu/vlba_calib/) is an ongoing project to provide phase-reference calibrators for *VLBA* experiments. When completed it will contain astrometric (∼1 mas) positions and 2.7 and 8.4 GHz images of the ∼3000 sources in the *JVAS* catalogue (§5.3).

The "Pearson-Readhead" ((*PR*) and "Caltech – Jodrell Bank" (CJ) imaging data base is a *VLBI* source archive at Caltech (`astro.caltech.edu/~tjp/cj/`). It offers images of over 300 *VLBI* sources at δ >35° observed in the (*PR* (Pearson & Readhead (1988)), CJ1 (Xu *et al.* (1995)), and CJ2 (Henstock *et al.* (1995)) surveys. Many of these sources are excellent calibrators for the *VLA* and *VLBA*. It has mostly 5 GHz (6 cm) and some 1.67 GHz (18 cm) *VLBI* images, as well as 1.4 and 5-GHz *VLA* images of extragalactic sources. Contour maps are publicly available as PostScript files.

A sample of 132 compact sources have been observed in "snapshot" mode with the *VLBA* at 15 GHz (2 cm; Kellermann *et al.* (1998)). At present it contains images at one epoch (`www.cv.nrao.edu/2cmsurvey/` and `www.mpifr-bonn.mpg.de/zensus/2cmsurvey/`), but it eventually will have multi-epoch sub-milliarcsecond data.

The "Radio Reference Frame Image Database" ((*RRFID*) is maintained at the U.S. Naval Observatory (*USNO*). Data obtained with the *VLBA* and the "Global Geodetic Network" at 2.3, 8.4, and 15 GHz (13, 3.6 and 2 cm) are publicly available as over 1400 images for more than 400 sources, at `maia.usno.navy.mil/rorf/rrfid.html`. In April 1998 first-epoch imaging of northern hemisphere radio reference frame sources was completed. Images at both 2.3 and 8.5 GHz now exist for ∼97 % of the "Radio Optical Reference Frame" (RORF) sources (Johnston *et al.* (1995)) north of δ=−20°, which is ∼90% of the "International Celestial Reference Frame" (ICRF) sources in this region of sky. A number of links are available from the (*RRFID* page, in particular to the RORF data base of sources (`maia.usno.navy.mil/rorf/rorf.html`).

The European *VLBI* Network (*EVN*; `www.nfra.nl/jive/evn/evn.html`) was formed in 1980 by the major European radio astronomy institutions, and is an array of radio telescopes spread over Europe, the Ukraine, and China. A catalogue of observations carried out so far can be retrieved from `terra.bo.cnr.it/ira/dira/vlbinet.dat`. The column explanation is available at `www.ira.bo.cnr.it/dira/vlbinet.doc`.

5.3. *Gravitational Lens Surveys*

About 2500 compact northern sources stronger than 200 mJy at 5 GHz have been mapped with the *VLA* at 8.4 GHz in the "Jodrell-Bank/*VLA* Astrometric Survey" (*JVAS*; `www.jb.man.ac.uk/~njj/glens/jvas.html`). The goal was (Patnaik *et al.* (1992)) to find phase calibrator sources for the *MERLIN* interferometer (`www.jb.man.ac.uk/merlin/`) and to search for gravitational lens candidates. If the redshift of both the parent object of the compact source and that of the intervening galaxy or cluster (causing the lensing effect) can be determined, and if in addition the compact source shows variability (not uncommon for such sources), the time delay between radio flares in the different images of the lensed object may be used to constrain the Hubble constant.

The "Cosmic Lens All-Sky Survey" (*CLASS*; `astro.caltech.edu/~cdf/class.html`), is a project to map more than 10 000 radio sources in order to create the largest and best studied statistical sample of radio-loud gravitationally lensed systems. Preliminary 8.4-GHz fluxes and positions are already available from `www.jb.man.ac.uk/~njj/glens/class.html` and `www.jb.man.ac.uk/~ceres1/list_pub.html`. The whole database will eventually be made public.

The "CfA – Arizona Space Telescope Lens Survey" (*CASTLeS*) provides information and data on gravitational lens systems at `cfa-www.harvard.edu/glensdata/`. It includes HST and radio images that can be downloaded via `ftp`. The service distinguishes between multiply imaged systems and binary quasars.

The "*VLBI* Space Observatory Program" (*VSOP* or *HALCA*; `www.vsop.isas.ac.jp/`)

has put an 8-m radio antenna into a highly elliptical Earth orbit so as to extend terrestrial interferometer baselines into space. Several hundred sources in the *VSOP* Survey Program (`www.vsop.isas.ac.jp/obs/Survey.html`) are listed, together with their observational status, at `www.ras.ucalgary.ca/survey.html`. These were selected to have 5-GHz flux above 1 Jy and a radio spectrum flatter than S$\sim \nu^{-0.5}$. Galactic masers are also being surveyed. An image gallery, including the first images ever made in Space-*VLBI*, may be viewed at `www.vsop.isas.ac.jp/general/Images.html`. Further images of *EVN-HALCA* observations are available at `www.nfra.nl/jive/evn/evn-vsop.html`. The same page will soon provide access to *VLBA* images of over 350 extragalactic sources observed with the *VLBA* at 5 GHz prior to the *VSOP* launch and some results of the pre-launch OH-maser survey.

5.4. *Variable Sources and Monitoring Projects*

Since 1997, about forty Galactic and extragalactic variable sources have been monitored with the Green Bank Interferometer (*GBI*) at 2.25 and 8.3 GHz (*HPBW* 11″ and 3″, resp.), under *NASA*'s *OSIRIS* project. The instrument consists of three 26-m antennas, and the targets are preferentially X-ray and γ-ray active. Radio light curves and tables of flux densities are provided at `info.gb.nrao.edu/gbint/GBINT.html`.

The "University of Michigan Radio Astronomy Observatory" (*UMRAO*) database (`www.astro.lsa.umich.edu/obs/radiotel/umrao.html`; Hughes *et al.* (1992)) contains the ongoing observations of the University of Michigan 26-m telescope at 5, 8.5 and 15 GHz. A number of strong sources are frequently (weekly) monitored and some weaker sources a bit less often. Currently the database offers flux densities, and polarisation percentage and angle (if available), for over 900 sources. For some objects the on-line data go back to 1965.

5.4.1. *Solar Radio Data*

An explanation of the types of solar bursts and a list of special events observed can be found at `www.ips.gov.au/culgoora/spectro/`. Daily flux measurements of the Sun at 2.8 GHz (10.7 cm) back to 1947 are at `www.drao.nrc.ca/icarus/www/sol_home.shtml`. The Ondrejov Solar Radio Astronomy Group (`sunkl.asu.cas.cz/~radio`) provides an archive of events detected with a 3.0 GHz continuum receiver and two spectrographs covering the range 1.0–4.5 GHz. The Metsahovi Radio Station in Finland offers solar radio data at `kurp.hut.fi/sun`, like e.g. radio images of the full Sun at 22, 37 and 87 GHz, a catalogue of flares since 1989, or "track plots" (light curves) of active regions of the solar surface. For further data, get in contact with the staff at `solar@hut.fi` or `Seppo.Urpo@hut.fi`.

The "National Geophysical Data Center" (*NGDC*) collects solar radio data from several dozen stations over the world at `www.ngdc.noaa.gov/stp/SOLAR/getdata.html`. This "Radio Solar Telescope Network" (*RSTN*) of 55 stations has now collected 722 station-years worth of data. Information about solar bursts, the solar continuum flux, and spectra from *RSTN* may be retrieved and displayed graphically at the *URL* `www.ngdc.noaa.gov:8080/production/html/RSTN/rstn_search_frames.html`.

Since July 1992, the Nobeyama Radio Observatory (NRO) has been offering (at `solar.nro.nao.ac.jp/`) a daily 17 GHz image of the Sun, taken with its Nobeyama Radio Heliograph. Also available are daily total flux measurements at five frequencies between 1 and 17 GHz observed since May 1994, as well as some exciting images of solar radio flares.

Cracow Observatory offers daily measurements of solar radio emission at six decimetric frequencies (410–1450 MHz) from July 1994 to the present (`www.oa.uj.edu.pl/sol/`).

Measurements with the Tremsdorf radio telescope of the Astrophysics Institute Potsdam (*AIP*), Germany, based on four solar sweep spectrographs (40–800 MHz) are available at `aipsoe.aip.de/~det`.

The Radio Astronomy Group (*RAG*) of the ETH Zürich offers the data from its various digital radio spectrometers and sweep spectrograph at `www.astro.phys.ethz.ch/rag`, and also hosts the homepage of the "Community of European Solar Radio Astronomers" (*CESRA*). The "Joint Organization for Solar Observations" (*JOSO*) offers a comprehensive list of links to solar telescopes and solar data centres at `joso.oat.ts.astro.it`.

A wide variety of solar data, including East-West scan images from the Algonquin Radio Observatory 32-element interferometer, and reports of ionospheric data, are provided by the University of Lethbridge, Alberta, Canada (`holly.cc.uleth.ca/solar`, or its `ftp` server at `ftp://ftp.uleth.ca/pub/solar/`).

Finally, there are many sites about solar-terrestrial processes and "Space Weather Reports", e.g. at `www.sel.noaa.gov` or `www.ips.gov.au/`. There are spacecraft solar radio data at `lep694.gsfc.nasa.gov/waves/waves.html` and `www-istp.gsfc.nasa.gov/`.

6. Raw Data, Software, Images and Spectral Line Data

6.1. *Radio Observatories and their Archives*

A list of 70 radio astronomy centres with direct links to their WWW pages has been compiled at `www.ls.eso.org/lasilla/Telescopes/SEST/html/radioastronomy.html`, and the *URL* `msowww.anu.edu.au/~anton/astroweb/astro_radio.html` presents links to 65 radio telescopes. Further WWW sites of radio observatories may be found on *AstroWeb* (`www.cv.nrao.edu/fits/www/astronomy.html`), searching for "Telescopes" or "Radio Astronomy".

A recent inquiry by E. Raimond of *NFRA* revealed the following (priv.comm., cf. also `www.eso.org/libraries/iau97/commission5.html`).

• Most major radio observatories saved their data, which does not necessarily mean that they are still usable. Only the larger institutions kept data readable by copying them to modern media.

• Data usually become available publicly after 18 months. Outside users can (sometimes) search a catalogue of observations and/or projects.

• Retrieval of archived data often requires the help of observatory staff. Support is offered in general.

• Radio observatories with a usable archive, and user support for those who wish to consult it, typically do not advertise this service well!

6.1.1. *Archives of Centimetre- and Metre-wave Telescopes/Arrays*

For the 305-m antenna at Arecibo (Puerto Rico; `www.naic.edu/`) all data were kept, some still on reel-to-reel magnetic tapes, but no catalogue is available remotely. A catalogue of projects is searchable with the help of the staff.

Data obtained at the Effelsberg 100-m telescope of the Max-Planck Institut für Radioastronomie (*MPIfR*) at Bonn, Germany (`www.mpifr-bonn.mpg.de/effberg.html`) were originally archived only for the five years prior to the overwriting of the storage medium. Presently data are archived on CD-ROMs, but these are not accessible to the outside world. There is no public observations catalogue on-line, and help from staff is required to work with data taken with the 100-m antenna.

The "Australia Telescope National Facility" (*ATNF*) has archived all raw data of its "Compact Array" (*ATCA*; `www.narrabri.atnf.csiro.au/`). Observations and project databases are available at `www.atnf.csiro.au/observers/data.html`.

The *ATNF* Parkes 64-m telescope, despite its 37 years of operation, has no data archive so far. The multibeam H I surveys (§6.4.1), just started, will be archived, and results will be made public.

At the "Dominion Radio Astronomy Observatory" (*DRAO*; `www.drao.nrc.ca`) raw and calibrated data are archived. Observatory staff will assist in searching the observations catalogue. Results of the Galactic Plane Survey (in progress, §8.1) will be made available publicly via the *CADC*.

The Molonglo Observatory (`www.physics.usyd.edu.au/astrop/most/`) has archived its raw data. In general, data can be retrieved via collaborations with staff. Results of recently started surveys (§3.7.4, §8.1) will be made publicly available.

The archive of *NRAO's Very Large Array* contains data from 1979 to the present (excluding 1987) and can be interrogated at `info.aoc.nrao.edu/doc/vladb/VLADB.html`. Data are reserved for the observing team for 18 months following the end of the observations. Archive data after this period must be requested from either the Assistant Director or the Scheduler at the *VLA*.

The *Netherlands Foundation for Research in Astronomy* (*NFRA*) keeps an archive of all the raw data ever taken with the Westerbork Synthesis Radio Telescope (*WSRT*), an aperture synthesis telescope of 14 antennas, of 25-m diameter, in the Netherlands. At *URL* `www.nfra.nl/scissor/` one may browse this archive by various criteria (*RA*, *DEC*, frequency, observation date, etc.) and even formulate a request to obtain the data. Note that you need to specify a username and password (both "guest") before you may query the database. However, special auxiliary files will be needed to reduce these data e.g. with *AIPS*. For the processed results of the *WENSS* survey, see §3.7.1.

The "Multi-Element Radio-Linked Interferometer" (*MERLIN*), in the UK, has all raw data archived since *MERLIN* became a National Facility in 1990. The catalogue of observations older than 18 months is searchable on position and other parameters (`www.jb.man.ac.uk/merlin/archive`). For the actual use of archived data, a visit to Jodrell Bank is recommended in order to get the processing done properly.

6.1.2. *Archives of (Very) Long Baseline Interferometers*

At the "Very Long Baseline Array" (*VLBA*; `www.nrao.edu/doc/vlba/html/VLBA.html`) in the USA, all correlated data are archived, and the observations catalogue may be retrieved from `www.nrao.edu/ftp/cumvlbaobs.txt`. (When I inquired about the latter *URL*, the reply came with a comment : "I'm not sure any *URL* is sufficiently permanent to be mentioned in a book.")

The "European *VLBI* Network" (*EVN*) has its correlated data archived at *MPIfR* Bonn, Germany, so far. A catalogue of observations is available (§5.2), but the data are *not* in the public domain. Once the correlator at the "Joint Institute for *VLBI* in Europe" (*JIVE*) becomes operational, archiving will be done at Dwingeloo (`www.nfra.nl/jive`). A general problem is that in order to re-analyse archived data, calibration data are also needed, and these are not routinely archived.

The US Naval Observatory (*USNO*) has made *VLBI* observations for geodetic and reference frame purposes for more than a decade. Among others, these lead to the latest estimates of the precession and nutation constants (`maia.usno.navy.mil/eo/`). This archive (§5.2) contains all of the images available from the Geodetic/Astrometric experiments, including the *VLBA*. Almost all of these *VLBA* observations have been imaged. However, the *USNO* Geodetic/Astrometric database is huge, and the large, continuing project of imaging has been only partially completed. Examples of archival use of the Washington *VLBI* Correlator database can be found in Tateyama *et al.* (1998) and references therein.

6.1.3. *Millimetre Telescopes and Arrays*

The "Berkeley-Illinois-Maryland Association" (*BIMA*, `bima.astro.umd.edu/bima/`, Welch *et al.* (1996)) maintains a millimetre array which has no formal archiving policy. However, there is a searchable observatory archive of raw data. To request authorisation for access, visit the page `bima-server.ncsa.uiuc.edu/bima/secure/bima.html`.

At the "Caltech Millimeter Array" (`www.ovro.caltech.edu/mm/main.html`) all observed data are in a database. Searching requires the help of the staff, because it is not easy to protect proprietary data in another way.

The "Institut de RadioAstronomie Millimétrique" (*IRAM*) maintains the 30-m dish at Pico Veleta (Spain, `ram.fr/PV/veleta.html`), and the "Plateau de Bure Interferometer" (PDBI; `iram.fr/PDBI/bure.html`). The raw data from the PDBI are archived, but so far the observation catalogues are made available only via Newsletter and e-mail. Web-access is being considered for the future. The P. de Bure archive is only accessible within `iram.fr`, the *IRAM* local network. The only external access to the P. de Bure archive is to pull up the observations list month by month (from 1990 to the present) from the page `iram.fr/PDBI/project.html`.

The "James-Clerk-Maxwell Telescope" (*JCMT*; `www.jach.hawaii.edu/JCMT/`) on Mauna Kea (Hawaii) has its raw and processed data archived at the Canadian Astronomy Data Center (*CADC*; `cadcwww.dao.nrc.ca/jcmt/`).

The "Caltech Sub-mm Observatory" (*CSO*; `www.cco.caltech.edu/~cso/`) is a 10.4-m sub-mm dish on Mauna Kea (Hawaii) in operation since 1988. Its archive can be reached via the *URL* `puuoo.submm.caltech.edu/doc_on_vax/html/doc/archive.html`.

The "Nobeyama Millimetre Array" (*NMA*; `www.nro.nao.ac.jp/NMA/nma-e.html`) and the Smithsonian Sub-mm Array (*SMA*; `sma2.harvard.edu/`) have their archives in the software development stage.

In conclusion, archiving in radio astronomy is far from optimal, but not too bad either. Most major radio observatories have an archive of some sort, and accessibility varies from excellent to usable. Some observatories could advertise their archives better, e.g. on their own Web-pages, or by registering it in *AstroWeb*'s links. Most importantly, these archives are **very little** used by astronomers, and would be well worth many (thesis) projects, e.g. to study source variability over more than a decade. The following section gives some hints on where to start searching when software is needed to reduce some of the raw data retrievable from the Internet.

6.2. *Software for Radio Astronomy*

As are the observing techniques for radio astronomy, its data reduction methods are much more diverse than those in optical astronomy. Although *AIPS* (`www.cv.nrao.edu/aips`) has been the dominating package for radio interferometer data, many other packages have been developed for special purposes, e.g. *GILDAS, GREG*, & *CLASS* at *IRAM* (`iram.fr/doc/doc/doc/gildas.html`), Analyz at *NAIC* (`ftp://naic.edu/pub/Analyz`), GIPSY at Groningen (`ftp://kapteyn.astro.rug.nl/gipsy/`), Miriad by the *BIMA* and *ATNF* staff (`www.atnf.csiro.au/computing/software/miriad/`), and Karma at *ATNF* (`ftp://ftp.atnf.csiro.au/pub/software/karma/`). A comprehensive compilation of links can be found from the *AstroWeb* at `www.cv.nrao.edu/fits/www/yp_software.html`. The "Astronomical Software and Documentation Service" (*ASDS*; `asds.stsci.edu/asds`) contains links to the major astronomical software packages and documentation. It allows one to search for keywords in all the documentation files available.

A complete rewrite of the *AIPS* package from Fortran to `C++` code, known as the

`AIPS++` project (`aips2.nrao.edu/aips++/docs/html/aips++.html`), has been under way since mid-1991.

A note on preparing radio-optical overlays with AIPS: With the public availability of 2-dimensional maps from radio (*NVSS*, *FIRST*, *WENSS*) and optical (*DSS*) surveys, it is relatively easy to prepare radio-optical overlays for identification or publication purposes. Radio and optical maps of similar size should be culled in *FITS* format from the WWW. To identify the co-ordinate system of a map, *AIPS* looks for the *FITS*-header keyword "*EPOCH*" rather than "*EQUINOX*" (which is one of the very few bugs in the *FITS* definition!). Maps from SkyView (e.g. *DSS* and *NVSS*) seem to lack the *EPOCH* keyword in their *FITS* header, thus *AIPS assumes*(!) them to be of equinox 1950.0. (Both *FIRST* and *NVSS* maps, when taken from their home institutions, *STScI* and *NRAO*, respectively, do have the *EPOCH* keyword properly set.) Thus, for *AIPS* to work correctly on SkyView maps, it is necessary to introduce the proper "*EPOCH*" value in the map headers. This may be done with `gethead` and `puthead` in *AIPS*. Then, one of the maps (usually the one with the coarser pixel size) has to be prepared for re-gridding to the grid of the map with the finer pixel size. This preparation may be done with `EPOSWTCH`, before the actual re-gridding is done with `HGEOM`. Finally the task `KNTR` permits one to plot one of the maps (usually the optical) in greyscale, and the other as contours (usually the radio map). However, for more sophisticated combined plots of greys and contours (including white contours), other software packages (§6.2) allow finer artwork to be produced.

6.3. *Radio Images on the Internet*

Here we have to distinguish between images extracted from large-scale surveys, and images of individual sources. Both types will be discussed in the following two subsections.

6.3.1. *Images from Large-scale Surveys*

I had already mentioned (§3.7) that the very large-scale radio surveys like *NVSS*, *FIRST*, and *WENSS* offer (or are in the process of developing) so-called "postage-stamp servers", i.e. WWW interfaces where desired pieces of the 2-dimensional maps may be extracted, either in `gif` format, or, if one needs to work with the data, in the (usually about 10 times larger) *FITS* format. For the retrieval of large lists of small images, typically for identification projects, it is worth noting that several sites offer scripts (mostly based on `perl`) which allow the retrieval of these maps "from the command line", i.e. without even opening a WWW browser! The source list and map sizes may be pre-edited locally within a sequence of commands which are run in background (e.g. during the night, if necessary), and which will save the requested maps as files with names of the user's choice. For the *NVSS*, these may be obtained from W. Cotton (`bcotton@nrao.edu`) or from `skyview.gsfc.nasa.gov/batchpage.html`. For *FIRST* images look at `www.ast.cam.ac.uk/~rgm/first/collab/first_batch.html`, or use the `lynx` browser from the command line (consult R. White at `rlw@stsci.edu` in case of doubt).

FIRST and *NVSS* have mirror sites for their data products at the "Mullard Radio Astronomical Observatory" (*MRAO*, Cambridge, UK; `www.mrao.cam.ac.uk/surveys`), to allow faster access from Europe. Presently only part of the *FIRST* maps (and not the *FIRST* source catalogue) are available from there. Make sure that the piece of information you need is included at this site before concluding that it has not been observed.

Several of large-scale radio surveys are accessible from *NASA*'s `SkyView` facility (`skyview.gsfc.nasa.gov/`):

• the 34.5 MHz survey (Dwarakanath & Udaya Shankar (1990)) with the *GEETEE* telescope in India

• the 408 MHz all-sky survey (Haslam *et al.* 1982);

• the 1.4 GHz Stockert 25-m dish surveys (Reich (1982), Reich & Reich (1986))

• *FIRST* and *NVSS* at 1.4 GHz (see §3.7)

• the 4.85 GHz surveys of 1986+87 (Condon *et al.* (1994)) with the Green Bank 300-ft telescope, as well as their southern counterparts made with the Parkes 64-m dish (*PMN*; Condon *et al.* (1993)).

A 4.85 GHz survey made with the *NRAO* 140-ft antenna (covering 0^h <RA<20^h, $-40°< \delta < +5°$) is also available (Condon *et al.* (1991)). Descriptions of the surveys accessible from `Skyview` can be found at *URL* `skyview.gsfc.nasa.gov/cgi-bin/survey.pl`.

An attempt to list some of the survey work at radio wavelengths in both hemispheres was made with the page "Radio Surveys of the Southern and Northern Sky" (`wwwpks.atnf.csiro.au/databases/surveys/surveys.html`). Links to the data from these surveys are included, where available.

Extractions from the large-scale surveys made at *MPIfR* Bonn can be retrieved interactively from the *URL* `www.mpifr-bonn.mpg.de/survey.html`, including polarisation maps (Stokes Q and U) of the Galactic plane at 2.7 GHz.

The *WSRT* has been used to survey a section of the Galactic plane at 327 MHz (Taylor *et al.* (1996)). The region $43°< \ell <91°$, $|b| <1.6°$ was covered with 23 overlapping fields. Each field was observed at two epochs, several years apart, to identify variable sources. Combined intensity maps from both epochs, having a sensitivity of typically a few mJy and angular resolution of $1'\times1' \csc(\delta)$, may be viewed or retrieved as *FITS* images from the *URL* `www.ras.ucalgary.ca/wsrt_survey.html`.

The Hartebeesthoek Radio Astronomy Observatory (*HartRAO*) has used its 26-m dish at 2.3 GHz to map 67% of the southern sky ($0^h < \alpha <12^h, -80°< \delta < +13°$; $12^h < \alpha <24^h, -83°< \delta < +32°$) with an angular resolution of $20'$ (Jonas *et al.* (1998)). Until now this is the highest frequency at which such large areas have been mapped, while preserving large-scale emission features. To see a combination with northern sky surveys, go to `www.ru.ac.za/departments/physics/physics.html`, and click on "Radio Astronomy Group". Survey maps are available at `ftp://phlinux.ru.ac.za/pub/survey` (or contact J. Jonas at `phjj@hippo.ru.ac.za`).

The southern Galactic plane has been surveyed with the Parkes 64-m dish at 2.42 GHz (Duncan *et al.* (1995)). The region $238°< \ell <5°$, $|b| <5°$ was mapped with $10.4'$ resolution. The polarisation data of that survey have been published in Duncan *et al.* (1997), and are accessible from `www.rp.csiro.au/~duncan/project.html`. With a noise level of ~17 mJy/beam in total intensity, and 5–8 mJy/beam in Stokes Q and U, it is currently the most sensitive southern Galactic plane survey. Its sensitivity to extended ($\gtrsim30'$) low surface brightness structures is 3–5 times better than a 12-h synthesis with *MOST* or *ATCA*. It is able to detect *SNRs* of up to 20° in size. Since the ratio of the size of the maximal structure detectable to angular resolution is the same as for the *MGPS* (§8.1), they are complementary surveys.

The *IAU* (Comm. 9) Working Group on Sky Surveys (formerly "Wide Field Imaging") offers a "butterfly" collection of links to sky surveys in radio and other wavebands at `www-gsss.stsci.edu/iauwg/survey_url.html`. Some images of outstanding radio sources in the sky are clickable from an all-sky radio map at `www.ira.bo.cnr.it/radiosky/m.html` (in Italian).

6.3.2. *Image Galleries of Individual Sources*

The *NED* database has radio images linked to some of their objects, typically bright radio galaxies from the 3C catalogue. The "Astronomy Digital Image Library" (*ADIL*; `imagelib.ncsa.uiuc.edu/imagelib`) at the National Center for Supercomputing Applications (*NCSA*) offers a search interface by co-ordinates, object name, waveband, and other criteria. However, it is not clear from the start how many and what kind of radio images one may expect.

In an attempt to more adequately describe the phenomenon of classical double radio sources, Leahy (1993) coined the term "Double Radiosource Associated with Galactic Nucleus" (*DRAGN*) for these objects (Fig. 8). A gallery of images of the 85 nearest *DRAGN*s is available in the interactive "Atlas of *DRAGN*s" (Leahy *et al.* (1998)) at `www.jb.man.ac.uk/atlas/`. Apart from high-resolution images, the Atlas gives extensive explanations and references on the physical processes involved. The gallery of icons which has the objects sorted by their radio luminosity is especially instructive as a demonstration of the well-established transition from "Fanaroff-Riley" class I (*FR* I) for low-luminosity objects to *FR* II for high-luminosity ones. The editors of the Atlas are planning to publish their work after reducing new data of objects for which the published maps are as yet inadequate. The maps may be downloaded in *FITS* format, which allows a number of analyses to be performed on them, at will.

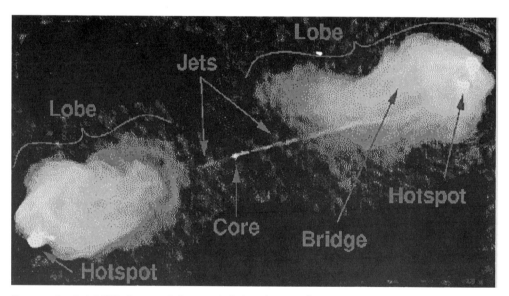

FIGURE 8. A 4.9 GHz image of Cygnus A (3C 405) at 0.4″ resolution, taken with all four configurations of the *VLA*, and showing the typical "ingredients" of a *DRAGN* (from Leahy et al. (1998), courtesy R. Perley, C. Carilli & J.P. Leahy). The overall size of the source is 2 arcmin, while the optical galaxy is ∼20″ in size on *DSS*, and coincides with the radio core.

More than a decade after its commissioning, the *VLA* has been equipped with low-frequency receivers at 74 and 330 MHz. At these frequencies, the field of view is so wide that the sky can no longer be approximated by a 2-dimensional plane, and 3-dimensional Fourier transforms are necessary to produce meaningful images. Some examples are given at `rsd-www.nrl.navy.mil/7214/weiler/4bandfarm.html`.

6.4. *Spectral Lines*

There is a general scarcity of WWW resources on spectral line data. Integrated source parameters may be obtained from selected catalogues in the *CDS* archive, most conveniently consulted via `vizier.u-strasbg.fr/cgi-bin/VizieR`.

A list of transition frequencies from 701 MHz to 3.43 THz, with references, has been published by Lovas (1992). A compilation of links to "Worldwide Molecular Astrophysics Resources" is offered at `www.strw.leidenuniv.nl/~iau34/links.html`. It provides pointers to several useful mm-wave line databases. The first of these links (dated June 1997) tells us that 114 molecules have now been detected in space, containing between two and *thirteen* atoms ($HC_{11}N$). One may search for known spectral lines by several parameters (frequency range, type of molecule, line strength, etc.) at the *URLs* `spec.jpl.nasa.gov/ftp/pub/catalog/catform.html` and `www.patnet.caltech.edu/~dmehring/find_lines.1.0.html`.

6.4.1. *Neutral Hydrogen (H I)*

By the 1950s, radio astronomers had already mapped the profiles of the hyperfine transition of the ground state of the hydrogen atom at 1420 MHz (21.1 cm). From this they could infer, using the formulae by Lindblad & Oort for differential Galactic rotation, the spiral structure of the H I distribution in the northern (Westerhout (1957)), and later in both Galactic hemispheres (e.g. Oort *et al.* (1958)).

Later these surveys were extended beyond the Galactic plane to the whole sky, and apart from their importance for the distribution and kinematics of Galactic H I, they are also a necessary tool for X-ray astronomers to estimate the Galactic absorption (from the H I column density) in X-ray spectra of extragalactic objects, near $\sim$0.1 keV. Many of these H I surveys are now available in the *CDS* archive (e.g. catalogues VIII/7, 8, 9, 10, 11, and 47). Curiously, some of these "treasures", e.g. the "Bell Laboratories H I Survey" by Stark *et al.* (1990; *CDS/ADC* #VIII/10) were never really published in an ordinary journal. Others, like VIII/47 (Westerhout *et al.* (1982)) are only now being prepared for integration into the *CDS* archive, 16 years after publication. Motivated by a poster displayed on the current winter school, the electronic version of the H I survey by Strong *et al.* (1982) was tracked down by the present author, and is now available as J/MNRAS/201/495/ from *ADC/CDS*! These authors used the Parkes 64-m dish to cover a Galactic plane region (ℓ=245°through 12°, $|b| <$10°), sampled every 0.5° in ℓ and 1° in b with an angular resolution of 15'.

One of the largest very recent H I surveys is indeed available electronically, but not in a public archive. Between 1988 through 1993 the sky north of δ=−30° was mapped with the Dwingeloo 25-m dish (*HPBW*=36') on a grid of 30' and with a velocity coverage of 850 km s^{-1}. The data resulting from this *Leiden-Dwingeloo H I Survey* are not presently available on the Internet, but were published as an Atlas and a CD-ROM (Hartmann & Burton (1997)). On the CD, there are colour GIF files for all images in the Atlas, as well as animations running through the data cube in velocity space.

Integrated H I fluxes and line widths of galaxies may be obtained from *LEDA*, accessible via telnet to `lmc.univ-lyon1.fr` (login as `leda`), or enter via the WWW at `www-obs.univ-lyon1.fr/leda`. Currently *LEDA* offers H I fluxes for $\sim$12,400 of a total of $\sim$170 000 galaxies in *LEDA*.

H I observations of external galaxies (mainly the gas-rich late-type ones) are important for deriving their rotation curves and detailed kinematics (e.g. Burton (1976)). These (flat) rotation curves have led to further evidence for the so-called "dark matter haloes" in galaxies. Numerous detailed studies have been published over the past three decades, very little of which has been preserved in electronic form, except for compilations of rotation curve parameters (e.g. Baiesi-Pillastrini *et al.* (1983), Persic & Salucci (1995) and Mathewson & Ford (1996)). Also, Martín (1998) gives a compilation of bibliographic references to HI maps of 1439 galaxies published between 1953 and 1995, as well as parameters drawn from these maps. A gallery of *ATCA* observations of HI in galaxies has been compiled by B. Koribalski at `www.atnf.csiro.au/~bkoribal/atca_obs.html`.

6.4.2. *Molecular and Recombination Lines, and Pulsars*

Carbon monoxide (CO) emits one of the most abundant molecular lines in space, e.g. the ^{12}CO(J=1-0) line at 115 GHz (λ=2.6 mm). These data allow one to infer the distribution of molecular hydrogen, e.g. in the plane of our Galaxy or throughout other galaxies. A composite survey of this line over the entire Milky Way (Dame *et al.* (1987)) at 8.7′ resolution is available as *ADC/CDS* catalogue #8039 and at *ADIL* (§6.3.2). It was made with two 1.2-m dishes, one in New York City, the other in Chile, and provides 720 latitude-velocity maps as *FITS* files, one for each 30′ of Galactic longitude, and a velocity-integrated map. The Five College Radio Astronomy Observatory (*FCRAO*) has mapped 336 deg^2 of the Galactic plane ($102.5°< \ell <141.5°$, $-3°< b < +5.4°$) in the ^{12}CO(1-0) line at 115 GHz with 45″ resolution (Heyer *et al.* (1998)), available at *ADIL* (§6.3.2; `imagelib.ncsa.uiuc.edu/document/97.MH.01`), and at *ADC*. A survey of the ^{12}CO(2-1) line at 230 GHz was made of the Galactic plane ($20°< \ell <60°$ and $|b| <1°$) with 9′ resolution, using the 60-cm sub-mm telescope at Nobeyama (Sakamoto *et al.* (1995)). It is available at *ADC* and *CDS* (#J/ApJS/100/125/).

The "Catalog of CO Observations of Galaxies" (Verter (1985), *ADC/CDS* #7064) is now largely outdated. The "*FCRAO* Extragalactic CO survey" has measured 1412 positions in 300 galaxies with the 14-m dish at *FCRAO* (*HPBW*=45″, Young *et al.* (1995)) in the ^{12}CO(1-0) line. It is apparently not available on the WWW, although its natural home appears to be the *URL* `donald.phast.umass.edu/~fcrao/library/`. The "Swedish *ESO* Submillimetre Telescope" (*SEST*) and the 20-m Onsala dish were used to search for CO in 168 galaxies (Elfhag *et al.* (1996)). The Nobeyama 45-m single dish has surveyed 27 nearby spirals in the ^{12}CO(1-0) line (Nishiyama & Nakai (1998)). Other comprehensive molecular line surveys and lists of transition frequencies have been published (not in electronic form) in Sutton *et al.* (1985), Turner (1989), and Schilke *et al.* (1997).

Radio recombination lines occur through transitions of electrons between two energy states with very high quantum number n. These lines are named after the atom, the destination quantum number and the difference in n of the transition (α for Δn=1, β for Δn=2, γ for Δn=3, etc.). Examples are H 157 α (transition from n=158 to n=157 in hydrogen) or He 109 β (transition from n=111 to n=109 in helium). They are mainly used to map departures from thermal equilibrium, and the velocity structure in HII regions or in planetary nebulae. To my knowledge there is no WWW site offering data on recombination lines, but see `www.hartrao.ac.za/spectra/SP_Hii.html` for an introduction to the research done with this type of observation. See Lilley & Palmer (1968) and Towle *et al.* (1996) for tables of radio recombination lines.

A successful search for pulsars over the entire southern sky with the Parkes 64-m dish, the so-called 70-cm Pulsar Survey or "Parkes Southern Pulsar Survey" detected 298 pulsars, 101 of them previously unknown (Lyne *et al.* (1998)).

7. Finding Literature, Addresses, and Proposal Forms on WWW

7.1. *Relevant Literature on the Internet*

As in other branches of astronomy, much of the most recent publications can be found on the WWW. The *LANL/SISSA* electronic preprint server (`xxx.lanl.gov`) has been described in some detail in my tutorial. Radio astronomy topics are becoming increasingly popular on this server, although some of the most productive radio astronomy institutions have not yet discovered its efficiency and cost savings for the distribution of their preprints. Some institutions offer at least the titles of their preprints (if not full versions) on the WWW. Those preprints still circulated only on paper may be found in the STEP- and RAPsheets of the *STScI* and *NRAO* (sesame.stsci.edu/lib/stsci-preprint-db.html and libwww.aoc.nrao.edu/aoclib/rapsheet.html). The comprehensive collection of astronomy newsletters at sesame.stsci.edu/lib/NEWSLETTER.htm, contains several items of interest to radio astronomers.

A few relevant proceedings volumes are also accessible on the WWW, e.g. the one on "Energy Transport in Radio Galaxies and Quasars" (Hardee *et al.* (1996)) discusses a wide variety of phenomena encountered in extragalactic radio sources, and papers from this volume are available as PostScript files from www.cv.nrao.edu/jetworks. Three other volumes (Cohen & Kellermann (1995), Zensus *et al.* (1995), Zensus *et al.* (1998)) bring together recent advances in high-resolution radio imaging of compact radio sources (see www.pnas.org/, www.cv.nrao.edu/vlbabook, and www.cv.nrao.edu/iau164/164book.html).

7.2. *Finding Radio Astronomers around the World*

Directories of astronomers in general have been described in section 7 of my tutorial for this winter school. Commission 40 ("Radio Astronomy") of the *IAU* offers a list of its 860 members (sma-www.harvard.edu/IAU_Com40/IAU_scroll.html), and 660 of them appear with their email address.

7.3. *Proposing Observations with Radio Telescopes*

Like in most other parts of the electromagnetic spectrum, proposals are accepted via email at most radio observatories. Many of them offer their proposal forms on the WWW, like e.g. for the various *NRAO* telescopes at www.nrao.edu/proposals.html, for the *ATNF* telescopes at www.atnf.csiro.au/observers/apply/form.html (including Parkes and *VLBI*), for *MERLIN* at www.jb.man.ac.uk/merlin/propsub/, for the Arecibo dish at www.naic.edu/vscience/proposal/proposal.htm, for the *BIMA* mm array at www.astro.uiuc.edu/~bima/call_for_proposals.html, or for the *JCMT* at www.jach.hawaii.edu/JCMT/pages/apply.html. *VLBI* proposal forms for the *VLBA* are found at the *NRAO* address above. *VLBI* proposal forms for the *EVN* can be found at www.nfra.nl/jive/evn/proposals/prop.html. No web forms were found e.g. for the *MRAO* or *IRAM* telescopes, nor for the *WSRT*.

Some institutions still require the proposals to be sent via regular mail, e.g. the *MPIfR* Bonn (www.mpifr-bonn.mpg.de/effelsberg/runkel/info_pke.html) for proposing time at the Effelsberg 100-m dish. It would be difficult here to give a comprehensive list of *URLs* for the many radio telescopes distributed over the globe.

8. The Near and Far Future of Radio Surveys and Telescopes

In this section I shall present some survey projects currently being carried out or planned, as well as some telescopes under construction or being designed. A good overview of current and planned radio astronomy facilities and recent research progress up to mid-1996 has been given in the latest Triennial Report of *IAU* Commision 40 ("Radio Astronomy"), at sma2.harvard.edu/IAU_Com40/c40rpt/c40report.html. The next 3-year report is to become available in late 1999 at www.iau.org/div10.html. Many such projects have also been described in the proceedings volume by Jackson & Davis (1997).

8.1. *Continuing or Planned Large-scale Surveys*

On the island of Mauritius a 151 MHz survey is being performed with the "Mauritius Radiotelescope" (MRT; Golap et al. (1998)), and may be regarded as the southern continuation of the *MRAO* 6C survey (cf. **Table 1**). This T-shaped array of helical antennas provides an angular resolution of $4' \times 4.6'$ csc(z), where z is the zenith distance. The aim is to map the sky between declinations $-10°$ and $-70°$ to a flux limit of $\lesssim 200$ mJy, including a map of the Galactic Plane and studies of pulsars. A catalogue of $\sim 10^5$ sources (icarus.uom.ac.mu/mrt2.html) can be expected after completion of the survey.

After the completion of *WENSS*, the *WSRT* started in late 1997 the "*WISH*" survey at 350 MHz (www.nfra.nl/nfra/projects/index.htm). The aim is to survey the region of effective overlap with *ESO*'s "Very Large Telescope" (www.eso.org/projects/vlt/), the "VLT", which is limited by the *WSRT* horizon and by the elongation of the synthesised beam. In order to have a minimum hour angle coverage of 4 h, the area $-30° \leq \delta \leq -10°$, $|b| > 10°$, or 5900 deg^2, will be covered. With an expected noise limit of about 3 mJy ($1\text{-}\sigma$), and a source density of 20 per deg^2, *WISH* should detect about 120 000 sources.

The *DRAO* Penticton aperture synthesis array is being used to survey the northern Galactic plane at 408 and 1420 MHz in the continuum, and at 1420 MHz in the H I line (www.drao.nrc.ca/web/survey.shtml). The area covered is $72° < \ell < 140°$, $-3° < b < +5°$, and the angular resolutions are $4'$ and $1'$. First results of this survey can be viewed at www.ras.ucalgary.ca/pilot_project.html.

At *MPIfR* Bonn a 1.4 GHz Galactic plane survey ($4° < |b| < 20°$), using the Effelsberg 100-m dish in both total intensity and polarisation, is under way (Uyanıker *et al.* (1998)). Examples of how this survey will be combined with the *NVSS*, and with polarisation data from Brouw & Spoelstra (1976), have been shown by Fürst *et al.* (1998).

The first-epoch "Molonglo Galactic Plane Survey" (*MGPS*-1; Green *et al.* (1999)) at 843 MHz, was obtained with the old, $70'$ field-of-view *MOST* and covers the region $245° < \ell < 355°$, $|b| < 1.5°$. The second-epoch Galactic plane survey (*MGPS*-2) is being made with the new, wide-field ($2.7°$) system at 843 MHz, and will cover the region $240° \leq \ell \leq 365°$, $|b| \leq 10°$. With an angular resolution of $43'' \times 43''$ csc δ and a noise level of 1–2 mJy/beam it is expected to yield over 80 000 sources above ~ 5 mJy (Green (1997)). As a part of *SUMSS* (§3.7), it is well under way, and its survey images can be viewed at www.physics.usyd.edu.au/astrop/MGPS/. Catalogues of sources will be prepared at a later stage.

The Hartebeesthoek Radio Astronomy Group (*HartRAO*) in South Africa, after having finished the 2.3 GHz southern sky survey (§6.3.1), is planning to use its 26-m dish for an 8.4 GHz survey of the southern Galactic plane in total intensity and linear polarisation (Jonas (1998)) at $\sim 6'$ resolution, and for deeper 2.3 GHz maps of interesting regions in the afore-mentioned 2.3 GHz survey.

8.2. *Very Recent Medium-Deep Multi-Waveband Source Surveys*

Between 1995 and 1997, the "*AT-ESP*" continuum survey was carried out at 1.4 GHz with *ATCA* (Prandoni *et al.* (1998)). This survey covers ~ 27 deg^2 near the South Galactic Pole with a uniform sensitivity of $\sim 70\,\mu$Jy ($1\,\sigma$). About 3 000 radio sources have been detected, one third of them being sub-mJy sources. Redshifts from the "*ESO* Slice Project" (*ESP*) redshift survey for 3342 galaxies (Vettolani *et al.* (1998)) down to $b_J \sim 19.4$ (boas5.bo.astro.it/~cappi/esokp.html) will allow studies of the population of low-power radio galaxies and of their 3-dimensional distribution.

The *VLA* has been used in C-configuration to carry out a sensitive 1.4 GHz survey of 4.22 deg^2 of the northern sky that have been surveyed also in the Far Infra-Red with the *ISO* satellite, as part of the "European Large Area *ISO* Survey" (*ELAIS*;

Ciliegi *et al.* (1998), `www.ast.cam.ac.uk/~ciliegi/elais/`). The 5σ flux limit of the survey ranges from 0.14 mJy (for 0.12 deg^2) to 1.15 mJy (for the entire 4.22 deg^2). A careful comparison of the catalogue of 867 detected radio sources with the *FIRST* and *NVSS* catalogues provided insights into the reliability and resolution-dependent surface brightness effects that affect interferometric radio surveys. Cross-identification with IR and optical objects is in progress.

The "Phoenix Deep Survey" (Hopkins *et al.* (1998)) has used the *ATCA* to map a 2° diameter region centred on $(\alpha,\delta)=$ J $01^h 14^m 12.2^s, -45° 44' 08''$. A total of 1079 sources were detected above ~ 0.2 mJy (`www.physics.usyd.edu.au/~ahopkins/cats`). Optical identifications were proposed for half of the sources, and redshifts were measured for 135 of these. A comparison with lower resolution 843 MHz *MOST* maps is in progress.

8.3. *Extending the Frequency Range of the Radio Window*

One of the very pioneers of radio astronomy, G. Reber, has been exploiting methods to observe cosmic radio emission at $\sim$2 MHz from the ground, even very recently. He quite successfully did so from two places in the world where the ionosphere appears to be exceptionally transparent (see Reber (1994) and Reber (1995)).

The lowest frequency observations regularly being made from the ground are done with the "Bruny Island Radio Spectrometer" (Erickson (1997)) on Bruny Island, south of Hobart (Tasmania). It is used for the study of solar bursts in the rarely observed frequency range from 3 to 20 MHz. Successful observations are made down to the minimum frequency that can propagate through the ionosphere. This frequency depends upon the zenith distance of the Sun and is usually between 4 and 8 MHz.

However, for many years radio astronomers have dreamt of extending the observing window to frequencies significantly below a few tens of MHz (where observations can be made more easily from the ground) to a few tens of kHz (just above the local plasma frequency of the interplanetary medium). Ionospheric absorption and refraction requires this to be done from space. The first radio astronomy experiment at kHz frequencies, and the first one from Space, was the "Radio Astronomy Explorer" (*RAE*; Kaiser (1987)), in the late 1960s and early 1970s. It consisted of a V-shaped antenna 450 m in extent, making it the largest man-made structure in space. It was equipped with radiometers for 25 kHz to 13.1 MHz. Although no discrete Galactic or extragalactic sources were detected, very crude all-sky maps were made, and solar system phenomena studied. Since then none of the various space projects proposed have been realised. Recent plans for developing low-frequency radio astronomy, both from the ground and from space, can be viewed at `rsd-www.nrl.navy.mil/7214/weiler` by following the links to Low Frequency Radio Astronomy (*LFRA*) and associated pages, but see also the *ALFA* project (§8.7). The proceedings volume by Kassim & Weiler (1990) is full of ideas on technical schemes for very low-frequency radio observatories, and on possible astrophysical insights from them.

Efforts to extend the radio window to very high frequencies have been much more serious and successful in the past two decades, and have led to a whole new branch of "mm-wave astronomy". The multi-feed technique (§2.1) has seen a trend moving away from just having a single receiver in the focal plane, towards having multiple receivers there, to help speed up the data collection (as e.g. in the Parkes H I multibeam survey, §8.4). By building big correlators, and taking the cross-products between the different beams, the complex field distribution in the focal plane of a dish may be mapped, and by transforming that one can correct for pointing, dish deformation, etc. Arrays of, say, 32 by 32 feeds are able to "image" the sky in real time (see e.g. the SEQUOIA system at the *FCRAO* 14-m dish, `donald.phast.umass.edu/~fcrao/instrumentation/`). This is only possible at mm wavelengths, where the equipment is small enough to fit into the focal plane. In perhaps three years such receivers should exist at $\sim$100 GHz (3 mm).

As mentioned in §3.4 there is a lack of large-area surveys at frequencies above $\sim$5 GHz. As Condon (1998) has pointed out, such surveys are made difficult since the beam solid angle of a telescope scales as ν^{-2} and system noise generally increases with frequency, so the time needed to survey a given area of sky rises very rapidly above 5 GHz. However, a 7-beam 15 GHz continuum receiver being built for the *GBT* (§8.6) could cover a 1-degree wide strip along the Galactic plane in one day, with an rms noise of $\sim$2 mJy. Repeating it several times would provide the first sensitive and systematic survey of variable and transient Galactic sources, such as radio stars, radio-emitting γ-ray sources, X-ray sources, etc.

More promising for the investigation of possible new source populations at these high frequencies (§3.4) may be the results from the new *CMB* satellites. One is the "Microwave Astronomy Probe" (*MAP*; map.gsfc.nasa.gov/html/web_site.html), expected to be launched by *NASA* in 2000. It will operate between 22 and 90 GHz with a 1.4×1.6 m diameter primary reflector, offering angular resolutions between 18′ and 54′. The other one is *PLANCK* (astro.estec.esa.es/SA-general/Projects/Planck/), to be launched by *ESA* in 2006 (possibly on the same bus as the "Far InfraRed and Submillimetre Telescope", *FIRST*, not to be confused with the *VLA FIRST* radio survey). *PLANCK* will have a telescope of 1.5 m aperture, and it will be used with radiometers for low frequencies (30–100 GHz; λ=3–10 mm), and with bolometers for high frequencies (100–857 GHz; λ 0.3–3.0 mm), with angular resolutions of $\sim$10′ at 100 GHz. Both the *MAP* and *PLANCK* missions should detect a fair number of extragalactic sources at 100 GHz (λ=3 mm). In fact, Tegmark & De Oliveira-Costa (1998) expect *PLANCK* to detect 40 000 discrete sources at 857 GHz. A highly important by-product will be the compilation of a much denser grid of calibration sources at mm wavelengths. The vast majority of the currently known mm-wave calibrators are variable anyway.

8.4. Spectral Line and Pulsar Surveys

The Australia Telescope National Facility (*ATNF*) has constructed and commissioned a 21-cm multi-feed system with 13 receivers at the prime focus of the Parkes 64-m telescope (Staveley-Smith (1997); wwwpks.atnf.csiro.au/people/multi/multi.html). The feeds are disposed to form beams with an angular resolution of 14′ and a distance of $\sim$28′ between neighbouring feeds. The on-line correlator measures flux density in all 13 channels and 2 polarisations simultaneously, with a spectral resolution of 16 km s^{-1} and a velocity range from -1200 km s^{-1} to $+12,700$ km s^{-1}. The Parkes multi-beam facility commenced regular observing in 1997, and a report on its status is regularly updated at www.atnf.csiro.au/research/multibeam/multibeam.html. Several major HI surveys are planned, including an "all-sky" survey ($\delta \lesssim +20°$) with a limiting sensitivity (5σ, 600 s) of $\sim$20 mJy per channel. The Zone of Avoidance (ZOA, $|b| <5°$) will be covered with the same velocity range and twice the sensitivity. It will be sensitive to objects with HI mass between 10^6 and 10^{10} M$_\odot$, depending on distance. This will be the first extensive "blind" survey of the 21-cm extragalactic sky. When scheduled, it is possible to watch the signal of all 13 beams almost *in real time* at wwwpks.atnf.csiro.au/people/multi/public_html/live/multibeam_live.html. An extension of this survey to the northern hemisphere ($\delta \gtrsim +20°$) will be performed with the Jodrell-Bank 76-m Mark I antenna, but only 4 receivers will be used.

This Parkes multi-beam system is also being used for a sensitive wide-band continuum search for pulsars at 1.4 GHz, initially limited to the zone $220° < \ell < 20°$ within $5°$ from the Galactic plane. A first observing run in August 1997 suggested that 400 new pulsars may be found in this survey, which is expected to take ~100 days of telescope time at Parkes, spread over two years (*ATNF* Newsletter 34, p. 8, 1998).

The "Westerbork observations of neutral Hydrogen in Irregular and SPiral galaxies" (*WHISP*; www.astro.rug.nl/~whisp) is a survey to obtain *WSRT* maps of the distribution and velocity structure of H I in 500 to 1000 galaxies, increasing the number of galaxies with well-studied H I observations by an order of magnitude. By May 1998 about 280 galaxies had been observed, and the data had been reduced for 160 of them. H I profiles, velocity maps, and optical finding charts are now available for 150 galaxies. Eventually the data cubes and (global) parameters of all galaxies will also be made available.

The Dwingeloo 25-m dish is currently pursuing the "Dwingeloo Obscured Galaxy Survey" (*DOGS*; Henning *et al.* (1998); www.nfra.nl/nfra/projects/dogs.htm) of the area $30° \le \ell \le 220°$; $|b| \le 5.25°$. This had led to the discovery of the nearby galaxy Dwingeloo 1 in August 1994 (Kraan-Korteweg *et al.* (1994)). After a shallow survey in the velocity range 0–4000 km s^{-1} with a noise level of 175 mJy per channel, a second, deeper survey is being performed to a noise level of 40 mJy. The latter has so far discovered 40 galaxies in an area of 790 deg^2 surveyed to date.

The first results of a dual-beam H I survey with the Arecibo 305-m dish have been reported in Rosenberg & Schneider (1998). In a 400 deg^2 area of sky 450 galaxies were detected, several of them barely visible on the Palomar Sky Survey.

Since 1990, the Nagoya University has been executing a ^{13}CO(1-0) survey at 110 GHz of the Galactic plane, with a 4-m mm-wave telescope. Since 1996, this telescope is operating at La Silla (Chile) to complete the southern Galactic plane (Fukui & Yonekura (1998)). The *BIMA* mm-array is currently being used to survey 44 nearby spiral galaxies in the ^{12}CO(1-0) line at 6″–9″ resolution (Helfer *et al.* (1998)).

8.5. *CMB and Sunyaev-Zeldovich Effect*

The cosmic microwave background (*CMB*) is a blackbody radiation of 2.73 K and has its maximum near ~150 GHz (2 mm). Measurements of its angular distribution on the sky are highly important to constrain cosmological models and structure formation in the early Universe, thus the mapping of anisotropies of the *CMB* has become one of the most important tools in cosmology. For a summary of current *CMB* anisotropy experiments, see Wilkinson (1998) and Bennett *et al.* (1997); the latter even lists the relevant *URLs* (cf. also brown.nord.nw.ru). As an example, the Cambridge "Cosmic Anisotropy Telescope" (CAT; Robson *et al.* (1993), www.mrao.cam.ac.uk/telescopes/cat/) has started to map such anisotropies, and it is the prototype for the future, more sensitive "Very Small Array" (*VSA*; §8.6.2).

The "Sunyaev-Zeldovich" (*SZ*) effect is the change of brightness temperature T_B of the *CMB* towards regions of "hot" (T ~10^7 K) thermal plasma, typically in the cores of rich, X-ray emitting clusters of galaxies. The effect is due to the scattering of microwave photons by fast electrons, and results in a diminution of T_B below ~ 200 GHz, and in an excess of T_B above that frequency. See Birkinshaw (1999) for a comprehensive review of past observations and the potential of these for cosmology. See Liang & Birkinshaw (1998) for the status and future plans for observing the Sunyaev-Zeldovich effect.

8.6. *Radio Telescopes: Planned, under Construction or being Upgraded*

8.6.1. *Low and Intermediate Frequencies*

The Arecibo observatory has emerged in early 1998 from a 2-year upgrading phase (www.naic.edu/techinfo/teltech/upgrade/upgrade.htm). Thanks to a new Gregorian reflector, the telescope has a significantly increased sensitivity.

The National Centre for Radio Astrophysics (*NCRA*) of the Tata Institute for Funda-mental Research (*TIFR*, India) is nearing the completion of the "Giant Metrewave Radio Telescope" (*GMRT*) at a site about 80 km north of Pune, India (information about the facility can be found at the *URL* www.ncra.tifr.res.in). With 30 fully steerable dishes of 45 m diameter, spread over distances of up to 25 km, it is the world's most powerful radio telescope operating in the frequency range 50–1500 MHz with angular resolutions between 50″ and 1.6″. In June 1998, all 30 dishes were controllable from the central electronics building. Installation of the remaining feeds and front ends is expected in summer 1998. The digital 30-antenna correlator, combining signals from all the anten-nas to produce the complex visibilities over 435 baselines and 256 frequency channels, is being assembled, and it will be installed at the *GMRT* site also in summer 1998. The entire *GMRT* array should be producing astronomical images before the end of 1999.

The *NRAO* "Green Bank Telescope" (*GBT*; www.gb.nrao.edu/GBT/GBT.html) is to re-place the former 300-ft telescope which collapsed in 1988 from metal fatigue. The *GBT* is a 100-m diameter single dish with an unblocked aperture, to work at frequencies from 300 MHz to ∼ 100 GHz, with almost continuous frequency coverage. It is finishing its construction phase, and is expected to be operational in 2000 (Vanden Bout (1998)).

The *VLA* has been operating for 20 years now, and a plan for an upgrade has been discussed for several years. Some, not very recent, information may be found at the *URL* www.nrao.edu/vla/html/Upgrade/Upgrade_home.shtml. Among other things, larger subre-flectors, more antennas, an extension of the A-array, a super-compact E-array for mosaics of large fields, and continuous frequency coverage between 1 and 50 GHz are considered.

An overview of current *VLBI* technology and outlooks for the future of *VLBI* have been given in the proceedings volume by Sasao *et al.* (1994).

For several years the need for and the design of a radio telescope with a collecting area of one square kilometre have been discussed. The project is known under different names: the "Square Kilometre Array Interferometer" (*SKAI*; www.nfra.nl/skai/; Brown (1996)); the "Square Kilometre Array" (*SKA*; www.drao.nrc.ca/web/ska/ska.html), and the "1-km teleskope" (1kT; www.atnf.csiro.au/1kT). A Chinese version under the name "Kilometer-square Area Radio Synthesis Telescope" (*KARST*; www.bao.ac.cn/bao/LT) was presented by Peng & Nan (1998), and contemplates the usage of spherical (Arecibo-type) natural depressions, frequently found in southwest China, by the placing of ∼ 30 passive spherical reflectors, of ∼ 300 m diamter, in each of them. A frequency coverage of 0.2–2 GHz is aimed at for such an array of reflectors.

A new design for a large radio telescope, based on several almost flat primary reflectors, has been recently proposed (Legg (1998)). The reflectors are adjustable in shape, and are of very long focal length. The receiver is carried by a powered, helium-filled balloon. Positional errors of the balloon are corrected either by moving the receiver feed point electronically, or by adjusting the primary reflector so as to move its focal point to follow the balloon. The telescope has the wide sky coverage needed for synthesis observations and an estimated optimum diameter of 100–300 m. It would operate from decimetre to cm-wavelengths, or, with smaller panels, mm-wavelengths.

8.6.2. *Where the Action is: Millimetre Telescopes and Arrays*

The "Smithsonian Submillimeter Wavelength Array" (*SMA*; sma2.harvard.edu/) on Mauna Kea (Hawaii) consists of eight telescopes of 6 m aperture, six of these provided by the Smithsonian Astrophysical Observatory (*SAO*) and two by the Astronomica Sinica Institute of Astronomy and Astrophysics (*ASIAA*, Taiwan). Eight receivers will cover all bands from 180 to 900 GHz (λ=1.7–0.33 mm). To achieve an optimised coverage of the uv plane, the antennas will be placed along the sides of Reuleaux triangles, nested in such a way that they share one side, and allow both compact and wide configurations. Baselines will range from 9 to 460 m, with angular resolutions as fine as 0.1″. The correlator-spectrometer with 92,160 channels will provide 0.8 MHz resolution for a bandwidth of 2 GHz in each of two bands. One of the *SMA* telescopes has had "first light" in spring 1998, and the full *SMA* is expected to be ready for observations in late 1999.

Since April 1998, the *ATNF* is being upgraded to become the first southern hemisphere mm-wave synthesis telescope (cf. *ATNF* Newsletter 35, Apr 1998). The project envisages the *ATNF* to be equipped with receivers for 12 and 3 mm (Norris (1998)).

The "Millimeter Array" (*MMA*; www.mma.nrao.edu/) is a project by *NRAO* to build an array of 40 dishes of 8–10 m diameter to operate as an aperture synthesis array at frequencies between 30 and 850 GHz (λ =0.35–10 mm). Array configurations will range from about 80 m to 10 km. It will most probably be placed in the Atacama desert in northern Chile at an altitude near 5000 m, a site rivalling the South Pole in its atmospheric transparency (Vanden Bout (1998)).

The "Large Southern Array" project (*LSA*) is co-ordinated by *ESO*, *IRAM*, *NFRA* and Onsala Space Observatory (*OSO*), and it anticipates the building of a large millimetre array with a collecting area of up to 10 000 m^2, or roughly 10 times the collecting area of today's largest millimetre array in the world, the *IRAM* interferometer at the Plateau de Bure with five 15-m diameter telescopes. With baselines foreseen to extend to 10 km, the angular resolution provided by the new instrument will be that of a diffraction-limited 4-m optical telescope. Current plans are to provide the collecting area equivalent to 50–100 dishes of between 11 and 16 m diameter, located on a plain above 3000 m altitude. Currently only site testing data are available on the WWW (puppis.ls.eso.org/lsa/lsahome.html).

A similar project in Japan, the "Large Millimeter and Submillimeter Array" (*LMSA*; www.nro.nao.ac.jp/LMSA/lmsa.html) anticipates the building of a mm array of 50 antennas of 10 m diameter each, with a collecting area of 3,900 m^2, to operate at frequencies between 80 and 800 GHz.

The *MMA*, *LSA* and *LMSA* projects will be so ambitious that negotiations to join the *LMSA* and *MMA* projects, and perhaps all three of them, are under way. The name "Atacama Array" has been coined for such a virtual instrument (see *NRAO* Newsletter # 73, p. 1, Oct. 1997). The *MMA* will also pose challenging problems for data archiving, and in fact will rely on a new data storage medium to enable archiving to be feasible (cf. www.mma.nrao.edu/memos/html-memos/abstracts/abs164.html).

A comparison of current and future mm arrays is given in Table 3.

The "Large Millimeter Telescope" (*LMT*) is a 50-m antenna to be built on the slopes of the highest mountain in Mexico in the Sierra Negra, ∼200 km east of Mexico City, at an elevation of 4500 m. It will operate at wavelengths between 90 and 350 GHz (λ=0.85–3.4 mm) achieving angular resolutions between 5″ and 20″ (see lmtsun.phast.umass.edu/).

There are plans for a 10-m sub-mm telescope at the South Pole (Stark *et al.* (1998); cfa-www.harvard.edu/~aas/tenmeter/tenmeter.html). The South Pole has been identified as the best site for sub-mm wave astronomy from the ground. The 10-m telescope will be suitable for "large-scale" (1 deg^2) mapping of line and continuum from sub-mm sources at mJy flux levels, at spatial resolutions from 4″ to 60″, and it will make arcminute scale *CMB* measurements.

The "Very Small Array" (www.jb.man.ac.uk/~sjm/cmb_vsa.htm; astro-ph/9804175), or "*VSA*", currently in the design phase, consists of a number of receivers with steerable horn antennas, forming an aperture synthesis array to work at frequencies around 30 GHz (λ=10 mm). It will be placed at the Teide Observatory on Tenerife (Spain) around the year 2000. The *VSA* will provide images of structures in the *CMB*, on angular scales

TABLE 3. Comparison of Current and Future mm Arrays [a]

Array	Completion Date	Wavelength Range (mm)	Sensitivity [b] at 3mm (Jy)	max baseline (km)
Nobeyama (6 × 10 m)	~1986	3.0, 2.0	1.7	0.36
IRAM (5 × 15 m)	~1988	3.0, 1.5	0.3–0.8	0.4
OVRO (6 × 10.4 m)	~1990 ?	3.0, 1.3	0.5	0.3
BIMA (9 × 6 m)	1996	3.0 (1.3)	0.7	1.4
SMA CfA (8 × 6 m)	1999 ?	1.7–0.33	-	0.46
ATCA (5 × 22 m)	2002 ?	12.0, 3.0	0.5 ?	3.0 (6.0 ?)
MMA USA (40 × 10 m?)	2010 ?	10.0–0.35	0.04 ?	10.0
LSA Europe (50 ? × 16 m?)	2010 ?	3.0, 1.3,...	0.02 ?	10.0
LMSA Japan (50 × 10m)	2010 ?	3.5–0.35	0.03?	10.0

a) adapted from Norris (1998), but see also Stark *et al.* (1998) for mm-wave single dishes
b) rms continuum sensitivity at 100 GHz to a point source observed for 8 hours

from $10'$ to $2°$. Such structures may be primordial, or due to the *SZ* effect (§8.5) of clusters of galaxies beyond the limit of current optical sky surveys.

The "Degree Angular Scale Interferometer" (*DASI*; `astro.uchicago.edu/dasi`) is designed to measure anisotropies in the *CMB*, and consists of 13 closely packed 20-cm diameter corrugated horns, using cooled High Electron Mobility Transistor (HEMT) amplifiers running between 26 and 36 GHz. It will operate at the South Pole by late 1999. A sister instrument, the "Cosmic microwave Background Interferometer" (CBI; `astro.caltech.edu/~tjp/CBI/`) will be located at high altitude in northern Chile, and it will probe the *CMB* on smaller angular scales.

8.7. *Space Projects*

Radioastron (Kardashev (1997); `www.asc.rssi.ru/radioastron/`) is an international space *VLBI* project led by the "Astro Space Center" of the Lebedev Physical Institute in Moscow, Russia. Its key element is an orbital radio telescope that consists of a deployable 10-m reflector made of carbon fibre petals. It will have an overall rms surface accuracy of 0.5 mm, and operate at frequencies of 0.33, 1.66, 4.83 and 22.2 GHz. It is planned to be launched in 2000–2002 on a Proton rocket, into a highly elliptical Earth orbit with an apogee of over 80 000 km.

The "Swedish-French-Canadian-Finnish Sub-mm Satellite" (*ODIN*) will carry a 1.1-m antenna to work in some of the unexplored bands of the electromagnetic spectrum, e.g. around 118, 490 and 560 GHz. The main objective is to perform detailed studies of the physics and the chemistry of the interstellar medium by observing emission from key species. Among the objects to be studied are comets, planets, giant molecular clouds and nearby dark clouds, protostars, circumstellar envelopes, and star forming regions in nearby galaxies (see `kurp-www.hut.fi/spectroscopy/space-projects.shtml`).

A new space *VLBI* project, the "Advanced Radio Interferometry between Space and Earth" (*ARISE*) has recently been proposed by Ulvestad & Linfield (1998). It consists in a 25-m antenna in an elliptical Earth orbit between altitudes of 5,000 and ~40,000 km, operating at frequencies from 5 to 90 GHz. The estimated launch date is the year 2008.

A space mission called "Astronomical Low-Frequency Array" (*ALFA*) has been proposed to map the entire sky between 30 kHz and 10 MHz. The project is in the development phase (`sgra.jpl.nasa.gov/html_dj/ALFA.html`) and no funding exists as yet.

The far side of the Moon has been envisaged for a long time as an ideal site for radio astronomy, due to the absence of man-made interference. Speculations on various kinds of radio observatories on the Moon can be found in the proceedings volumes by Burns & Mendell (1988), Burns *et al.* (1989), and Kassim & Weiler (1990). Since then, the subject has been "dead" as there has been no sign of interest by the major space agencies in returning to the Moon within the foreseeable future.

8.8. *Nomenclature and Databases*

More and more astronomers rely on databases like *NED, SIMBAD & LEDA*, assuming they are complete and up-to-date. However, researchers should make life easier for the managers of these databases, not only by providing their results and data tables directly to them, but also by making correct references to astronomical objects in their publications. According to *IAU* recommendations for the designation of celestial objects outside the solar system (`cdsweb.u-strasbg.fr/iau-spec.html`), existing names should neither be changed nor truncated in their number of digits. Acronyms for newly detected sources, or for large surveys, should be selected carefully so as to avoid clashes with existing acronyms. The best way to guarantee this is to register a new acronym with the *IAU* at `vizier.u-strasbg.fr/cgi-bin/DicForm`. For example, in D. Levine's lectures for this winter school the meaning of *FIRST* is very different from that in the present paper (cf. §8.3). Together with the Task Group on Designations of Commission 5 of the *IAU*, the author is currently involved in a project to allow authors to check their preprints for consistency with current recommendations. This should *not* be seen just as a further obstacle for authors, but as an offer to detect possible non-conforming designations which are likely to lead to confusion when it comes to the ingestion of these data into public databases.

9. Summary of Practicals

Two afternoons of three to four hours were set aside for exercises using the WWW facilities listed in these lectures. In the first practical, the students were offered the names of five very extended ($\sim 20'$) radio galaxies from the 3C catalogue, and asked to find out the positions of one or more of them from *NED* or *SIMBAD*, to obtain an optical finding chart from one of the various *DSS* servers, and to plot these with sky co-ordinates along the margins. The next task was to extract a 1.4 GHz radio image from the *NVSS* survey, and a list of sources from the *NVSS* catalogue of the same region. A comparison of the two gave an idea of how well (or less well) the catalogued sources (or components) represent the real complex structures of these sources. The students were also asked to look at higher-resolution images of these sources in the "Atlas of *DRAGNs*". A further exercise was to find out where the many names under which these sources were known in *NED* or *SIMBAD* come from, by looking up the acronyms in the On-line Dictionary of Nomenclature. The optical object catalogues like *APM, APS* and *COSMOS* were then queried for the same regions of sky, which allowed the object classification as star or galaxy to be checked by comparison with the optical charts from the *DSS* server. Eventually, radio images from the *FIRST* survey were extracted in order to see to what extent the large-scale structure of the radio galaxies could still be recognised.

In the second practical, the students were given a chance to discover a new, optically bright radio-loud quasar, a radio galaxy, a starburst galaxy, or even a radio star! Each of the roughly two dozen participating students was assigned a region of sky of the size of a Palomar plate ($6.5° \times 6.5°$) in the zone $08^h < \text{RA} < 16^h$, $+22° < \delta < +42°$ (the region covered by the *FIRST* 1.4 GHz survey at that time). Each student was asked to extract all bright objects (10 mag $<$ B $<$ 17 mag) from the *USNO* A1.0 catalogue (cf. §3 of my tutorial) in the region assigned to them. This was done by remote interrogation of the

USNO site, using the command `findpmm` from the *CDS* client software, which had been installed for the school on the computers at the *IAC*. The resulting object list (typically 3 000–10 000 objects per student, depending on the Galactic latitude of the assigned field) was reformatted so as to serve as input for interrogation of the *NVSS* source catalogue, which I had installed at the *IAC* for the winter school. It had 1.67 million sources in Nov. 1997. The FORTRAN program *NVSS*list, publicly available from *NRAO*, was used to search a circle of radius $10''$ around each optical object in the *NVSS* catalogue, limited to 1.4 GHz fluxes greater than 10 mJy so as to assure the positional accuracy of the radio sources. The students were asked to estimate the chance coincidence rate, and they found that between 0.5 and 2 matches were to be expected by chance. Actually each student had between three and 18 "hits" and was asked to concentrate on the optically or radio-brightest objects to find out whether the identification was correct, whether it was new, and what was known previously about the object. For promising candidates, a search in the *FIRST* 1.4 GHz image database was suggested, as well as an extraction of a *DSS* image. Some students even managed a radio-optical overlay, and one of them found that the overlay facility in SkyView had a bug when the pixel sizes of the overlaid images (like e.g. *NVSS* and *DSS*) was not identical. This was later reported to, acknowledged and fixed by the SkyView team. Unfortunately the *FIRST* image server went "out of service" right during the practical.

With the 23 participating students, a sky area of 800 deg^2 had been covered, and a cross-identification of altogether ∼80 000 optical objects with ∼12,500 radio sources was accomplished. Fifteen students sent me their results, of which only the most spectacular will be mentioned here. The 13.9 mag IRAS galaxy NGC 3987 almost filled the $3' \times 3'$ *DSS* image (which the students were asked to extract) and gave a splendid appearance with its edge-on orientation and strong dust lane. It was found to coincide with a 58 mJy *NVSS* source extended along the disk of the galaxy, while *FIRST* clearly shows a strong compact source (*AGN*?) and weak radio emission along the disk. A subsequent search in *NED* turned up a few other detailed radio studies (Burns *et al.* (1987), Condon & Broderick (1986b), and Jaffe *et al.* (1986)). Comparison of the 609 MHz flux from the latter reference shows that the compact central source has an inverted spectrum (rising with frequency), apparently not noticed before in literature. Another student came across UGC 5146 (Arp 129), an interacting pair. While the *FIRST* image server was unavailable, the *FIRST* catalogue showed it to be a very complex source, aligned along the connecting line between the pair, and very extended. A third student "rediscovered" the well-known Seyfert galaxy NGC 4151 with its 600 mJy nuclear point source. At first sight no bright radio star had been discovered, not surprising given the more thorough searches for radio stars now available (§3.6).

10. Conclusions

This was a most unusual and rewarding winter school, and exhausting for lecturers, students and organisers. The organisers are to be congratulated for the excellent planning and running of the school, and the *IAC* for its vision and courage to choose such a topic, which, at least when it comes to requests for funding, is often claimed to lack merit, and to be regarded as *not scientific*. The school has clearly proven that scientific expertise is a prerequisite for constructing and maintaining data archives, databases, and WWW interfaces, so as to make them user-friendly and reliable at the same time. Too much effort is often spent on fancy user interfaces, rather than on the content or its adequate documentation in an archive or database.

I have tried to show that more concerted effort is necessary to avoid duplication of similar WWW facilities, and the deterioration of WWW pages with outdated information. Much effort is being spent by individuals, without an institutional support or obligation, in providing useful WWW pages. The advantage is that these are often highly motivated and qualified researchers, but also with the disadvantage that the service will likely be discontinued with personal changes.

I also hope that my lectures will stimulate the use of archives both for advanced research as well as for thesis projects. Nevertheless, the warnings and pointers to possible pitfalls I have tried to strew about my lectures cannot replace a sound observational experience during the first years of research.

Clearly, the educational possibilities of the Internet have not been fully exploited. In fact, this winter school, gathering 50 students in one place and putting them in front of 25 computer terminals, to go and try what they had been taught during the lectures, may have been an interim between a classical school without hands-on exercises and a fully distributed and interactive one, where students would follow lectures over the WWW and perform exercises at home.

I am grateful to the organisers for inviting me to give these lectures, and for their financial support. The persistence and excitement of the students during the practicals was truly impressive. These practicals would have been impossible without the excellent computing facilities prepared for the school by R. Kroll and his team of the "Centro de Calculo" of the *IAC*. I would like to thank the many people who provided useful information, enriching and completing these lecture notes: D. Banhatti, E. Brinks, S. Britzen, J. Burns, J.J. Condon, W.R. Cotton, G. Dulk, W. Erickson, L. Feretti, G. Giovannini, Gopal Krishna, L. Gurvits, S.E.G. Hales, R.W. Hunstead, S. J. Katajainen, K. Kingham, N. Loiseau, V. Migenes, R. Norris, E. Raimond, W. Sherwood, S. A. Trushkin, K. Weiler, and Rick L. White. Thanks also go to all authors of useful WWW pages with compilations of links which I came across while surfing the WWW for this contribution. This paper has made use of *NASA*'s Astrophysics Data System Abstract Service. A. Koekemoer provided help to print Figure 1 successfully, E. Tago kindly helped to produce Figure 6, and special thanks also go to A.C. Davenhall, A. Fletcher, S. Kurtz, and O.B. Slee for their careful reading of the manuscript at the very last moment. All of them strengthened my belief that most of the information given here must have been correct at least at some point in time. Last, but not least, the Editors of this volume are thanked for their eternal patience with the delivery of this report.

REFERENCES

ABELL, G. O., CORWIN, H. G. JR, & OLOWIN, R. P. 1989, *ApJS* **70**, 1–138

ACKER, A., STENHOLM, B., & VÉRON, P. 1991, *A&AS* **87**, 499–506

ACKER, A., & STENHOLM, B. 1990, *A&AS* **86**, 219–225

ALTENHOFF, W. J., DOWNES, D., PAULS, T., & SCHRAML, J. 1979, *A&AS* **35**, 23–54

ALTSCHULER, D. R. 1986, *A&AS* **65**, 267–283

AMIRKHANYAN, V. R., GORSHKOV, A. G., LARIONOV, M. G., KAPUSTKIN, A. A., KONNIKOVA, V. K., LAZUTKIN, A. N., NIKANOROV, A. S., SIDORENKOV, V. N., & UGOL'KOVA, L. S. 1989, The Zelenchuk Survey of Radio Sources between Declinations 0° and +14°, MIR Publ., Moscow, ISBN 5-211-01151-1

ANDERNACH, H. 1989, *Bull. Inf. CDS* **37**, 139–141

ANDERNACH, H. 1990, *Bull. Inf. CDS* **38**, 69–94

ANDERNACH H. 1992, in *Astronomy from Large Data Bases – II*, eds. A. Heck & F. Murtagh, ESO Conf. & Workshop Proceedings, **43**, 185–190, ESO, Garching

ANDERNACH, H., FERETTI, L., GIOVANNINI, G., KLEIN, U., ROSSETTI, E., & SCHNAUBELT, J. 1992, *A&AS* **93**, 331–357

BAADE, W., & MINKOWSKI, R. 1954, *ApJ* **119**, 206–214

BAARS, J. W. M., GENZEL, R., PAULINY-TOTH, I. I. K., & WITZEL, A. 1977, *A&A* **61**, 99–106

BAIESI-PILLASTRINI, G. C., PALUMBO, G. G. C., & VETTOLANI, G. 1983, *A&AS* **53**, 373–381

BALDWIN, J. E., BOYSEN, R. C., HALES, S. E. G., JENNINGS, J. E., WAGGETT, P. C., WARNER, P. J., & WILSON, D. M. A. 1985, *MNRAS* **217**, 717–730, microfiche

BALEISIS, A., LAHAV, O., LOAN, A. J., & WALL, J. V. 1998, *MNRAS* **297**, 545–558

BECKER, R. H., WHITE, R. L., & EDWARDS, A. L. 1991, *ApJS* **75**, 1–229

BECKER, R. H., WHITE, R. L., HELFAND, D. J., & ZOONEMATKERMANI, S. 1994, *ApJS* **91**, 347–387

BENNETT, C. L., LAWRENCE, C. R., BURKE, B. F., HEWITT, J. N., & MAHONEY, J. 1986, *ApJS* **61**, 1–104

BENNETT, C. L., TURNER, M. S., & WHITE M. 1997, *Physics Today*, **50**, 32–38

BIRKINSHAW, M. 1998, *Phys. Rep.*, **310**, 97–195 www.star.bris.ac.uk/mb1/Export.html

BISCHOF, O. B., & BECKER, R. H. 1997, *AJ* **113**, 2000–2005

BOLLER, T., BERTOLDI, F., DENNEFELD, M., & VOGES, W. 1998, *A&AS* **129**, 87–145

BOLTON, J. G., STANLEY, G., & SLEE, O. B. 1949, *Nature*, **164**, 101–102

BOLTON, J. G., GARDNER, F. F., & MACKEY, M. B. 1964, *Aust. J. Phys.* **17**, 340

BOZYAN, E. P. 1992, *ApJS* **82**, 1–92

BRAUDE, S. IA., MEGN, A. V., SOKOLOV, K. P., TKACHENKO, A. P., & SHARYKIN, N. K. 1979, *Ap&SS* **64**, 73–126

BRAUDE, S. Y., SOKOLOV, K. P., SHARYKIN, N. K., & ZAKHARENKO, S. M. 1995, *Ap&SS* **226**, 245–271

BREMER, M., JACKSON, N., & PÉREZ-FOURNON, I. (EDS.) 1998, *Observational Cosmology with the New Radio Surveys*, Kluwer Acad. Publ., ASSL vol. 226

BRINKMANN, W., & SIEBERT, J. 1994, *A&A* **285**, 812–818

BRINKS, E., & SHANE, W. W. 1984, *A&AS* **55**, 179–251

BROTEN, N. W., MacLEOD, J. M., & VALLÉE J.P. 1988, *Ap&SS* **141**, 303–331

BROUW, W. N., & SPOELSTRA, T. A. TH. 1976, *A&AS* **26**, 129–144

BROWN, R. L. 1996, in *The Westerbork Observatory, Continuing Adventure in Radio Astronomy*, Kluwer Acad. Publ., Dordrecht, p. 167–183

BURN, B. J. 1966, *MNRAS* **133**, 67–83

BURNS, J. O., FEIGELSON, E. D., & SCHREIER, E. J. 1983, *ApJ* **273**, 128–153

BURNS, J. O., HANISCH, R. J., WHITE, R. A., NELSON, E. R., MORRISETTE, K. A., & WARD, M. J. 1987, *AJ* **94**, 587–617

BURNS, J. O., & MENDELL, W. W. (EDS.) 1988, Future Astronomical Observatories on the Moon, NASA Conf. Publ. 2489, Proc. of Workshop held in Houston, TX, USA, Jan. 1986

BURNS, J. O., JOHNSON, S., JOHNSON, S., & TAYLOR, G. J. (EDS.) 1989, A Lunar Far-Side Very Low Frequency Array, NASA Conf. Publ. 3039, Proc. of Workshop held in Albuquerque, NM, USA, Feb. 1988

BURSOV, N. N., LIPOVKA, N. M., SOBOLEVA, N. S., TEMIROVA, A. V., GOL'NEVA, N. E., PARIJSKAYA, E. YU., & SAVASTENYA, A. V. 1997, *Bull. SAO* **42**, 5

BURTON, W. B. 1976, *ARA&A* **14**, 275–306

BYSTEDT, J. E. V., BRINKS, E., DE BRUYN, A.G., ISRAEL, F. P., SCHWERING, P. B. W., SHANE, W. W., & WALTERBOS, R. A. M. 1984, *A&AS* **56**, 245–280

CHRISTIANSEN, W. N. & HÖGBOM, J. A. 1985, *Radio Telescopes*, 2nd ed., Cambridge University Press, Cambridge, UK

CILIEGI, P., McMAHON, R. G., MILEY, G., GRUPPIONI, C., ROWAN-ROBINSON, M., CE-

SARSKY, C., DANESE, L., FRANCESCHINI, A., GENZEL, R., LAWRENCE, A., LEMKE, D., OLIVER, S., PUGET, J.-L., ROCCA-VOLMERANGE, B. 1999, **302**, 222–244

CIOFFI, D. F., & JONES, T. W. 1980, *AJ* **85**, 368–375

CLARKE, M. E., LITTLE, A. G., & MILLS, B. Y. 1976, *Aust. J. Phys. Ap. Suppl.* **40**, 1–71

CLARKE, J. N., KRONBERG, P. P., & SIMARD-NORMANDIN, M. 1980, *MNRAS* **190**, 205–215

COHEN, M. H., & KELLERMANN, K. I. 1995, *Quasars and Active Galactic Nuclei: High Resolution Radio Imaging*, Proc. Natl. Acad. Sci., USA, **92**, 11339–11450

COLLA, G., FANTI, C., FANTI, R., FICARRA, A., FORMIGGINI, L., GANDOLFI, E., GRUEFF, G., LARI, C., PADRIELLI, L., ROFFI, G., TOMASI, P., & VIGOTTI, M. 1970, *A&AS* **1**, 281–317

COLLA, G., FANTI, C., FANTI, R., FICARRA, A., FORMIGGINI, L., GANDOLFI, E., LARI, C., MARANO, B., PADRIELLI, L., & TOMASI, P. 1972, *A&AS* **7**, 1–34

COLLA, G., FANTI, C., FANTI, R., FICARRA, A., FORMIGGINI, L., GANDOLFI, E., GIOIA, I., LARI, C., MARANO, B., PADRIELLI, L., & TOMASI, P. 1973, *A&AS* **11**, 291–325

COLLINS, M. 1977, *Astronomical catalogues 1950–1975*, Inspec Bibliography Series no. 2, IN-SPEC, ISBN 0 95296 440 4

COMBI, J. A., ROMERO, G. E., & ARNAL, E. M. 1998, *A&A* **333**, 298–304

CONDON, J. J. 1974, *ApJ* **188**, 279–286

CONDON, J. J., & BRODERICK, J. J. 1986a, *AJ* **91**, 1051–1057

CONDON, J. J., & BRODERICK, J. J. 1986b, *AJ* **92**, 94–102

CONDON, J. J., BRODERICK, J. J., & SEIELSTAD, G. A. 1989, *AJ* **97**, 1064–1073

CONDON, J. J., & LOCKMAN, F. J. 1990, Large-Scale Surveys of the Sky, Proc. NRAO Green-Bank Workshop No. 20, held Sept. 1987, publ. by NRAO, Charlottesville, WV, USA

CONDON, J. J., BRODERICK, J. J., & SEIELSTAD, G. A. 1991, *AJ* **102**, 2041–2046

CONDON, J. J., GRIFFITH, M. R., & WRIGHT, A. E. 1993, *AJ* **106**, 1095–1100

CONDON, J. J., BRODERICK, J. J., SEIELSTAD, G. A., DOUGLAS, K., & GREGORY, P. C. 1994, *AJ* **107**, 1829–1833

CONDON, J. J., ANDERSON, E., & BRODERICK, J. J. 1995, *AJ* **109**, 2318–2354

CONDON, J. J. 1997, *PASP* **109**, 166–172

CONDON, J. J., KAPLAN, D. L., & YIN, Q. F. 1997, *BAAS* **29**, 1231

CONDON, J. J. 1998, IAU Symp. **179**, 19–25, eds. B.J. McLean *et al.* , Kluwer Acad. Publ., Dordrecht

CONDON, J. J., COTTON, W. D., GREISEN, E. W., YIN, Q. F., PERLEY, R. A., TAYLOR, G. B., & BRODERICK, J. J. 1998, *AJ* **115**, 1693–1716; www.cv.nrao.edu/~jcondon/nvss.html

CONDON, J. J., & KAPLAN, D. L. 1998, *ApJS*, **117**, 361–385

COORAY, A. R., GREGO, L., HOLZAPFEL, W. L., JOY, M., & CARLSTROM, J. E. 1998, *AJ* **115**, 1388–1399

CORNWELL, T. J. 1988, *A&A* **202**, 316–321

CORNWELL, T. 1989, in *Synthesis Imaging in Radio Astronomy*, eds. Perley, R. A., Schwab, F. R. & Bridle, A. H., ASP Conf. Ser. **6**, ASP, San Francisco, p. 277–286

CORNWELL, T. J. & PERLEY, R. A. (EDS.) 1991, *Radio Interferometry: Theory, Techniques, and Applications*, ASP Conf. Ser. **19**, ASP, San Francisco

CRAF HANDBOOK FOR RADIO ASTRONOMY, 2ND ED. 1997, Committee on Radio Astronomy Frequencies (*CRAF*), Netherlands Foundation for Research in Astronomy (*NFRA*), Dwingeloo, The Netherlands

CRAWFORD, T., MARR, J., PARTRIDGE, B., & STRAUSS, M. A. 1996, *ApJ* **460**, 225–243

DAME, T. M., UNGERECHTS, H., COHEN, R. S., DE GEUS, E. J., GRENIER, I. A., MAY, J., MURPHY, D. C., NYMAN, L.-A., & THADDEUS, P. 1987, *ApJ* **322**, 706–720

DAVIES, I. M., LITTLE, A. G., & MILLS, B. Y. 1973, *Aust. J. Phys. Ap. Suppl.* **28**, 1–59

DE BRUYN, A. G., & SIJBRING, D. 1993, chap. 2 of Sijbring, D., PhD Thesis, Groningen, 1993

DE RUITER, H. R., ARP, H. C., & WILLIS, A. G. 1977, *A&AS* **27**, 211–293

DIXON, R. S. 1970, *ApJS* **20**, 1–503

DOUGLAS, J. N., BASH, F. N., BOZYAN, F. A., TORRENCE, G. W., & WOLFE, C. 1996, *AJ* **111**, 1945–1963

DREHER, J. W., CARILLI, C. L., & PERLEY, R. A. 1987, *ApJ* **316**, 611–625

DRESSEL, L. L., CONDON, J. J. 1978, *ApJS* **36**, 53–75

DUNCAN, A. R., STEWART, R. T., HAYNES, R. F., & JONES, K. L. 1995, *MNRAS* **277**, 36–52

DUNCAN, A. R., HAYNES, R. F., JONES, K. L., & STEWART, R. T. 1997, *MNRAS* **291**, 279–295

DURDIN, J. M., PLETICHA, D., CONDON, J. J., YERBURY, M. J., JAUNCEY, D. L., & HAZARD, C. 1975, The NAIC 611-MHz Multi-Beam Sky Survey Source List, NAIC Report **45**, March 1975; `cats.sao.ru/doc/NAIC.html`

DWARAKANATH, K. S., & UDAYA SHANKAR, N. 1990, *J. Ap. Astron.* **11**, 323–410

EDGE, D. O., & MULKAY, M. J. 1976, *Astronomy Transformed: The Emergence of Radio Astronomy in Britain*, John Wiley & Sons Inc., New York, NY, USA

EGGER, R.J., & ASCHENBACH, B. 1995, *A&A* **294**, L25–L28

ELFHAG, T., BOOTH, R. S., HÖGLUND, B., JOHANSSON, L. E. B., & SANDQVIST, AA. 1996, *A&AS* **115**, 439–468

EMERSON, D. T., KLEIN, U., & HASLAM, C. G. T. 1979, *A&A*, **76**, 92–105

EMERSON, D. T., & GRÄVE, R. 1988, *A&A*, **190**, 353–358

EMERSON, D. T. & PAYNE, J. M. (EDS.) 1995, *Multi-Feed Systems for Radio Telescopes*, ASP Conf. Ser. **75**, ASP, San Francisco

ERICKSON, W. C. 1997, *Proc. ASA* **14**, 278–282

FANTI, C., FANTI, R., FICARRA, A., & PADRIELLI, L. 1974, *A&AS* **18**, 147–156

FERETTI, L., GIOVANNINI, G., & BÖHRINGER, H. 1997, *New Astron.* **2**, 501–515

FICARRA, A., GRUEFF, G., & TOMASSETTI, G. 1985, *A&AS* **59**, 255–347

FOMALONT, E. B., KELLERMANN, K. I., RICHARDS, E. A., WINDHORST, R. A., & PATRIDGE R. B. 1997, *ApJ* **475**, L5–L8

FÜRST E., REICH, W., REICH, P., & REIF, K. 1990, *A&AS* **85**, 805–811

FÜRST, E., REICH, W., REICH, P., UYANIKER, B., & WIELEBINSKI, R. 1998, *IAU Symp.* **179**, 165–171, eds. B.J. McLean *et al.* , Kluwer Acad. Publ., Dordrecht

FUKUI, Y., & YONEKURA, Y. 1998, *IAU Symp.* **179**, 165–171, eds. B.J. McLean *et al.* , Kluwer Acad. Publ., Dordrecht

GARDNER, F. F., WHITEOAK, J. B., & MORRIS, D. 1966, *ARA&A* **4**, 245–292

GIOVANNINI, G., FERETTI, L., & STANGHELLINI, C. 1991, *A&A* **252**, 528–537

GOLAP, K., UDAYA SHANKAR, N., SACHDEV, S., DODSON, R., SASTRY, CH. V. 1998, *J. Ap&A* **19**, 35–53

GOWER, J. F. R., SCOTT, P. F., & WILLS, D. 1967, *MmRAS* **71**, 49–144

GRAHAM, I. 1970, *MNRAS* **149**, 319–339

GRAHAM-SMITH, F. 1974, *Radio Astronomy*, 4th edition, Penguin Books Ltd., UK

GRAY, A. D. 1994a, *MNRAS* **270**, 822–834

GRAY, A. D. 1994b, *MNRAS* **270**, 861–870

GRAY, A. D., LANDECKER, T. L., DEWDNEY, P. E., & TAYLOR, A. R. 1998, *Nature*, **393**, 660

GREEN, A. J. 1997, *Proc. ASA* **14**, 73–76

GREEN, A. J., CRAM, L. E., & LARGE, M. I. 1998, *ApJ*, in press

GREGORY, P. C., & TAYLOR, A. R. 1986, *AJ* **92**, 371–411

GREGORY, P. C., & CONDON, J. J. 1991, *ApJS* **75**, 1011–1291

GREGORY, P. C., VAVASOUR, J. D., SCOTT, W. K., & CONDON, J. J. 1994, *ApJS* **90**, 173–177

GREGORY, P. C., SCOTT, W. K., DOUGLAS, K., & CONDON, J. J. 1996, *ApJS* **103**, 427–432

GREGORY, P. C., SCOTT, W. K., & POLLER, B. 1998, in IAU Coll. 164, *Radio Emission from Galactic and Extragalactic Compact Sources*, ASP Conf. Ser. **144**, 283–284, eds. J. A. Zensus, G. B. Taylor, & J. M. Wrobel, ASP, San Francisco.

GRIFFITH, M., LANGSTON, G., HEFLIN, M., CONNER, S., LEHÁR J., & BURKE, B. 1990, *ApJS* **74**, 129–180

GRIFFITH, M., LANGSTON, G., HEFLIN, M., CONNER, S., & BURKE, B. 1991, *ApJS* **75**, 801–833

GRIFFITH, M. R., WRIGHT, A. E., BURKE, B. F., & EKERS, R. D. 1994, *ApJS* **90**, 179–295

GRIFFITH, M. R., WRIGHT, A. E., BURKE, B. F., & EKERS, R. D. 1995, *ApJS* **97**, 347–453

GUBANOV, A. G., & ANDERNACH, H. 1997, *Balt. Astron.* **6**, 263–266

HACKING, P., CONDON, J. J., HOUCK, J. R., & BEICHMAN, C. A. 1989, *ApJ* **339**, 12–26

HAIGH, A. J., ROBERTSON, J. G., & HUNSTEAD, R. W. 1997, *Proc. ASA* **14**, 221–229

HALES, S. E. G., BALDWIN, J. E., & WARNER, P. J. 1988, *MNRAS* **234**, 919–936

HALES, S. E. G., MASSON, C. R., WARNER, P. J., & BALDWIN, J. E. 1990, *MNRAS* **246**, 256–262

HALES, S. E. G., MAYER, C. J., WARNER, P. J., & BALDWIN, J. E. 1991, *MNRAS* **251**, 46–53

HALES, S. E. G., MASSON, C. R., WARNER, P. J., BALDWIN, J. E., & GREEN, D. A. 1993a, *MNRAS* **262**, 1057–1061

HALES, S. E. G., BALDWIN, J. E., & WARNER, P. J. 1993b, *MNRAS* **263**, 25–30

HALES, S. E. G., WALDRAM, E. M., REES, N., & WARNER, P. J. 1995, *MNRAS* **274**, 447–451

HARDEE, P. E., BRIDLE, A. H., & ZENSUS, A. (EDS.) 1996, *Energy Transport in Radio Galaxies and Quasars*, ASP Conf. Ser. **100**, ASP, San Francisco, (www.cv.nrao.edu/jetworks)

HARRIS, D. E., & MILEY, G. K. 1978, *A&AS* **34**, 117–128

HARRIS, D. E., GRANT C. P. S., & ANDERNACH, H. 1995, *Astronomical Data Analysis Software and Systems – IV*, ASP Conf. Ser. **77**, 48–51, eds. R. A. Shaw, H. E. Payne, J.J.E. Hayes; ASP, San Francisco, astro-ph/94011021

HARTMANN, D. & BURTON, W. B. 1997, *Atlas of Galactic Neutral Hydrogen*, Cambridge University Press, ISBN 0-521-47111-7 (no *URL* available)

HASLAM, C. G. T. 1974, *A&AS* **15**, 333–338

HASLAM, C. G. T., KLEIN, U., SALTER, C. J., STOFFEL, H., WILSON, W. E., CLEARY, M. N., COOKE, D. J., & THOMASSON, P. 1981, *A&A* **100**, 209–219

HASLAM, C. G. T., STOFFEL, H., SALTER, C. J., & WILSON, W. E. 1982, *A&AS* **47**, 1–142

HAUSCHILDT, M. 1987, *A&A* **184**, 43–56

HAYNES, R. F., CASWELL, J. L., & SIMONS, L. W. J. 1979, *Aust. J. Phys. Ap. Suppl.* **48**, 1–30

HAYNES, RAYMOND; HAYNES, ROSLYNN; MALIN, D., & McGEE, R. 1996, *Explorers of the Southern Sky: A History of Australian Astronomy*, Cambridge University Press, Cambridge, UK

HELFAND, D. J., ZOONEMATKERMANI, S., BECKER, R. H., & WHITE, R. L. 1992, *ApJS* **80**, 211–255

HELFAND, D. J., SCHNEE, S., BECKER, R. H., WHITE, R. L., & McMAHON, R. G. 1997, *BAAS* **29**, 1231

HELFER, T. T., THORNLEY, M. D., REGAN, M. W., SHETH, K., VOGEL, S. N., WONG, T., BLITZ, L., & BOCK, D. 1998, *BAAS* **30**, 928

HENNING, P. A., KRAAN-KORTEWEG, R. C., RIVERS, A. J., LOAN, A. J., LAHAV, O., & BURTON, W. B. 1998, *AJ* **115**, 584–591

HENSTOCK, D. R., BROWNE, I. W. A., WILKINSON, P. N., TAYLOR, G. B., VERMEULEN, R. C., PEARSON, T. J., & READHEAD, A. C. S. 1995, *ApJS* **100**, 1–36

HERBIG, T., & READHEAD, A. C. S. 1992, *ApJS* **81**, 83–66

HEWITT, A., & BURBIDGE, G. 1989, *ApJS* **69**, 1–63

HEY, J. S. 1971, *The Radio Universe*, 1st edition, Pergamon Press Ltd, UK

HEY, J. S. 1973, *The Evolution of Radio Astronomy*, 1st edition, Neale Watson Academic Publications, New York, NY 10010, USA

HEYER, M. H., BRUNT, C., SNELL, R. L., HOWE, J. E., SCHLOERB, F. P., & CARPENTER, J. M. 1998, *ApJS* **115**, 241–258

HOOPER, E. J., IMPEY, C. D., FOLTZ, C. B., & HEWETT, P. C. 1996, *ApJ* **473**, 746–759

HOPKINS, A. M. 1998, Ph. D. thesis, University of Sydney

HOPKINS, A. M., MOBASHER, B., CRAM, L., & ROWAN-ROBINSON, M. 1998, *MNRAS* **296**, 839–846

HUGHES, V. A., & MACLEOD, G. C. 1989, *AJ* **97**, 786–800

HUGHES, P. A., ALLER, H. D., & ALLER, M. F. 1992, *ApJ* **396**, 469–486

HUNSTEAD, R. W., CRAM, L. E., & SADLER, E. M. 1998, Proc. Observational Cosmology with the New Radio Surveys, eds. M. Bremer, N. Jackson & I. Pérez-Fournon, Kluwer Acad. Publ., p. 55–62

HUTCHINGS, J. B., DURAND, D., & PAZDER, J. 1991, *PASP* **103**, 21–25

JACKSON, N. & DAVIS, R. J. (EDS.) 1997, *High Sensitivity Radio Astronomy*, Cambridge University Press, Cambridge, UK

JAFFE, W., GAVAZZI, G., & VALENTIJN, E. 1986, *AJ* **91**, 199–203

JAUNCEY, D. L. (ED.) 1977, *Radio Astronomy and Cosmology*, IAU Symp. **74**, 398 pp., D. Reidel, Dordrecht

JOHNSTON, K. J., FEY, A. L., ZACHARIAS, N., RUSSELL, J. L., MA, C., DE VEGT, C., REYNOLDS, J. E., JAUNCEY, D. L., ARCHINAL, B. A., CARTER, M. S., CORBIN, T. E., EUBANKS, T. M., FLORKOWSKI, D. R., HALL, D. M., McCARTHY, D. D., McCULLOCH, P. M., KING, E. A., NICOLSON, G., & SHAFFER, D. B. 1995, *AJ* **110**, 880–915

JONAS, J. L., DE JAGER, G., & BAART, E. E. 1985, *A&AS* **62**, 105–128

JONAS, J. L. 1998, *IAU Symp* **179**, 95–96, eds. B.J. McLean *et al.* , Kluwer Acad. Publ., Dordrecht

JONAS, J. L., BAART, E. & NICOLSON, G. 1998, *MNRAS* **297**, 977–989

JONCAS, G., DURAND, D., & ROGER, R. S. 1992, *ApJ* **387**, 591–611

JUNKES, N., FÜRST, E., & REICH, W. 1987, *A&AS* **69**, 451–464

JUNKES, N., HAYNES, R. F., HARNETT, J. I., & JAUNCEY, D. L. 1993, *A&A* **269**, 29–38

KAISER, M. L. 1987, in Radio Astronomy from Space, Proc. Workshop #18 held at NRAO Green Bank, WV, USA, Oct. 1986, ed. K. W. Weiler, p. 227–238, publ. by NRAO

KALLAS, E., & REICH, W. 1980, *A&AS* **42**, 227–243

KAPLAN, D. L., CONDON, J. J., ARZOUMANIAN, Z., & CORDES, J. M. 1998, *ApJS*, **119**, 75–82

KARDASHEV, N. S. 1997, *Exp. Astron.* **7**, 329–343

KASSIM, N. E. 1988, *ApJS* **68**, 715–733

KASSIM, N. E., & WEILER, K. W. (EDS.) 1990, Low Frequency Astrophysics from Space, Proc. of Workshop held in Crystal City, VA, USA, Jan. 1990, Lecture Notes in Physics, Springer Verlag, Berlin

KATZ-STONE, D. M., & RUDNICK, L. 1994, *ApJ* **426**, 116–122

KELLERMANN, K. I., & SHEETS, B., EDS. 1984, *Serendipitous discoveries in radio astronomy*, Proc. NRAO Workshop 7, Green Bank, WV, USA, May 1983; publ. by National Radio Astronomy Observatory

KELLERMANN, K. I., VERMEULEN, R. C., ZENSUS, J. A., & COHEN, M. H. 1998, *AJ* **115**, 1295–1318

KLEIN, U., & EMERSON, D. T. 1981, *A&A* **94**, 29–44

KLEIN, U., WIELEBINSKI, R., HAYNES, R. F., & MALIN, D. F. 1989, *A&A* **211**, 280–292

KLEIN, U. & MACK, K.-H. 1995, in *Multi-Feed Systems for Radio Telescopes*, eds. Emerson, D. T. & Payne, J. M., ASP Conf. Ser. **75**, ASP, San Francisco, p. 318–326

KLEIN U., VIGOTTI M., GREGORINI L., REUTER H.-P., MACK K.-H., & FANTI R. 1996, *A&A* **313**, 417–422

KOHOUTEK, L. 1997, *AN* **318**, 35–44

KOLLGAARD, R. I., BRINKMANN, W., MCMATH CHESTER, M., FEIGELSON, E. D., HERTZ, P., REICH, P., & WIELEBINSKI, R. 1994, *ApJS* **93**, 145–159

KOUWENHOVEN, M., BERGER, M., DEICH, W., & DE BRUYN, G. 1996, Pulsars: Problems & Progress, eds. S. Johnston, M. A. Walker, & M. Bailes, ASP Conf. Ser. **105**, 15–16, ASP, San Francisco

KRAAN-KORTEWEG, R. C., LOAN, A. J., BURTON, W. B., LAHAV, O., FERGUSON, H. C., HENNING, P. A., & LYNDEN-BELL, D. 1994, *Nature* **372**, 77–79

KRAAN-KORTEWEG, R. C., WOUDT, P. A., & HENNING, P. A. 1997, *Proc. ASA* **14**, 15–20

KUCHAR, T. A., & CLARK, F. O. 1997, *ApJ* **488**, 224–233

KÜHR, H., NAUBER, U., PAULINY-TOTH, I. I. K., & WITZEL, A. 1979, A Catalogue of Radio Sources, MPIfR preprint no. 55

KÜHR, H., WITZEL, A., PAULINY-TOTH, I. I. K., & NAUBER, U. 1981, A&AS, **45**, 367–430

KURTZ, S., CHURCHWELL, E., & WOOD, D. O. S. 1994, *ApJS* **91**, 659–712

LANDECKER, T. L., CLUTTON-BROCK, M., & PURTON, C. R. 1990, *A&A* **232**, 207–214

LANDECKER, T. L., ANDERSON, M. D., ROUTLEDGE, D., & VANELDIK, J. F. 1992, *A&A* **258**, 495–506

LANGER, W. D., VELUSAMY, T., KUIPER, T. B. H., LEVIN, S., OLSEN, E., & MIGENES, V. 1995, *ApJ* **453**, 293–307

LANGSTON, G. I., HEFLIN, M. B., CONNER, S. R., LEHÁR J., CARRILLI, C. L., & BURKE, B. F. 1990, *ApJS* **72**, 621–691

LARGE, M. I., MILLS, B. Y., LITTLE, A. G., CRAWFORD, D. F., & SUTTON, J. M. 1981, *MNRAS* **194**, 693–704

LARGE, M. I., CRAM, L. E., & BURGESS, A. M. 1991, *The Observatory* **111**, 72–75

LARIONOV, M. G. 1991, *Soobshch. Spets. Astrof. Obs.* **68**, 14–46

LAURENT-MUEHLEISEN, S. A., KOLLGAARD, R. I., RYAN, P. J., FEIGELSON, E. D., BRINKMANN, W., & SIEBERT J. 1997, *A&AS* **122**, 235–247

LAWRENCE, C. R., BENNETT, C. L., GARCIA-BARRETO, J. A., GREENFIELD, P. E., & BURKE, B. F. 1983, *ApJS* **51**, 67–114

LEAHY, J. P. 1993, in Jets in Extragalactic Radio Sources, Lecture Notes in Physics **421**, p. 1–13, eds. H.-J. Röser & K. Meisenheimer, Springer Verlag, Berlin

LEAHY, J. P., BRIDLE, A. H., & STROM, R. G. (EDS.) 1998, An Atlas of DRAGNs, see *URL* www.jb.man.ac.uk/atlas/

LEDLOW, M. J., OWEN, F. N., & KEEL, W. C. 1998, *ApJ* **495**, 227–238

LEGG, T. H. 1998, *A&AS* **130**, 369–379

LEWIS, J. W. 1995, Astronomical Data Analysis Software and Systems – IV, eds. R. A. Shaw, H. E. Payne, & J. J. E. Hayes, ASP Conf. Ser. **77**, p. 327–330, ASP, San Francisco

LIANG, H. & BIRKINSHAW, M. 1999, "Looking Deep in the Southern Sky", Sydney, eds. R. Morganti & W. J. Couch, Springer-Verlag, Berlin, p. 159–166

LILLEY, A. E., & PALMER, P. 1968, *ApJS* **16**, 143–173

LORIMER, D. R., JESSNER, A., SEIRADAKIS, J. H., LYNE, A. G., D'AMICO, N., ATHANA-SOPOULOS, A., XILOURIS, K. M., KRAMER, M., & WIELEBINSKI, R. 1998, *A&AS* **128**, 541–544

LORTET, M.-C., BORDE, S., & OCHSENBEIN, F. 1994, *A&AS* **107**, 193–218

LOVAS, F. J. 1992, *J. Phys. Chem. Ref. Data* **21**, 181

LYNE, A. G., MANCHESTER, R. N., LORIMER, D. R., BAILES, M., D'AMICO, N., TAURIS, T. M., JOHNSTON, S., BELL, J. F., NICASTRO, L. 1998, *MNRAS* **295**, 743–755

MACHALSKI, J. 1978, *Acta Astron.* **28**, 367–440

MACK, K.-H., KLEIN U., O'DEA, C. P., & WILLIS A. G. 1997, *A&AS* **123**, 423–444

MALOFEEV, V. M. 1996, Pulsars: Problems & Progress, eds. S. Johnston, M. A. Walker, & M. Bailes, ASP Conf. Ser. **105**, 271–277, ASP, San Francisco

MARSALKOVA, P. 1974, *Ap&SS* **27**, 3–110

MARTÍN, M. C. 1998, *A&AS* **131**, 73–76

MASLOWSKI, J. 1972, *Acta Astron.* **22**, 227–260

MATHEWSON, D. S., & FORD, V. L. 1996, *ApJS* **107**, 97–102

MCADAM, W. B. 1991, *Proc. ASA* **9**, 255–256

MCGILCHRIST, M. M., BALDWIN, J. E., RILEY, J. M., TITTERINGTON, D. J., WALDRAM, E. M., & WARNER, P. J. 1990, *MNRAS* **246**, 110–122

MCKAY, N. P. F. & MCKAY, D. J. 1998, in *ADASS VII*, ASP Conf. Ser. **145**, p. 240 (`www.stsci.edu/stsci/meetings/adassVII`)

MILEY, G. K., PEROLA, G. C., VAN DER KRUIT, P. C., & VAN DER LAAN, H. 1972, *Nature* **237**, 269–272

MILEY, G. K. 1980, *ARA&A* **18**, 165–218

MORAN, E. C., HELFAND, D. J., BECKER, R. H., & WHITE, R. L. 1996, *ApJ* **461**, 127–145

NICHOLLS, P. N. 1987, *J. Amer. Soc. Information Science*, **38**, 443

NISHIYAMA, K., & NAKAI, N. 1998, *IAU Symp.* **184**, in press; ADS `1997IAUS..184E.132N`

NODLAND, B., & RALSTON, J. P. 1997, *Phys. Rev. Lett.* **78**, 3043; `astro-ph/9704196`

NOORDAM, J. E., & DE BRUYN. A. G. 1982, *Nature* **299**, 597–600

NORMANDEAU, M., JONCAS, G., & GREEN, D. A. 1992, *A&AS* **92**, 63–83

NORMANDEAU, M., TAYLOR, A. R., DEWDNEY, P. E. 1996, *Nature* **380**, 687–689

NORRIS, R. P. 1999, "Looking Deep in the Southern Sky", eds. R. Morganti & W. Couch, Springer-Verlag, Berlin, p. 140–145

NOTNI, P., & FRÖHLICH, H.-E. 1975, *AN* **296**, 197–219

O'DEA, C. P. 1998, *PASP* **110**, 493–532

OORT, J. H., KERR, F. J., & WESTERHOUT, G. 1958, *MNRAS* **118**, 379–389

OORT, M. J. A., & VAN LANGEVELDE, H. J. 1987, *A&AS* **71**, 25–38

OTRUPCEK, R. E., & WRIGHT, A. E. 1991, *Proc. ASA* **9**, 170 (PKSCAT90; `ftp://ftp.atnf.csiro.au/pub/data/pkscat90/`)

OTT, M., WITZEL, A., QUIRRENBACH, A., KRICHBAUM, T. P., STANDKE, K. J., SCHALINSKI, C. J., & HUMMEL, C. A. 1994, *A&A* **284**, 331–339

PACHOLCZYK, A. G. 1970, *Radio Astrophysics*, W. H. Freeman & Co., San Francisco

PACHOLCZYK, A. G. 1977, *Radio Galaxies*, Pergamon Press, Oxford

PADRIELLI, L., & CONWAY, R. G., 1977, *A&AS* **27**, 171–180

PARIJSKIJ, YU. N., BURSOV, N. N., LIPOVKA, N. M., SOBOLEVA, N. S., & TEMIROVA, A. V. 1991, *A&AS* **87**, 1–32

PARIJSKIJ, YU. N., BURSOV, N. N., LIPOVKA, N. M., SOBOLEVA, N. S., TEMIROVA, A. V., & CHEPURNOV, A. V. 1992, *A&AS* **96**, 583–592

PARIJSKIJ, YU. M., BURSOV, N. M., LIPOVKA, N. M., SOBOLEVA, M. S., TEMIROVA, A. V., & CHEPURNOV, A. V. 1993, *A&AS* **98**, 391–392

PATNAIK, A. R., BROWNE, I. W. A., WILKINSON, P. N., & WROBEL, J. M. 1992, *MNRAS* **254**, 655–676

PEARSON, T. J., & READHEAD, A. C. S. 1988, *ApJ* **328**, 114–142

PENG, B., & NAN, R. 1998, IAU Symp. **179**, 93–94, eds. B.J. McLean *et al.* , Kluwer Acad. Publ., Dordrecht

PERLEY, R. 1989, in *Synthesis Imaging in Radio Astronomy*, eds. Perley, R. A., Schwab, F. R. & Bridle, A. H., ASP Conf. Ser. **6**, ASP, San Francisco, p. 287–313

PERLEY, R. A., SCHWAB, F. R. & BRIDLE, A. H. (EDS.) 1989, *Synthesis Imaging in Radio*

Astronomy ASP Conf. Ser. **6**, ASP, San Francisco

PERSIC, M., & SALUCCI, P. 1995, *ApJS* **99**, 501–541

PILKINGTON, J. D. H., & SCOTT, P. F. 1965, *MmRAS* **69**, 183–224

PORCAS, R. W., URRY, C. M., BROWNE, I. W. A., COHEN, A. M., DAINTREE, E. J., & WALSH, D. 1980, *MNRAS* **191**, 607–614

POUND, M. W., GRUENDL, R., LADA, E. A. & MUNDY, L. 1997, Star Formation Near and Far, eds. S. Holt & L. Mundy, p. 395–397

PRANDONI, I., GREGORINI, L., PARMA, P., DE RUITER, R. H., VETTOLANI, G., WIERINGA, M. H., & EKERS, R. D. 1999, "Looking Deep in the Southern Sky", eds. R. Morganti & W. J. Couch, Springer-Verlag, Berlin, p. 114–119

PURTON, C., & DURRELL, P. 1991, *Program SURSEARCH: SEARCHing radio continuum source SURveys for overlap with a given area of sky*, FORCE software package available from C. R. Purton (DRAO); see also `cats.sao.ru/doc/SURSEARCH.html`

PURVIS, A., TAPPIN, S. J., REES, W. G., HEWISH, A., & DUFFETT-SMITH, P. J. 1987, *MNRAS* **229**, 589–619

QUINIENTO, Z. M., CERSOSIMO, J. C., & COLOMB, F. R. 1988, *A&AS* **76**, 21–34

REBER, G. 1994, *J.Roy.Astr.Soc. Canada*, **88**, 297–302

REBER, G. 1995, *Ap&SS* **227**, 93–96

REES, N. 1990a, *MNRAS* **243**, 637–639

REES, N. 1990b, *MNRAS* **244**, 233–246

REFERENCE DATA FOR RADIO ENGINEERS 1975, publ. by Howard W. Sams & Co., ITT, ISBN 0-672-21218-8, p. 1-3

REICH, W. 1982, *A&AS* **48**, 219–297

REICH, W., FÜRST, E., HASLAM, C. G. T., STEFFEN, P., & REIF, K. 1984, *A&AS* **58**, 197–199

REICH, P., & REICH, W. 1986, *A&AS* **63**, 205–288

REICH, P., & REICH, W. 1988, *A&AS* **74**, 7–23

REICH W. 1991, *IAU Symp.* **144**, 187–196, ed. H. Bloemen, Kluwer Acad. Publ., Dordrecht

REICH, W., REICH, P., & FÜRST E. 1990, *A&AS* **83**, 539

REICH, P., REICH, W., & FÜRST E. 1997, *A&AS* **126**, 413–435

RENGELINK, R. B., TANG, Y., DE, BRUYN, A. G., MILEY, G. K., BREMER, M. N., RÖTTGERING H. J. A., & BREMER, M. A. R. 1997, *A&AS* **124**, 259–280

RICHTER, G. A. 1975, *AN* **296**, 65–81

RIGHETTI, G., GIOVANNINI, G., & FERETTI, L. 1988, *A&AS* **74**, 315–324

ROBERTSON, P. 1992, *Beyond Southern Skies: Radio Astronomy and the Parkes Telescope*, Cambridge University Press, Cambridge, UK

ROBSON, M., YASSIN, G., WOAN, G., WILSON, D. M. A., SCOTT, P. F., LASENBY, A. N., KENDERDINE, S., & DUFFETT-SMITH, P. J. 1993, *A&A* **277**, 314–320

ROGER, R. S., COSTAIN, C. H., LANDECKER, T. L., & SWERDLYK, C. M. 1999, *A&AS*, in press, `astro-ph/9902213`

RÖTTGERING H. J. A., WIERINGA, M. H., HUNSTEAD, R. W., & EKERS, R. D. 1997, *MNRAS* **290**, 577–584

ROHLFS, K. & WILSON, T. L. 1996 *Tools of Radio Astronomy*, 2nd ed., Springer Verlag, Berlin

ROSENBERG, J. L., & SCHNEIDER, S. E. 1998, *BAAS* **30**, 914

SADLER, E. M. 1999, "Looking Deep in the Southern Sky", eds. R. Morganti & W. J. Couch, Springer-Verlag, Berlin, p. 103–109

SAKAMOTO, S., HASEGAWA, T., HAYASHI, M., HANDA, T., & OKA, T. 1995, *ApJS* **100**, 125–131

SALTER, C. J. 1983, *Bull. Astron. Soc. India* **11**, 1–142

SALTER, C. J., & BROWN, R. L. 1988, in *Galactic and Extragalactic Radio Astronomy*, eds.

G. L. Verschuur & K. I. Kellermann, 2nd. edn., Springer-Verlag, Berlin

SASAO, T., MANABE, S., KAMEYA, O., & INOUE, M. (EDS.) 1994, *VLBI Technology*, Terra Science Publ. Company, Tokyo

SCHILKE, P., GROESBECK, T. D., BLAKE G. A., & PHILLIPS, T.G 1997, *ApJS* **108**, 301–337

SCHOENMAKERS, A. P., MACK, K.-H., LARA, L., RÖTTGERING, H. J. A., DE BRUYN, A.G., VAN DER LAAN, H., GIOVANNINI, G. 1998, *A&A*, **336**, 455–478

SEIRADAKIS, J. H., REICH, W., SIEBER, W., SCHLICKEISER, R. & KÜHR, H. 1985, *A&A* **143**, 478–480

SHAKESHAFT, J. R., RYLE, M., BALDWIN, J. E., ELSMORE, B., & THOMSON, J. H. 1955, *MmRAS* **67**, 106–154

SHARPLESS, S. 1959, *ApJS* **4**, 257–279

SHRAUNER, J. A., TAYLOR, J. H., WOAN, G., 1998, *ApJ* **509**, 785–792

SIEBER, W., HASLAM, C. G. T., & SALTER, C. J. 1979, *A&A* **74**, 361–368

SIJBRING, D., DE BRUYN, A. G. 1998, *A&A* **331**, 901–915

SIMARD-NORMANDIN, M., KRONBERG, P. P., & BUTTON, S. 1981, *ApJS* **45**, 97–111

SLEE, O. B., & HIGGINS, C. S. 1973, *Aust. J. Phys. Ap. Suppl.* **27**, 1–43

SLEE, O. B., & HIGGINS, C. S. 1975, *Aust. J. Phys. Ap. Suppl.* **36**, 1–60

SLEE, O. B. 1977, *Aust. J. Phys. Ap. Suppl.* **43**, 1–123

SLEE, O. B. 1995, *Aust. J. Phys.* **48**, 143–186

SNELLEN, I. A. G., SCHILIZZI, R. T., RÖTTGERING, H. J. A., & BREMER, M. N. (EDS.) 1996, *Second Workshop on GPS and CSS Radio Sources*, Leiden Observatory

SOVERS, O. J., FANSELOW, J. L., & JACOBS, C. S. 1998, *Rev. Mod. Phys.* **70**, 1393–1454

STARK, A. A., *et al.* 1990, The Bell Laboratories H I Survey, "Preliminary Draft", available as ADC/CDS catalogue #8010

STARK, A. A., CARLSTROM, J. E., ISRAEL, F. P., MENTEN, K. M., PETERSON, J. B., PHILLIPS, T. G., SIRONI, G., & WALKER, C. K. 1998, SPIE, in press (`astro-ph/9802326`)

STAVELEY-SMITH, L. 1997, *Proc. ASA* **14**, 111–116

STRONG, A. W., RILEY, P. A., OSBORNE, J. L., & MURRAY, J. D. 1982, *MNRAS* **201**, 495–501

SULLIVAN III, W. T. 1982, *Classics in Radio Astronomy*, Reidel, Dordrecht, The Netherlands

SULLIVAN III, W. T. (ED.) 1984, *The Early years of radio astronomy: reflections fifty years after Jansky's discovery*, Cambridge University Press, Cambridge, UK

SUTTON, E. C., BLAKE G. A., MASSON, C. R., & PHILLIPS T.G 1985, *ApJS* **58**, 341–378

TABARA, H., & INOUE, M. 1980, *A&AS* **39**, 379–393

TASHIRO, M., KANEDA, H., MAKISHIMA, K., IYOMOTO, N., IDESAWA, E., ISHISAKI, Y., KOTANI, T., TAKAHASHI, T., & YAMASHITA, A. 1998, *ApJ* **499**, 713–718

TATEYAMA, C. E., KINGHAM, K. A., KAUFMANN, P. PINER, B. G., DE LUCENA, A. M. P., & BOTTI, L. C. L. 1998, *ApJ* **500**, 810–815

TASKER, N., WRIGHT, A., McCONNELL, D., SAVAGE, A., KESTEVEN, M., TROUP, E., & GRIFFITH, M. 1993, *Proc. ASA* **10**, 320–321

TAYLOR, J. H., MANCHESTER, R. N., & LYNE, A. G. 1993, *ApJS* **88**, 529–568

TAYLOR, A. R., GOSS, W. M., COLEMAN, P. H., VAN LEEUWEN, J., & WALLACE, B. J. 1996, *ApJS* **107**, 239–254

TEGMARK, M., & DE OLIVEIRA-COSTA, A. 1998, *ApJ* **500**, L83–L86

TOSCANO, M., BAILES, M., MANCHESTER, R. N., SANDHU, J. S., 1998, *ApJ* **506**, 863–867

TOWLE, J. P., FELDMAN, P. A., & WATSON, J. K. G. 1996, *ApJS* **107**, 747–760

TRIMBLE, V., & McFADDEN, L.-A. 1998, *PASP* **110**, 223–267

TRUSHKIN, S. A. 1996, *A&ATr* **11**, 225–233

TRUSHKIN, S. A. 1997, *Bull. SAO* **41**, p. 64–79

TRUSHKIN, S. A. 1998, SAO Preprint, N131, 30 pp.; (ftp://cats.sao.ru/SNR_spectra/)

TURNER, B. E. 1989, *ApJS* **70**, 539–622

ULVESTAD, J. S., & LINFIELD, R. P. 1998, in IAU Coll. 164, *Radio Emission from Galactic and Extragalactic Compact Sources*, ASP Conf. Ser. **144**, 397–398, eds. J. A. Zensus, G. B. Taylor, & J. M. Wrobel, ASP, San Francisco.

UYANIKER, B., FÜRST, E., REICH, W., REICH, P., & WIELEBINSKI, R. 1998, *A&A*, **132**, 401–411

VANDEN BOUT, P. A. 1998, SPIE Proc. 3357, NRAO preprint 98/038

VERKHODANOV, O.V., TRUSHKIN, S.A., ANDERNACH, H., & CHERNENKOV, V.N. 1997, ASP Conf. Ser., **125**, 322–325, eds. G. Hunt & H.E. Payne, ASP, San Francisco (astro-ph/9610262)

VÉRON-CETTY, M. P., & VÉRON, P. 1983, *A&AS*, **53**, 219–221; ADC/CDS catalogue # 7054

VÉRON-CETTY, M. P., & VÉRON, P. 1989, *Catalogue of Quasars and Active Galactic Nuclei*, 4th Edition, *ESO Sci. Rep.* **7**; ADC/CDS catalogue # 7126

VÉRON-CETTY, M. P., & VÉRON, P. 1998, *Catalogue of Quasars and Active Galactic Nuclei*, 8th Edition, *ESO Sci. Rep.* **18**; ADC/CDS catalogue # 7207

VERSCHUUR, G. L. & KELLERMANN, K. I. (eds.) 1988, *Galactic and Extragalactic Radio Astronomy*, 2nd. edn., Springer-Verlag, Berlin

VERTER, F. 1985, *ApJS* **57**, 261–285, ADC/CDS # 7064

VESSEY, S. J., & GREEN, D. A. 1998, *MNRAS* **294**, 607–614

VETTOLANI, G., ZUCCA, E., MERIGHI, R., MIGNOLI, M., PROUST, D., ZAMORANI, G., CAPPI, A., GUZZO, L., MACCAGNI, D., RAMELLA, M., STIRPE, G. M., BLANCHARD, A., CAYATTE, V., COLLINS, C., MACGILLIVRAY, H., MAUROGORDATO, S., SCARAMELLA, R., BALKOWSKI, C., CHINCARINI, G., & FELENBOK, P., 1998, A&AS **130**, 323–332

VISSER, A. E., RILEY, J. M., RÖTTGERING H. J. A., & WALDRAM, E. M. 1995, *A&AS* **110**, 419–439

WALDRAM, E. M., YATES, J. A., RILEY, J. M., & WARNER, P. J. 1996, *MNRAS* **282**, 779–787, erratum in *MNRAS* **284**, 1007

WALSH, A. J., HYLAND, A. R., ROBINSON, G., & BURTON, M. G. 1997, *MNRAS* **291**, 261–278

WALTERBOS, R. A. M., BRINKS, E., & SHANE, W. W. 1985, *A&AS* **61**, 451–471

WARDLE, J. F. C., PERLEY, R. A., & COHEN, M. H. 1997, *Phys. Rev. Lett.* **79**, 1801

WELCH, W. J., THORNTON, D. D., PLAMBECK, R. L., WRIGHT, M. C. H., LUGTEN, J., URRY, L., FLEMING, M., HOFFMAN, W., HUDSON, J., LUM, W. T., FORSTER, J. R., THATTE, N., ZHANG, X., ZIVANOVIC, S., SNYDER, L., CRUTCHER, R., LO, K. Y., WAKKER, B., STUPAR, M., SAULT, R., MIAO, Y., RAO, R., WAN, K., DICKEL, H. R., BLITZ, L., VOGEL, S. N., MUNDY, L., ERICKSON, W., TEUBEN, P. J., MORGAN, J., HELFER, T., LOONEY, L., DE GUES, E., GROSSMAN, A., HOWE, J. E., POUND, M., & REGAN, R. 1996, *PASP* **108**, 93–103

WENDKER, H. J. 1995, *A&AS* **109**, 177–179

WESTERHOUT, G. 1957, *Bull. Astron. Inst. Netherlands*, **13**, 201–246

WESTERHOUT, G., MADER, G. L. & HARTEN, R. H. 1982, *A&AS* **49**, 137–141

WHITE, G. L. 1984, *Proc. ASA* **5**, 290–340

WHITE, R. L., & BECKER, R. H. 1992, *ApJS* **79**, 331–467

WHITE, R. L., BECKER, R. H., HELFAND, D. J., & GREGG, M. D. 1997, *ApJ* **475**, 479–493; see also http://sundog.stsci.edu

WIERINGA, M. H. 1991, Ph. D. thesis, Leiden University

WIERINGA, M. H. 1993a, *Bull. Inf. CDS* **43**, 17; and PhD Thesis, Leiden Univ. (1991)

WIERINGA, M. H., DE BRUYN, A. G., JANSEN, D., BROUW, W. N., & KATGERT P. 1993b, *A&A* **268**, 215–229

WILKINSON D. 1998, *Proc. Natl. Acad. Sci. USA* **95**, 29–34; www.pnas.org/all.shtml

WILLS, B. J. 1975, *Aust. J. Phys. Ap. Suppl.* **38**, 1–65

WOOD, D. O. S., & CHURCHWELL, E. 1989, *ApJ* **340**, 265–272

WRIGHT, A. E., GRIFFITH, M. R., BURKE, B. F., & EKERS, R. D. 1994, *ApJS* **91**, 111–308

WRIGHT, A. E., GRIFFITH, M. R., HUNT, A. J., TROUP, E., BURKE, B. F., & EKERS, R. D. 1996, *ApJS* **103**, 145–172

XU, W., READHEAD, A. C. S., PEARSON, T. J., POLATIDIS, A. G., & WILKINSON, P. N. 1995, *ApJS*, **99**, 297–348

YOUNG, J. S., XIE, S., TACCONI, L., KNEZEK, P., VISCUSO, P., TACCONI-GARMAN, L., SCOVILLE, N., SCHNEIDER, S., SCHLOERB, F. P., LORD, S., LESSER, A., KENNEY, J., HUANG, Y.-L., DEVEREUX, N., CLAUSSEN, M., CASE, J., CARPENTER, J., BERRY, M., & ALLEN, L. 1995, *ApJS* **98**, 219–257

ZENSUS, J. A., DIAMOND, P. J. & NAPIER, P. J. (EDS.) 1995, *Very Long Baseline Interferometry and the VLBA*, ASP Conf. Ser. **82**, ASP, San Francisco; (http://www.cv.nrao.edu/vlbabook/)

ZENSUS, J. A., TAYLOR, G. B., & WROBEL, J. M. (EDS.) 1998, *Radio Emission from Galactic and Extragalactic Compact Sources*, Proc. IAU Coll. 164, ASP Conf. Ser. **144**, ASP, San Francisco; (http://www.cv.nrao.edu/iau164/164book.html)

ZHANG, X., ZHENG, Y., CHEN, H., WANG, S., CAO, A., PENG, B., & NAN, R. 1997, *A&AS* **121**, 59–63

ZOONEMATKERMANI, S., HELFAND, D. J., BECKER, R. H., WHITE, R. L., & PERLEY, R. A. 1990, *ApJS* **74**, 181–224

Science With Infrared Surveys

By CHARLES M. TELESCO

Department of Astronomy, University of Florida, Gainesville, Florida 32611, USA

This article gives an initial overview of the infrared range and the observational techniques which are used in it. This is followed by a detailed examination of the applications of infrared surveys to astronomical problems, such as Galactic evolution and the search for brown dwarfs.

1. Introduction

A survey – the systematic observation of a large number of objects and/or a large region of the sky – is almost always both time consuming and expensive. Therefore, the justification for a survey must be well formulated, the rationale must be clear, before one can expect to be awarded the required funding and telescope time. The rationale is, of course, the anticipated scientific value of the survey, i.e., how the survey will add to our knowledge.

In this series of lectures we will try to get a feeling for the kinds of astronomical research programs for which infrared (IR) surveys have proven to be particularly useful. First, I will give a technical overview of the IR regime: general definitions, atmospheric properties and limitations, observational techniques, etc. Then we will look at how IR emission is used as a diagnostic tool; i.e., what does the IR emission from astronomical sources look like and what can it tell us? Finally, we will consider scientific highlights of several successful IR surveys and look at some anticipated results from selected future programs.

2. Overview of the Infrared Spectral Region and Techniques

The infrared (IR) spectral region spans the wavelength range between about $1\mu m$ and $300\mu m$. Generally, it is sub-divided into the near-IR (1-$5\mu m$), the mid-IR (5-$30\mu m$), and the far-IR (30-$300\mu m$) regions, with longer wavelengths referred to as the sub-millimeter ($300\mu m$ - 1 mm) and millimeter regions. These are useful designations, because they convey to an experienced astronomer information about the technical characteristics (and challenges) of the observations: the types of detectors and the associated support equipment, the atmospheric properties, and the required observing sites or platforms.

In **Figure 1** I show transmission of the Earth's atmosphere between $1\mu m$ and several millimeters for a high, dry site like Mauna Kea (McLean (1997)). The atmospheric transmission is a strong function of wavelength. Water vapour and CO_2 are especially important absorbers throughout the IR, with water vapour making the atmosphere completely opaque over large wavelength ranges from all groundbased sites. Other absorbers, such as ozone, cause more limited (in wavelength) trouble, but can be important for specific scientific programs. In addition, the atmosphere emits molecular lines and a very broad ("thermal") continuum, which at each wavelength is proportional to the product of the emissivity and the Planck function evaluated at the temperature of the atmosphere (i.e., integrated over the full range of temperatures along the line of sight through the atmosphere). The thermal emission from the observer's astronomical telescope also contributes significantly to, and often dominates, the total thermal continuum emission at wavelengths longer than a few microns.

2.1. *The Near-Infrared Spectral Region*

The near-IR spectral region extends from 1μm to 5μm. The atmosphere is transparent over several reasonably well defined spectral regions, or "windows", for which there are a set of widely accepted standard filters designated J (1.25μm), H (1.65μm), K (2.2μm), L (3.6μm), and M (4.8μm). Slight variants on some of these filters (e.g., the K' and L' filters) have also been devised for special purposes, particularly to improve sensitivity by more optimally sampling the atmospheric windows. As we will see below, the colour information provided by these filters can be a useful discriminant of a broad range of physical processes, and one can thereby often very conveniently classify large numbers of astronomical objects. A fairly large gap in the atmospheric transmission, due entirely to absorption by water occurs between 5 and 8μm, which partly accounts for the treatment of the near-IR as a distinct spectral region. That designation has been reinforced by the development of indium antimonide (InSb) detectors (both single and in arrays) that are very responsive throughout this region but which cut-off sharply at 5μm.

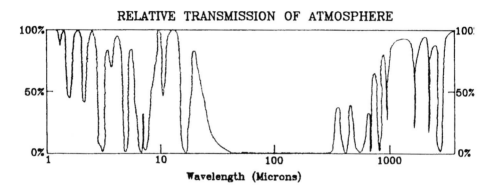

FIGURE 1. The atmospheric transmission in the IR.

Blackbody (i.e., thermal) continuum emission from "room" temperature ($\sim$300 K) sources becomes prominent in the near-IR at $\geq 3\mu$m. What this means is that radiation emitted by the telescope and the atmosphere (especially the atmosphere's water vapour) along the line of sight to an astronomical source is an often strong background against which the source must be detected. Some of the atmospheric windows are very transparent, and therefore, by Kirchhoff's Law, not very emissive; at those wavelengths, such as in the high-quality 10μm window discussed below, the IR emission from the telescope optics dominates the background. As long as the detector noise is very small, it is the photon shot noise (i.e., the fluctuations in the number of collected photons that result from the discreteness of the photon stream) associated with this background that is the principal source of noise for the IR observations; such observations are said to be "background-limited", and one always wants the detectors to be so good that it is the background shot noise (proportional to the square root of the collected photons), rather than the detector noise, that dominates the observational noise. This IR spectral region beyond about 3μm where the background "heat" emission becomes so obvious is called the "thermal IR". Because of the large thermal background (and therefore the large associated photon shot noise), observations at thermal-IR wavelengths are generally much less sensitive than those at 1-3μm. It is partly for this reason that most of thground-baseded near-IR observations are made in the J (1.25μm), H (1.65μm), and K (2.2μm) bands. Indeed, mercury-cadmium-telluride (HgCdTe) detector arrays (such as the NICMOS detectors for the Hubble Space Telescope), which have high responsivity only at $<2.5\mu$m, are optimised for operation in the region spanned by the J, H, and K bands. The thermal continuum radiation is relatively weak in the 1-3μm region, but other types of atmospheric emission play a significant role there. In particular, a forest of OH lines is emitted by the atmosphere as a result of its interaction with sunlight (see, e.g., McLean (1997)). This line emission, which can vary significantly in minutes, provides the fundamental background against which earth-based 1-3μm observations are made. O_2 atmospheric emission bands are also present in the 1-3μm region, but they are much less important than are the OH lines.

There are two main types of detectors now used in the near-IR: Indium Antimonide (InSb) devices that are sensitive out to 5μm, and mercury cadmium telluride (HgCdTe) devices sensitive out to 2.5μm. Generally, one wants to use a detector that is sensitive only out to the longest wavelength necessary for one's scientific program, since the longer the wavelength cut-off (and therefore the narrower the band-gap across which photoelectrons must be excited), the greater is the detector susceptibility to thermally generated electrons (dark current). Recently, these detectors have become available as arrays with over 10^6 pixels, in a 1024×1024 pixel format. This is an impressive advance over the first such arrays, with 3600 pixels (58×62 pixels), used for astronomy back in 1986.

2.2. *The Mid-Infrared Spectral Region*

The mid-IR region extends from about 5μm to the atmospheric cut-off near 30μm, the latter being the longest IR wavelength observable from the ground from the best sites such as Mauna Kea. The best mid-IR atmospheric window extends from about 8μm to 14μm, with another important, but much weaker, window (actually a collection of narrow transmission regions) extending from about 16μm to somewhat above 25μm. The broadband filter spanning the entire 8-14μm window (referred to as "the 10 micron window") is usually called the N-band filter, and the broadband filter spanning the most transmissive part of the 18-25μm window ("the 20μm window") is usually called the Q-band filter. Numerous narrower-band filters (typically with $\Delta\lambda \approx$1-2μm) usually called the "silicate filters" for reasons described below, are also available for use in these windows. At the best sites (i.e., the highest and driest), like Mauna Kea, the atmosphere is somewhat transmissive even out to 30μm.

Atmospheric CO_2 causes a deep absorption around 15μm. Water vapour is responsible for the short-wavelength cut-off of the mid-IR region near 8μm and for absorption lines throughout the mid-IR region. The water content of the atmosphere is a usefully strong function of altitude. In fact, its scale height is much less than that of most other atmospheric constituents. The atmospheric pressure drops roughly exponentially with altitude; i.e.,

$$P_z = P_0 \exp^{(-z/H)} \qquad (2.1)$$

where z and H are the altitude and scale height, respectively. The scale height of air is about 8000 m, whereas the scale height of water is about 3000 m. Thus, at a high observing site like Mauna Kea (4200 m), the observer is above 40% of the air but above 80% of the atmospheric water vapour, which greatly improves observing at mid-IR (and thermal near-IR) wavelengths. Note that the previously mentioned OH lines, which are a special problem for 1-3μm observations, originate at an altitude of about 90 km, so observing from even the highest ground-based sites does not reduce their interference to the observer.

The entire mid-IR region is in the thermal-IR regime; the Planck spectral intensity B_λ

for a 300 K "room-temperature" object peaks at 17μm. Depending on the telescope's IR emission properties, the telescope thermal emission may dominate over the atmosphere's thermal emission at wavelengths where the atmosphere is especially transparent and therefore less emissive; the 10μm window is such a region, but emission in the 20μm window, which is much less transparent, is dominated by the atmosphere. The mid-IR thermal emission from the telescope and sky is generally so large that IR astronomers have developed special techniques, called "chopping" and "nodding" to cope with it; these techniques are also used for broadband thermal near-IR and far-IR observations. It is not uncommon for the background emission from the telescope and atmosphere to be 10^5 - 10^6 times larger that the detected astronomical signal! How is this amazing signal extraction accomplished? The telescope secondary mirror is oscillated, or "chopped", at 5-10 Hz so that the detector's view is rapidly switched between the celestial source and an adjacent piece of sky. Subtracting those two views effectively removes the sky background emission as well as electronic drifts and offsets that vary less rapidly than the chop frequency. However, there is a residual IR background emission, called "radiative offset", from the telescope, which results from the fact that at each of the secondary-mirror chop positions the detector sees a slightly different telescope thermal geometry. That is removed by "nodding", or "beam switching", the whole telescope about every 10 s. The nodding consists of interchanging which of the two secondary mirror chop positions directs the celestial source's image onto the detector. Subtracting the images seen at the two different telescope nod positions removes the radiative offset, which, for a good telescope and a well-behaved atmosphere, leaves only the flux from the celestial source. Under the best conditions, the noise associated with that flux is the shot noise associated with the background emission; that noise, which results from the discreteness of the photon stream, is fundamental and cannot be removed.

Until the mid-1980's the principal detector for ground-based mid-IR astronomy was the germanium bolometer invented by Frank Low (Low (1961)). Although these detectors are still in use, the state-of-the-art detectors are the blocked-impurity band *BIB*), also called impurity band conduction *IBC*), photoconductors available from Boeing (from a division previously part of Rockwell) and Hughes Santa Barbara Research Center. The principal mid-IR detector in use now is the Boeing arsenic-doped silicon 128×128 pixel array, but during 1998, both 256×256 pixel arrays (Boeing) and 320×240 pixel arrays (Hughes) are expected to become available.

2.3. *The Far-Infrared Spectral Region*

The far-IR spectral region stretches from 30μm to about 300μm. Primarily because of water vapour, the Earth's atmosphere above sites even as good as Mauna Kea is completely opaque to far-IR radiation (although Antarctica may permit observing out to 60μm). The atmosphere is transparent at $\lambda > 300\mu$m, which marks the short-wavelength end of the sub-millimetre region. Far-IR astronomy must be carried out using airborne, balloon-borne, or space-based telescopes. That astronomers have been willing to go through so much trouble to make far-IR observations is a strong testament to the great diagnostic value of the numerous spectral lines and the continuum emission available there.

During its 21-year lifetime (1974-1995), *NASA*'s Kuiper Airborne Observatory (*KAO*), a 0.95-m telescope in a Lockheed C-141 Starlifter jet aircraft, was the most prominent and most productive sub-orbital far-IR observatory. The *KAO* is soon (2001) to be replaced with the US/German Stratospheric Observatory for Infrared Astronomy (*SOFIA*), which will consist of a 2.5-m telescope in a Boeing 747 aircraft. These planes must fly at altitudes of 41 000 - 45 000 ft (12.5 - 13.7 km) to be above most of the water vapour.

When one's altitude increases from that of even the highest ground-based IR site (Mauna Kea) to those of the *KAO* and *SOFIA*, the atmosphere goes from being totally opaque to nearly transparent in the far-IR. The primary limitation to far-IR sensitivity from these airborne observatories is the IR emission from the telescope and the atmosphere. Enormous gains in sensitivity are achievable by going to a cooled, orbiting IR telescope, although the costs are also enormous. We will explore this issue more in the next section.

Once you are observing from an altitude sufficiently high to open up the far-IR region, there are no well-defined atmospheric transmission windows, and so, unlike the situation in the near-IR and mid-IR, there has been no standard set of photometric filters for the far-IR. Far-IR observers have therefore used spectral filters selected on the basis of the specific science they want to do and on the availability of filter materials. Both photoconductive and bolometric detectors are in wide use in the far-IR, with the choice being made on the basis of the specific sensitivity requirements (e.g., spectral bandwidth) and operational constraints (e.g., detector temperature) of the program.

2.4. *Ground-based, Airborne, & Space-based IR Astronomy: Trade-offs*

The relatively low cost, the long lifetimes, the large telescope apertures, and the virtually unconstrained access to, and complexity of, the focal-plane instrumentation are the primary advantages of being able to do IR astronomy from the ground. Consider the following:

• The cost is about US$60-100M to construct one of the Keck 10-m telescopes or one of the Gemini 8-m IR-optimised telescopes, whereas space IR telescopes like *ISO* and *SIRTF* are an order of magnitude more expensive;

• Major ground-based IR-optimised telescopes like those of Gemini are expected to be key facilities in astronomy for 50 years before they are significantly surpassed by other technologies (although the Palomar 5-m telescope is over 50 years old and still doing outstanding astronomy thanks to constant technology upgrades); orbiting IR telescopes, which must have at least some actively cooled (usually with liquid helium) components such as the detectors, have much more restricted lifetimes of a few years at most;

• Whereas we have reached an era of 8-to-10-m IR-optimised ground-based telescopes, orbiting IR telescopes like *SIRTF* are still in the 1-m class. Although larger ones are on the drawing board, the enormous expense of developing, launching, and operating them may continue to restrict the size of these facilities. Thus, at least for the near future, the angular resolution of the largest ground-based telescopes will be superior to that of space facilities as long as development and implementation of adaptive-optics techniques for ground-based telescopes continues to progress;

• Although in theory one can upgrade the focal plane instruments on low-Earth-orbit IR telescopes, the short lifetimes of these missions usually make that impractical. In addition, the facilities are often (e.g., *ISO*, *SIRTF*) in inaccessible orbits, so one must use the same suite of modest IR instruments for the duration of the mission. The necessarily limited size of these instruments is usually reflected in their much more limited capabilities: each of the complex IR imagers and spectrometers on Gemini weighs 2000 kg and is several meters in size, whereas the *ISO* and *SIRTF* instruments are much more modest since all instruments for each of these missions must fit into an envelope smaller than a meter. The more generous weight and size restrictions and the ability to conveniently improve instruments over time make ground-based telescopes dynamic laboratories with very long lifetimes.

However, orbiting IR observatories have one major advantage that justifies the enormous expense and effort: they are extremely sensitive. This very high sensitivity is achieved by (1) being above the IR-absorbing and IR-emitting atmosphere, and (2) cooling the telescope so that its IR emission is negligible. A comparison of IR sensitivities for several ground-based, airborne, and orbiting IR observatories is given in **Figure 2**, taken from Thronson *et al.* (1995). The *KAO* and *SOFIA* are interesting "intermediate" solutions in that they permit one the advantages of long life, constant access, larger instruments, 1-2.5 m telescopes, and, of course, they permit observation in the far-IR region where the atmosphere is completely opaque from the ground. However, even though the telescope temperatures are lower than Earth-based ones, the thermal emission of the telescope and atmosphere still greatly restricts the sensitivity, as indicated in **Figure 2** (adapted from Thronson *et al.* 1995).

3. Infrared Diagnostics

3.1. *Emission from Gas & Stars*

IR continuum emission from gas is either free-free (bremsstrahlung) or bound-free radiation from fairly low-density gas or from stellar photospheres. For more exotic objects, such as active galactic nuclei and supernovae, we may also see non-thermal (synchrotron) IR emission. The spectral energy distribution from gas at temperatures higher than 1000 K peaks at wavelengths shorter than $5\mu m$, so stars, especially red ones, are bright in the near-IR. In the left panel of **Figure 3** (adapted from Doyon *et al.* (1990)), I show the loci of JHK colours for giant and dwarf stars. Red supergiant stars have near-IR colours that are barely distinguishable from those of red giants. The vectors in the figure indicate how the near-IR colours of red stars are modified by extinction (A_V) or by adding light from blue stars (A0), ionised gas (H II), or hot dust. Late-type stars have a photospheric CO absorption band at $\lambda > 2.3\mu m$. The depth of the CO band increases with decreasing stellar effective temperature and with increasing luminosity. The depth of the CO feature, as represented by a CO index, can be used to distinguish between red giants and red supergiants, as indicated in the right panel of **Figure 3** (Doyon *et al.* (1990)).

The IR spectral region is rich in emission lines from gas. IR fine-structure lines (also called forbidden transitions) occur between levels with excitation temperatures of a few hundred degrees, and so the atoms and ions are easily excited collisionally in many environments of astrophysical interest. The populations of the levels depend on the collision rates and on the radiative transition probabilities from the relatively long-lived excited states, and thus the ratios of various line strengths indicate gas densities and temperatures (see, e.g., Stacey (1989); Spinoglio & Malkan (1992); Voit (1992)). In addition, the energy distribution of the exciting source can be deduced by determining which ionisation states and gaseous components are present. Because the interstellar extinction is so much lower than in the optical, IR emission lines are incisive probes of even the most heavily obscured regions. The number of fine-structure lines that one expects to see throughout the IR is very large, and the potential for characterising the interstellar medium and the sources that power it in other galaxies is enormous. Spinoglio & Malkan (1992) give the ionisation potentials and critical densities appropriate for numerous IR fine structure lines. (The critical density is that density at which the collisional de-excitation and radiative de-excitation are equal; for densities below the critical density, collisional de-excitation is negligible, and radiative transitions dominate.) Not all of these lines have been observed in astrophysical environments.

The strengths of recombination lines are determined primarily by the rate at which gas is ionised. In a star-forming region, for example, the photoionization rate is a useful direct measure of the number of massive stars. The near-IR recombination lines of hydrogen at $4.05\mu m$ (Brackett α) and $2.17\mu m$ (Brackett γ) are strong, easily observable from the

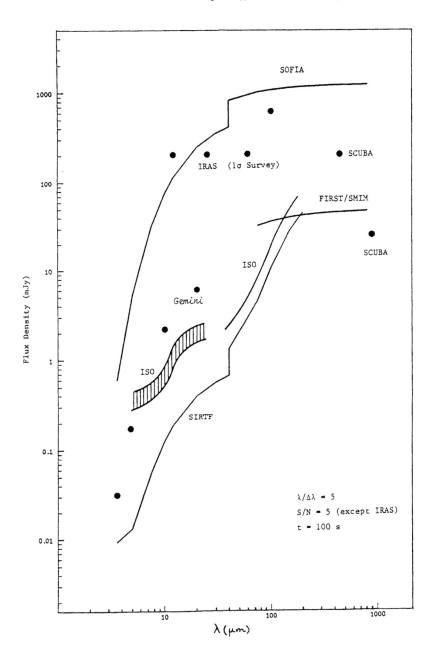

FIGURE 2. Comparison of photometric sensitivities of selected ground-based, airborne, and space-based IR observatories.

ground, and, as noted above for all IR lines, usually suffer only modestly from the effects of interstellar extinction. Since $A_{2.2\mu m} \approx 0.1A_V$, the ratio of these two lines is a useful measure of the extinction to regions with very high visual obscuration, for which the Balmer decrement is virtually useless as a measure of the extinction. The H I line at $1.28\mu m$ (Paschen β) and the He I line at $2.06\mu m$ are also widely used.

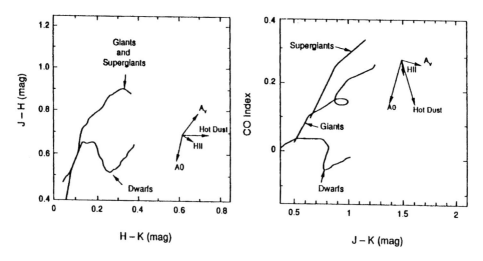

FIGURE 3. Near-IR colours and the CO index for stars.

The fine structure and recombination lines discussed so far originate in neutral and ionised atoms. However, there are astrophysically significant molecules, particularly H_2, that have been detected in their IR emission lines. Numerous H_2 lines due to transitions between vibration-rotation states are emitted in the 2-4μm region in Galactic star-formation regions (e.g., Beckwith *et al.* (1983)), with several of these, including the v = 2-1 S (1) line at 2.248μm and the v = 1-0 S (1) line at 2.122μm, being strong enough to observe in other galaxies. The H_2 is thought to be excited either collisionally in shocks or by absorption of 912-1100Å photons followed by fluorescent de-excitation. Depending on which lines are observed, the observations can be used to determine values of the interstellar extinction, shock velocities, and UV energy densities.

3.2. *Emission from Dust*

Dust and radiation are everywhere, so IR emission from dust heated by that radiation is also everywhere. The IR energy distribution from astrophysical dust is a broad bump, with the maximum flux density occurring somewhere between 10μm and 250μm, depending on the energy density of the radiation field at the location of those particles. Since dust particles evaporate at temperatures above 1500-2000 K, the IR emission from the hottest dust is often detected at K (2.2μm), but it is usually weak at J (1.25μm) and H (1.65μm). I show in **Figure 4** the IR energy distributions for various Galactic sources that illustrate the range of continua that IR astronomers observe. For most IR sources, the main body of the broad emission bump closely resembles a blackbody emission spectrum that is modified by a wavelength-dependent emission efficiency of the form λ^n, where the value of "n" is usually taken to be in the range -1 to -2. IR emission in excess of this blackbody on the short-wavelength side of the bump is usually present and indicates that there is dust at temperatures higher than that characterising most of the bump.

The region of the Orion nebula is the nearest (450 pc) site of active star formation (see review in Genzel & Stutzki (1989)). In **Figure 4**, I show the energy distribution for the 1′-diameter region, corresponding to 0.13 pc, centred on the IR cluster (Werner *et al.* (1976)). This cluster of young, massive stars emits $10^5 L_\odot$ of which 90% emerges at ($\lambda > 30\mu$m. It is buried in a dense, molecular cloud that is subject to complex radiative transport effects which include the absorption by outlying dust in the molecular cloud of near-IR and mid-IR radiation emitted by dust near the inner edges of the cluster cavity, resulting in the steep decline of flux at shorter wavelengths. This IR energy distribution peaking at 60-100μm is characteristic of H II regions containing OB stars still buried in dense molecular clouds where nearly all of the dust is cooler than 100 K (e.g., Churchwell *et al.* (1990)). OB stars that have dispersed their shroud of molecular material can still heat the remaining cloud from the outside, with the Trapezium stars and their visible companions in Orion being a notable example.

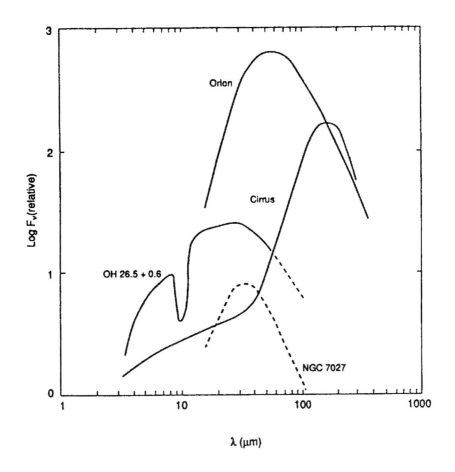

FIGURE 4. Illustrative IR spectral energy distributions for Galactic sources.

A unique discovery of the *IRAS* survey was that the Galaxy is laced with a vast filamentary structure that resembles cirrus clouds, which is discussed in more detail below. This cirrus emission is most evident at 60-100μm, but is also clearly present at 12-25μm. About 80% of the cirrus is emitted by dust mixed with the extensive distribution of low density H I, with the remainder mixed with diffuse H II (Boulanger & Pérault (1988)). The interstellar radiation field (ISRF) heats the cirrus-emitting dust. In the solar neighbourhood, the ISRF mean intensity $4\pi J = U_c = 2.17 \times 10^{-2}$ ergs cm^{-2} s^{-1}, and the energy density is U = 0.45 eV cm^{-3} (Mathis *et al.* (1983)). Only $\sim$7% of the ISRF luminosity is

UV radiation at $\lambda < 0.25\mu$m (Mathis *et al.* (1983)), but the efficient absorption by dust of the UV accounts for about 50% of the cirrus emission (Boulanger & Pérault (1988)). This UV is emitted by OB stars, and so stars younger than 10^8 yr. account for about half of the cirrus emission in the solar neighbourhood, with the remaining half of the cirrus luminosity being powered by older stars. As we will discuss later, the IR spectral energy distributions of galaxies usually resemble those of star-forming regions or cirrus emission.

When intermediate mass stars (2-8M$_\odot$) ascend the asymptotic giant branch, they rapidly lose mass and may become completely obscured by dust formed in the ejecta. These red giant and supergiant stars heat their dust shells to several hundred degrees. I show in **Figure 4** the IR energy distribution for the OH/IR dust-shell star OH 26.5+0.6 (Werner *et al.* (1980); Olnan *et al.* (1984)). The IR energy distribution is significant throughout the mid-IR and the shorter far-IR wavelengths.

I also show in **Figure 4** the IR energy distribution for the planetary nebula NGC 7027. I have included this object because it is the most extreme example I can find of a Galactic H II region powered by a hot star. The central star has a temperature of 1.7×10^5 K (see references in Telesco & Harper (1977)), and, although there is a massive neutral shell surrounding the H II region (Bieging *et al.* (1991)), nearly all of the IR emission appears to originate from dust within the H II region (Telesco & Harper (1977)). NGC 7027 is useful as a template for an ionised gas ball energised by a relatively hard UV radiation field; the average energy for a Lyman continuum photon in NGC 7027 is 50 eV (Telesco & Harper (1977)), compared to 18 eV for an O8.5 star (Telesco *et al.* (1989)). Such a gas ball might resemble in some ways an active galactic nucleus where dust is heated by the hard UV/x-ray spectrum generated near a massive black hole.

4. Survey Science in the Near-IR

Often the primary diagnostic value of surveys at 1-5μm derives from the fact that stellar photospheres are the most prominent emission sources at these wavelengths, and the extinction is low compared to that in the visible. For example, studies of the stellar initial mass function and galactic structure benefit greatly from the reduced effects of interstellar extinction in the IR, since in both cases you must be able to count stars that are either deeply buried in their parent molecular clouds or, because they are located at great distances, must be viewed through a long column of interstellar matter. In the following discussion we will look first at near-IR surveys of relatively limited parts of the sky that have been carried out to answer specific scientific questions about star formation and galactic evolution. Then, we will consider "all sky" surveys, which, by their observationally comprehensive nature, provide a database that can address a broad range of scientific questions.

4.1. *The Locations and Mass Function(s) of Young Stars*

The development of near-IR detector arrays with many pixels has provided astronomers with powerful tools to examine fundamental astronomical problems that were previously out of reach. The determination of where stars are born and how they are distributed in space and by mass is one such problem. Not only are the youngest stars deeply buried in molecular clouds, where they are visually obscured, but the parent cloud complexes are enormous, tens to hundreds of parsecs in size, which corresponds to many degrees for the closest star-forming regions, at $\sim$400 pc. The panoramic near-IR detector arrays, such as the pioneering SBRC 58$\times$62 pixel Indium Antimonide device, have lifted the obscuring veil from these regions, permitting practical surveys to be carried out over such large regions for the first time.

To illustrate what we can learn from ground-based near-IR surveys of star forming regions, consider the study by Lada *et al.* (1991) of the molecular cloud L1630 (Orion B). They used the above-mentioned array on the Kitt Peak 1.3-m telescope to image 2800 fields, each about $1'$ in size. The near-IR fields were chosen to span the primary molecular cloud regions. Lada *et al.* (1991) estimate the survey to be complete to a K magnitude of about 13, which, at the distance of this cluster (400 pc), corresponds to a main sequence star with a mass of $0.6 M_\odot$. They detected a total of 1185 sources. Using images taken in parts of the adjacent sky free of molecular clouds, Lada *et al.* (1991) estimate that about 50% of the stars seen toward L1630 are actually background, or field, stars observed through, or in front of, the cloud.

Although Lada *et al.* (1991) see sources everywhere throughout the surveyed region (this is at least partly due to the contamination by background stars), the source distribution is clustered rather than uniform, with three or four clusters being obvious. The central stellar number densities, i.e., the densities near the regions spanned by each cluster's central-most 50 stars, range from 440 stars pc^{-3} up to 4000 stars pc^{-3}. For comparison, the central part of the Trapezium cluster in the Orion nebula has a stellar density of about 14 000 stars pc^{-3}, whereas star cluster densities in the ρ Oph dark cloud appear to be comparable to those in L1630. Showing that these young stars form in clusters is the main result of the Lada *et al.* (1991) study, in clear disagreement with long-held notions that lower-mass stars tend to form in much smaller and isolated groups more uniformly distributed throughout dark clouds. Knowing how the number of stars varies with K magnitude permitted Lada *et al.* (1991) to also examine the stellar initial mass function (IMF), which is defined to be the distribution of stellar masses at birth. This is a fundamental property of star formation, and variations (or lack thereof) of the IMF throughout the Galaxy must hold important clues to the formation processes of stars in clusters. The IMF, often defined as the number of stars per logarithmic mass interval, is usually represented as a power law:

$$dN/d\log M) \propto M^{-\alpha} \tag{4.2}$$

with

$$\alpha = 1.7 \pm 0.5 \tag{4.3}$$

describing the IMF for solar neighbourhood stars (Miller & Scalo (1979)). A famous early estimate, called the Salpeter function, which is still consistent with available observations, has the value $\alpha = 1.35$ derived by Salpeter (1955) for solar neighbourhood stars. Lada *et al.* (1991) compare their cluster K distributions to those expected for clusters having a solar neighbourhood IMF: the results are only suggestive, but the cluster IMFs are consistent with that inferred for the solar neighbourhood. Clearly, here is an area to which near-IR surveys will make an enormous contribution.

4.2. Galactic Evolution

The near-IR spectral region, especially at K, provides several advantages for deep surveys of galaxies that extend out to large redshifts (see, e.g., Cowie & Songaila (1993)). Generally the K light from galaxies is dominated by the light from solar mass, red giant stars with lifetimes comparable to the lifetime of the galaxy; thus, modelling their evolution is easier, since one is trying to characterise long-term "passive" evolution rather than short-term, essentially stochastic events (such as starbursts). Also, the optical and near-IR energy distributions are fairly flat and similar for a wide range of galaxy types, so that the corrections to the flux detected in the spectral band to account for the redshift of the *SED* (the so-called K correction, this K being unrelated to the near-IR K band) are relatively small; this means that real evolutionary effects should be much more evident rather than being submerged by the various modelling uncertainties. Finally, the flat optical/near-IR *SED* for most galaxies results in the K band being much brighter out to large redshifts than are the optical bands.

A productive approach has been to make deep K-band and optical surveys, followed by optical spectroscopy to determine redshifts of the discovered sources. These data can then be used to infer galaxy evolution as a function of look-back time. An interesting example is the study by Songaila, Hu, & Cowie (see Cowie & Songaila (1993)). The left panel of **Figure 5** shows the redshift-magnitude distribution for their sample of 230 K-band-selected galaxies, which have a median redshift of 0.588. In the right panel of **Figure 5**, the median redshift is indicated for each magnitude bin. The observed distribution is compared to models for galaxies with no evolution and $q_0 = 0.5$ (solid line), 0.02 (dashed line), and mild evolution with $q_0 = 0.5$ (dot-dash); looking at the mild-evolution case specifically, we see that at a given redshift (e.g., z = 1), the models predict that a galaxy will be about a magnitude brighter there compared to the no-evolution case. However, the best fit of the faint-galaxy data is obtained (for $q_0 = 0.5$) for models that invoke a factor of four increase in galaxy number density coupled with a half magnitude dimming. These characteristics are most consistent with a scenario in which a large number of galaxies merged to form larger and therefore brighter ones near a redshift of 0.6, as proposed by Carlberg (1992); this is indicated by the presence of galaxies fainter than K = 18 falling below the no-evolution curve (i.e., they correspond to negative evolution). One might actually be seeing the formation, through mergers, of spiral galaxies (Cowie & Songaila (1993)).

4.3. All-Sky Surveys and the Structure of the Milky Way Galaxy.

Surveys that cover a large fraction of the sky have special appeal, because they usually provide a huge data base that permits statistically satisfying studies of a broad range of scientific problems, many of which are not fully defined or appreciated when the survey is conceived and executed. In this section we will consider all-sky surveys at near-IR wavelengths, illustrating their utility by considering how such surveys can be used to determine the structure of the Milky Way galaxy.

The first all-sky IR survey was the Two Micron Sky Survey (*TMSS*) carried out by Neugebauer & Leighton (1968) in the K band. The adopted limit of the survey was K = 3, the magnitude for which the survey was considered complete and reliable, but in fact the detection limit was K≈4.5. Most of the 5612 sources in the *TMSS* are late-type stars with the colours expected for normal photospheric emission, in particular oxygen-rich red giants (see, e.g., Jura (1993)). However, there are a small number of early-type stars and luminous supergiants, some of the latter possessing IR-emitting dust shells that have been expelled from the stars. The *TMSS* was a pioneering survey that gave us our first introduction to the IR sky.

DENIS and *2MASS*, two very ambitious near-IR all-sky surveys that are similar in scope, are now in progress. The Deep Near Infrared Southern Sky Survey (*DENIS*) will use a 1-m telescope to survey the southern sky in the three passbands I (0.8μm), J, and K_S with an angular resolution of 1″-3″ (see, e.g., Epchtein (1997)) and an average source positional accuracy of 1″. Initial observations imply 3σ limiting magnitudes of I = 18.5, J = 16.3, and K_S = 14, with the detectors saturating on sources brighter than I = 9.5, J = 8.5, and K_S = 6.5. The Two Micron All Sky Survey (*2MASS*) survey will use two

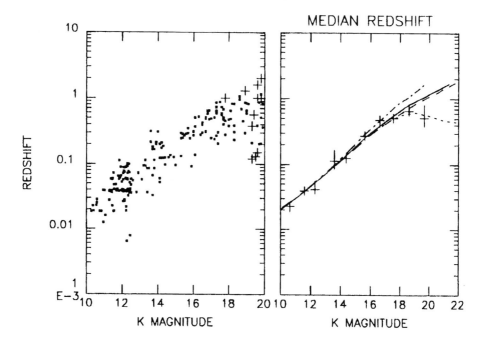

FIGURE 5. Redshift-magnitude diagram for 219 galaxies with spectroscopic redshifts (filled squares) and 11 galaxies with colour-estimated redshifts (crosses). The right panel shows the fit, using median redshifts, of the data to a variety of different evolutionary models.

1.3 m telescopes to image the entire sky in the J, H, and K_S bands with 3σ limiting magnitudes of 17.1, 16.4, and 14.3, respectively (Skrutskie *et al.* (1997)). The *2MASS* pixels subtend $2''$, and point source positions will be determined to better than $0.5''$. A fundamentally important technical feature of both the *DENIS* and *2MASS* surveys is that all three bandpasses for each survey are observed simultaneously, which insures accurate registration of multi-wavelength images of each field and thus accurate colours for the objects.

To put these survey limits into perspective, we note that the *2MASS* K-band limit corresponds to the detection of a K0 main-sequence star at a distance of 1 kpc and a K0 red giant at nearly 14 kpc. Thus, the determination of the Galactic structure using the near-IR emission from giant stars is a potentially powerful application of both *2MASS* and *DENIS*.

The near-IR surveys discussed above are detecting almost the entire range of types of astronomical source: old and young field stars and star clusters, H II regions and planetary nebulae, shocked gas, and near and distant galaxies. However, as already noted, the detection of the old, especially giant, stars is of particular value in determining the structure of our galaxy. The fact that the interstellar extinction is nearly ten magnitudes lower at K $(2.2\mu m)$ than at V $(0.5\mu m)$ means that near-IR star counts in a given direction, especially in the Galactic plane, can much more accurately reflect the true number of stars. Galaxy model builders use these star counts along with assumptions about the stellar luminosity functions and populations to infer the structure of the Galaxy. As discussed by, e.g., Robin (1997), these models can be static, in which the match to the observations is made using stellar properties at a fixed (i.e., current) evolutionary stage, or evolving, in which initial models of the gas distribution are evolved into a galaxy. Let's look in more detail at how this modelling is done for the static case.

Although there are somewhat different approaches to the problem of determining Galactic structure from star counts, they all assume that the Galaxy is composed of several distinct components each of which has a different spatial distribution (see, e.g., Wainscoat *et al.* (1992)). The basic idea is that by observing the numbers of stars as a function of magnitude in many different directions, one should be able to constrain the structure of each of the Galactic structural components. Wainscoat *et al.* (1992), and references therein consider the Galaxy to be composed of five geometrical components: the exponential disk, the bulge, the stellar halo, the spiral arms, and the molecular ring. Although these components are defined geometrically (i.e., by shape, or spatial distribution), their existence as relatively distinct entities is also supported by their distinctive kinematics and metallicities, and by examination of other galaxies thought to be similar to our own galaxy. For example, the spiral arms are represented in the Wainscoat *et al.* (1992) models as consisting primarily of massive young stars, and the "molecular ring" has both massive and lower mass stars. Where possible, specific analytic expressions are used to represent the shapes of each of the components (see Wainscoat *et al.* (1992)). An additional complication is that extinction, while much smaller in the IR than in the visible, must still be taken into account in the near IR.

An important part of the procedure is to use as much prior information to establish the general properties of the Galaxy, and then use the near-IR star counts to test that picture and make major or minor adjustments to it. A source list must be constructed that gives the relative numbers of all stellar types (or other types of astrophysical source) that are thought to contribute to the content of each of the Galactic structural components (see, e.g., Table 2 in Wainscoat *et al.* (1992)). One must first construct a Galactic model using each of the structural components with its appropriate source content. Then one considers the appearance that this composite model has along any line of sight originating at the Earth. One actually integrates the properties along the line of sight to infinity, collecting the counts at each position for each class of star in each of the five Galactic components into magnitude bins at each relevant wavelength (or passband). Wainscoat *et al.* (1992) carried out such a procedure to match the source counts in the V, K, and 12, 25, and 60μm *IRAS* bands. However, because the *DENIS* and *2MASS* surveys will provide a particularly powerful data base for Galactic structure research, I show in **Figure 6** a comparison between the observed star counts at K and those predicted by the model of Ruphy (1997) for a representative line of sight in the Galactic plane (l = 303°, b = 0.4°); the agreement is reasonably good.

5. Survey Science in the Mid-IR and Far-IR

The Infrared Astronomical Satellite (*IRAS*) has had the greatest impact of any IR survey so far. It has really changed our view of how the sky appears at IR wavelengths, as is evident from the enormous number of scientific publications based on *IRAS* results. What I want to do here is to describe some key scientific results of *IRAS*, and then examine briefly how several more recent surveys have expanded, or will expand, on them. I also refer you to the reviews of *IRAS* results by Beichman (1987), Soifer *et al.* (1987), and Sanders & Mirabel (1996).

IRAS mapped 96% of the sky at 12, 25, 60 and 100μm during 10 months in 1983. The angular resolution was about 0.76'×4.5' at 12 and 25μm, 1.5'×4.75' at 60μm, and 3'×5' at 100μm, with the smallest dimension being along the scan direction. Survey sensitivities were typically 0.3 Jy, although co-addition of many scans in some regions of the sky

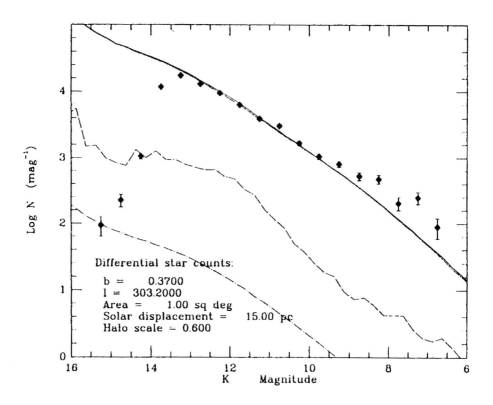

FIGURE 6. *DENIS* observed differential star counts at K compared to model predictions. Diamonds, observed star counts; solid line, total model prediction; dotted line (nearly coincident with solid line), disk; long-dashed line, spiral arms; dash-dot line, halo.

provided sensitivity that was about an order of magnitude better than this. Long before *IRAS* was launched, mid-IR astronomy was being done from the ground, and far-IR astronomy was being done from the Kuiper Airborne Observatory (*KAO*), which began operation in 1974. As a result, we knew that all kinds of astronomical sources are bright at these wavelengths. For example, early ground-based and *KAO* mapping of H II regions such as the Orion nebula showed that most of the luminosity of these star-forming regions emerges in the IR (e.g., Werner *et al.* 1976). Small samples of more evolved objects such as planetary nebulae also proved bright in the IR (Telesco & Harper (1977); Moseley (1980)). Likewise, ground-based, Learjet, and *KAO* observations of other galaxies such as M82 and NGC 253 demonstrated that at least a few of them are IR-luminous (Rieke & Low (1972); Harper & Low (1973); Telesco & Harper (1980)). At the time, these detections of large mid-IR and far-IR fluxes were very surprising and, indeed, very exciting, and I remember how hard we worked to interpret the observations of what was clearly a relatively small sample of objects; by 1980, we had detected only about a dozen galaxies in the far-IR. The points I want to make here are that (1), even before *IRAS*, we were very aware of the fact that many types of astronomical sources are bright in the mid- and far-IR, but (2) the samples of objects, at least in the far-IR, were relatively small. Both of these points provided a strong motivation for *IRAS*. Also worth noting is that, before *IRAS*, we essentially chose well-known examples of different classes of celestial objects and observed them at mid and far-IR wavelengths. The *IRAS* survey was unbiased, and so *IRAS* made some really unexpected significant discoveries. Here I want to explore three particularly unique contributions of *IRAS*.

5.1. *Infrared Cirrus*

One of the first startling results of the *IRAS* survey was the discovery that the sky at high Galactic latitudes ($|b| \geq 10°$) is covered in a vast complex of patchy, often filamentary emission that is superposed on smoother emission for which the brightness distribution is roughly proportional to $csc(|b|)$. This emission looks like cirrus clouds in the Earth's sky and so is referred to as "infrared cirrus" (Low *et al.* (1984)). The cirrus, which is evident in all four *IRAS* bands, is emission from interstellar dust particles. Some of the cirrus is associated with faint filaments previously identified at visible wavelengths (e.g., Sandage (1976)). It appears that the cirrus-emitting dust detected by *IRAS* has optical depths corresponding to visual extinctions down to values as low, or lower than, $A_V = 0.01$ mag, which implies that *IRAS* was very sensitive to low-surface-brightness material. The structure of the cirrus is, as the name implies, very complex. On the *IRAS* images, one can trace individual filaments over tens of degrees, and fine structure is evident down to the survey's resolution limit of several arcminutes. The cirrus is almost certainly emission from dust grains, because it correlates well with extinction to background stars and the scattering of Galactic light (e.g., De Vries & Le Poole (1985)), and warm dust emits IR radiation.

What is the relationship of this cirrus to other constituents of the interstellar medium? If you actually compare pixel-by-pixel the *IRAS* all-sky maps to H I, H II, and CO maps, you find that there is a very good correlation between the IR and H I distributions. What one finds is that, across the sky, there is a variation of about a factor of three in the ratio of the far-IR (i.e., 60 and 100μm) emission per H I atom, with there being more far-IR emission per H I atom in the vicinities of star forming regions (e.g., Boulanger & Pérault (1988)). The obvious interpretation of this is that the large-scale distribution of interstellar dust in the Galaxy is heated by the general interstellar radiation field, with there being an increase in this heating near OB associations. A significant fraction, about 20%, of the gas outside of H II regions and dense clouds is ionised; i.e., ionised hydrogen accounts for 20% of the total amount of neutral plus ionised hydrogen. On a large scale, these two components are reasonably well mixed, and therefore, at least approximately, dust mixed with the diffuse H II appears to account for about 20% of the far-IR cirrus emission. Molecular clouds also contribute a small amount to the cirrus emission, since there appears to be a good correlation between CO emission and the cirrus clouds. Molecular clouds are heated internally by young stars (this IR emission is not strictly cirrus emission) and externally by the interstellar radiation field, as described below.

The spectral energy distribution (*SED*) of the cirrus-emission holds some important clues to the nature of the dust grains. **Figure 7** shows the normalised *SED* derived by Boulanger & Pérault (1988) for the cirrus at $|b| \geq 10°$ (see also, Fich & Terebey (1996)). This *SED* clearly does not look like that of an H II region or, for that matter, any of the other canonical sources that we are familiar with (see **Figure 4**); cirrus has a very distinctive, very peculiar *SED* characterised by a "cool" far-IR slope (high 100μm-to-60μm ratio) and a "hot" mid-IR slope (high 12μm-to-25μm ratio). The dust temperature derived from the typical far-IR flux ratio is 20-30 K, assuming the dust emission efficiency varies with frequency as $\nu^{1.5}$. The 12 and 25μm fluxes imply temperatures in the range

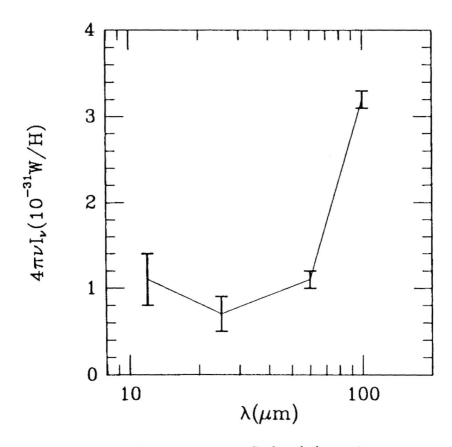

FIGURE 7. Cirrus spectrum normalised per hydrogen atom.

80-500 K (Gautier (1986)). There seems to be hot dust and cold dust and not much in between. The emission between 7 and 35μm comprises about 40% of the total cirrus emission emerging shortward of 120μm and about 25% of the total starlight absorbed by the dust (Boulanger & Pérault (1988)). In fact, this *SED* implies that there are essentially two distinct populations of grains (Section 2): particles large enough (roughly 0.01μm) to be in radiative equilibrium with the ambient heating radiation field, and very small particles ($(0.001\mu$m) each of which, because of its small size and therefore its small heat capacity, experiences a large increase in temperature when it absorbs a single photon (Andriesse (1978); Sellgren (1984)). Because the radiation field heating the cirrus-emitting dust is of relatively low energy density, the larger grains, which are in equilibrium with it, are very cool, thus giving rise to the extremely red far-IR colour. On the other hand, the temperature of a small grain, after it absorbs a UV photon, is independent of the radiative energy density. Even if a large population of very small grains is present in an H II region, which they certainly are, their *SED* is not as distinct as that seen in the cirrus, because the IR emission from the larger grains in the higher energy density environment of the H II region is much more prominent at the shorter wavelengths.

Based on a group of emission features in the 3-12μm region observed in Galactic reflection nebulae and other sources, polycyclic aromatic hydrocarbon molecules (*PAHs*), or some variation thereof, are thought to constitute one component of this population of small grains (Léger & Puget (1984)). A large fraction of the cirrus emission in the *IRAS* 12μm band may arise in these bands, but one or more additional components of small grains is needed to account for the 25μm and at least part of the 60μm emission (Draine & Anderson (1985)). Absorption of UV photons appears to be important in heating the small grains and *PAH*s (e.g., Puget & Léger (1989)), but recent studies (Sellgren *et al.* (1990)) have shown that visual photons may also play a role.

The general interstellar radiation field (ISRF) heats the cirrus-emitting dust. The ISRF is the interstellar field that results from the superposition of the radiation "leaked" into the ISM from stars of all ages and spectral types. The *SED* of the ISRF is obviously to some extent a function of position in the Galaxy. In the Galactic bulge, which is dominated by red stars, the ISRF is undoubtedly "soft" compared to the ISRF in one of the Galactic spiral arms where there are large numbers of young, blue stars. Boulanger & Pérault (1988) have done a thorough analysis of the types of stars that contribute to the heating of the cirrus-emitting dust in the solar neighbourhood and of the types of interstellar gas that those dust particles are mixed with. Let me summarise their results.

The cirrus emission in the solar neighbourhood originates from dust mixed with atomic gas (9 $L_\odot/pc^2$), molecular clouds (1 $L_\odot/pc^2$), and H II regions (4 $L_\odot/pc^2$). These values represent emission per unit surface area of the Galactic plane within about 500 pc of the sun. Dust associated with atomic and ionised gas far away from star-forming regions accounts for two-thirds of the cirrus emission; that dust is heated by the ISRF. Boulanger & Pérault (1988) then used the interstellar extinction curve and the *SED* for the ISRF presented by Mathis *et al.* (1983) to estimate that UV (0.0912μm $< \lambda <$ 0.346μm), optical (0.346μm $< \lambda <$ 0.8μm), and infrared ($\lambda >$ 0.8μm) photons account for 50%, 30%, and 20%, respectively, of the heating of the dust by the ISRF. Thus, about half of the cirrus emission actually arises from young stars, since the UV photons in the ISRF originate from B-type stars. Adding this to the cirrus emission identified with very extended, diffuse H II implies that about two-thirds of the cirrus emission is due to heating of the dust by stars younger than 10^8 yr. This conclusion has special significance for the interpretation of the IR emission observed in other galaxies and discussed below.

5.2. *Infrared-Luminous Galaxies*

IRAS neither discovered that galaxies are bright in the mid-IR and far-IR, nor did it provide the basis for the basic interpretation of the origin of the emission (see the review in Telesco (1988)). However, by virtue of the enormous number of galaxies observed in the *IRAS* survey, it permitted the first real statistical examination of these properties, which in turn allowed us to begin to appreciate how each galaxy fits into the broader context of this phenomenon.

The first direct indication that IR observations could provide unique information about other galaxies came in the late 1960's when ground-based observations by Low and others showed that the 2 - 22μm emission from a small sample of quasars and Seyfert galaxies is well above an interpolation between their optical and radio spectra (e.g., Low & Keinmann (1968), and references therein) and that the flux must be even higher at inaccessible wavelengths beyond about 30μm where the Earth's atmosphere is opaque from the ground. Conservative estimates of the IR luminosities based on those observations implied values much higher than expected to originate from the old stars in a normal galaxy. The urgent question became: what is the source of that large luminosity?

The next logical step after these exploratory ground-based observations was to carry out ground-based surveys of galaxies to see how many galaxies are strong IR emitters, and to look for trends, for patterns, in the distributions in luminosity, size, host-galaxy type, shape of the spectral energy distributions, etc. The first IR surveys of galaxies, carried out primarily through the good atmospheric window at 10μm, showed that extragalactic IR emission is ubiquitous (Kleinmann & Low (1970); Rieke & Low (1972)), and there was reason to conclude that the radiation in most, although maybe not all, of these objects is emitted by warm dust grains (Burbidge & Stein (1970)), so the mystery became, What heats the dust? By considering a reasonably large number of normal and active (Seyferts and quasars) galaxies, the 10μm survey by Rieke & Low (1972), in particular, demonstrated how the observations of a large sample could be used to try to relate the emission phenomenon to the type of galaxy and the source of power. At about this time, exploratory observations by Harper & Low (1973), using the *NASA* 30-cm airborne (Learjet) telescope, showed that at least for the two nearby galaxies M82 and NGC 253, the IR emission peaked near 80μm, and they demonstrated the great potential value of using airborne observations to extend the extragalactic research to beyond the atmospheric cut-off near 30μm.

Throughout the 1970's and early 1980's this potential began to be realised as the new Kuiper Airborne Observatory (*KAO*) permitted the first systematic studies of extragalactic IR emission (Telesco & Harper (1980); Rickard & Harvey (1984); and references therein). These studies actually represented small surveys. Those first far-IR surveys of numerous galaxies indicated that high IR luminosities, those greater than $10^{10}L_\odot$, are probably common and that the emission from many (but not all) of the galaxies originates from dust heated by young stars. It was shown, for example, that the shapes of the 10 - 200μm spectral energy distributions are like those of Galactic star-forming regions, and that the amount of far-IR luminosity is correlated with the amount of CO emission, thus linking the magnitude of extragalactic star formation to the amount of fuel available to form stars. *KAO* Observations, albeit limited ones, of Seyfert galaxies, particularly the Seyfert 2 archetype NGC 1068, showed that they may harbour two IR emission components, a compact one associated with the active nucleus and an extended one associated with intense star formation. These *KAO* and ground-based IR observations of galaxies formed the backdrop, and indeed an important part of the justification, for the launch of the Infrared Astronomical Satellite (*IRAS*).

IRAS detected about 25 000 galaxies in the mid and far-IR (see Soifer *et al.* (1987) for a more complete review of the *IRAS* extragalactic sample). Nearly all of these are late-type spirals, with there being very few ellipticals or lenticulars. At Galactic latitudes greater than 30° over 75% of the 60μm sources in the *IRAS* Point Source Catalogue are galaxies. Such a large data base has permitted us to gain an understanding of the range of luminosities and IR *SEDs* for galaxies and how those correlate with their other properties such as their radio-continuum and molecular emission and their optical morphology. For example, *IRAS* detected galaxies with IR luminosities spanning at least six orders of magnitude (see Soifer *et al.* (1987)). For the typical galaxy less luminous than $10^{11}L_\odot$, about 20% of the total luminosity emerges in the mid and far-IR, whereas nearly all of the luminosity emerges in the IR for more luminous galaxies. Although they receive a lot of attention because of their spectacular rate of energy production, galaxies more luminous than about $10^{12}L_\odot$, the so-called ultraluminous galaxies, are very rare; there are 10 000 times fewer galaxies emitting $10^{12}L_\odot$ than are emitting $10^{10}L_\odot$.

The *SEDs* exhibit a range of shapes, a fact manifested nicely in the *IRAS* 2-colour plot shown in **Figure 8** Helou (1986). The distribution of galaxy colours in this plot follows a locus that Helou (1986) first interpreted as a mixing line: the location of a galaxy's colours in the plot is determined by the relative contribution from star-forming regions and cirrus regions, as also indicated by the models. The IR expected from these

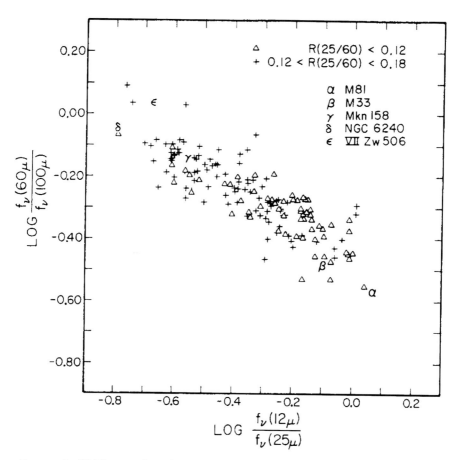

FIGURE 8. *IRAS* two-colour diagram for mostly normal (non-*AGN*) galaxies.

two types of regions was discussed in Section 3.2. It appears that the dust emitting the IR emission from most galaxies is heated by stars of various ages; those in which young stars dominate the heating are called starburst galaxies, whereas those in which the IR luminosity is mainly cirrus emission are "normal" galaxies. This picture is supported by the determination that the IR luminosity is correlated with CO and radio-non-thermal luminosities, both of which are expected from large ensembles of young stars (see Telesco & Harper (1980); Telesco (1988)); the CO is part of the clouds from which stars form, and the radio non-thermal emission is synchrotron radiation associated with the supernova remnants resulting from the deaths of massive stars. Galaxies, such as Seyferts, with active galactic nuclei (*AGNs*) can also emit very large IR luminosities (Telesco *et al.* (1988)). It is very clear that at least part of the mid and far-IR emission from these active galaxies arises from heating of dust by radiation produced directly or indirectly by the exotic central object (black hole). *KAO* observations showed this for the first time (Telesco & Harper (1980)), but *IRAS* demonstrated this as a general phenomenon (De Grijp *et al.* (1987)). In fact, the *AGNs* and starbursts apparently coexist in enough IR-luminous galaxies that an evolutionary connection between the two has been hypothesised.

I have said very little about surveys being carried out with the Infrared Space Observatory (*ISO*), which, during this Winter School, celebrates its second successful year in orbit. This omission results mainly from the fact that few of the *ISO* surveys have been completed, and the data that have been obtained are being analysed. The next year or so will see a dramatic increase in the dissemination of impressive *ISO* survey results. Nevertheless, I would like to mention an *ISO* survey that I am participating in (Telesco *et al.* , in preparation) that demonstrates the continuing value of the *IRAS* extragalactic database. These results are preliminary, and I do not plan on being held accountable for what I say here!

Deep surveys at radio wavelengths followed by visual and near-IR imaging and spectroscopy of the radio sources has proven to be very effective for finding distant galaxies. Interestingly, as one considers progressively fainter radio fluxes and higher redshifts, one finds an increasingly larger fraction of these galaxies are very blue and evidently experiencing intense star formation; blue radio galaxies constitute one-third of the 1.4 GHz sample at flux densities above 10 mJy but over two-thirds of the sample below 3 mJy Windhorst *et al.* (1985). These blue radio galaxies appear to be the excess population of 1.4 GHz radio sources discovered by Windhorst *et al.* (1985). Complex optical morphology of many of these galaxies implies that they may be interacting or merging, which may account for the apparent episodes of star formation. Most of these galaxies for which redshifts have been measured have $z < 1$. Our *ISO* program has focused on the detection of these galaxies at 15, 60, and $100\mu m$ to permit us to estimate their luminosities, and to see how those luminosities compare to known classes of galaxies. These observations should also permit an assessment to be made of their *SED*s and characteristic dust temperatures.

Our total sample consists of 51 galaxies distributed among five different fields. *IRAS* had previously detected only about three of these galaxies. Our preliminary reduction of the data indicates that we have detected about 50% of the galaxies, with our 3σ detection limit being about 90 mJy at 60 and $100\mu m$. For the small subset of the galaxies for which we have found published redshifts, we estimate IR luminosities in the range of roughly 10^{10}- $10^{11}L_\odot$. I wish to emphasise here that our knowledge of the *IRAS* extragalactic database permits us to say that galaxies in this luminosity range are not unusual, at least by virtue of their luminosity. Indeed, many of the archetypal starburst galaxies, such as M82 and NGC 253, fall right in the middle of this range. However, if it turns out that most of these galaxies are, for example, dwarfs, then the IR luminosity may prove to be unusually high for that class. *IRAS* provides the backdrop, the frame of reference, permitting us to initially assess how these galaxies compare to other more familiar ones.

5.3. *Vega-Type Debris Disks*

Before *IRAS*, certain types of stars were known to have IR emission above that expected from the photosphere: (1) relatively young stars, still associated with the dusty molecular clouds from which they formed; (2) massive stars with ionised stellar winds; (3) cool stars with dust shells due to mass loss during the asymptotic giant branch phase. However, the detection of a large mid and far-IR excesses from the main sequence A stars α Lyrae, β Pictoris, α Piscis Austrinis, and ϵ Eridani, which are not expected to undergo significant mass loss, was a big surprise (Aumann *et al.* (1984)). These stars show IR excesses extending from the mid-IR (12-$25\mu m$) out to the 60 and $100\mu m$ where the *SED*s actually peak. Since the original discovery of these so-called "Vega-type debris disks", over 100 additional such stars have been discovered by careful examination of the original *IRAS* data base (see Backman & Paresce (1993)).

The possibility that disks surround these stars was given dramatic support by the coronagraphic imaging at $0.89\mu m$ of β Pic by Smith & Terrile (1984). Their image shows what is most straightforwardly interpreted as an edge-on disk. The inner few hundred AU of the disk have been resolved in the mid-IR in studies by Telesco *et al.* (1988) and Backman *et al.* (1992). The former study has permitted a direct estimate of the dust temperature at 80 AU from the star, which shows that the dust is much hotter (160 K) than the value of 50 K expected for a blackbody at that distance. This result implies that the dust grains are very small, inefficient emitters: indeed, a more thorough analysis implies grain sizes of about 0.2μm (Telesco *et al.* (1988)), that has been confirmed by the detection of the silicate feature (Telesco & Knacke (1991); Knacke *et al.* (1993)), a feature which is not expected to be observable if the grains are large (e.g., bigger than about 10μm). The interesting point here is that grains of that size will lose orbital angular momentum due to the Poynting-Robertson drag; they will spiral into the star in 3×10^{5} years. Since this is a much shorter time scale than the lifetime of the star, we conclude that the dust is being replenished. Replenishment of the dust by comets or asteroids is currently considered possible. In addition, a central clearing of the dust has been detected by direct mid-IR imaging by Lagage & Pantin (1994). They point out that the clearing may be caused by gravitational perturbation of a planet. We see that disks may hold important clues to the origin and evolution of planetary systems. *IRAS* opened up this important area of research, with future ground-based and space-based surveys in the IR poised to make additional important discoveries.

5.4. *Galactic Structure at Longer Wavelengths*

The extraordinarily successful Cosmic Background Explorer (COBE) was launched in November 1989 to study the diffuse IR and microwave background radiation thought to be the 2.7 K remnant of the primeval explosion, the Big Bang. COBE has three essentially independent instruments contained within a common envelope/shield structure. These instruments are the Differential Microwave Radiometer (*DMR*), the Far-Infrared Absolute Spectrophotometer (*FIRAS*), and the Diffuse Infrared Background Experiment *DIRBE*). The *DMR* experiment determined the degree of isotropy of the cosmic background at 3.3, 5.7, and 9.6 mm. The *FIRAS* experiment determined the spectral energy distribution of the cosmic background radiation to unprecedented accuracy over the wavelength range 100μm to 5 mm. Both *DMR* and *FIRAS* have angular resolutions of about 7° and operate in the submillimeter and microwave regions. The third COBE experiment, *DIRBE*, operates only at IR wavelengths, 1.2 to 240μm, and I would like to look at some of the *DIRBE* results here.

The *DIRBE* experiment mapped the entire sky simultaneously in ten different passbands with a 0.7° field of view: 1.2, 2.2, 3.5, 4.9, 12, 25, 60, 100, 140, and 240μm. The primary purpose of *DIRBE* is to map out the cosmic background at these wavelengths, to search for diffuse light from the earliest, most distant galaxies and stars. However, as we have already discussed, the nearby universe emits heavily at those same wavelengths, and the goal of the *DIRBE* observers has been to extract the cosmic emission from this much stronger, and very complex, emission arising in the solar system, the Galaxy, and in nearby galaxies. However, those non-cosmic data are very interesting in themselves.

An intriguing result from radio astronomy in the late 1950s was that the H I layer of the outer Milky Way is warped (Burke (1957)). Actually, we see warping in the disks of many galaxies, and such a result for the Milky Way may provide us with special insight on the Galactic structure and its evolution. Djorgovski & Sosin (1989) showed that the warp is also evident in the distribution of asymptotic giant branch (AGB) stars, as determined from the *IRAS* 12μm point source catalogue. Since the AGB stars are old stars that, it is thought, trace the general stellar distribution, this latter result implies that the warp is not restricted to the youngest components of the Galaxy represented by the H I. *DIRBE*'s broad wavelength coverage has provided the opportunity to examine this warp systematically as a function of wavelength, and, thus, as a function of Galactic population. Freudenreich *et al.* (1994) present contour plots and longitude scans of *DIRBE* images of the Galactic plane which confirm beautifully the previously inferred warp, and offer the opportunity to analyse it in much greater detail than previously possible. For example, the warp is evident at both 3.5μm and 240μm, but it is noticeably smaller at 3.5μm ($\sim 1°$) than it is at 240μm ($\sim 2°$). The 240μm emission comes from cool dust that follows closely the distribution of the H I, whereas, the 3.5μm emission follows the distribution of primarily K giant stars. One interpretation of the smaller warp in the giant-star disk than in the cool-dust disk is that the stellar disk is truncated, so that we are actually seeing a much larger disk in the dust and H I (Djorgovski & Sosin (1989)). Indeed, it is not uncommon to see galaxies in which the H I distribution is much more extended than the optically evident starlight. Among the explanations for the warp is that it results from the Galaxy's having an aspherical, massive dark halo (see Djorgovski & Sosin (1989), and references therein). Such an aspherical halo, which would severely perturb the disk, could be a product of merging of galaxies in the early history of the Galaxy.

6. Glimpses of the Future: SIRTF

*N*ASA's Space Infrared Telescope Facility *SIRTF*), which is scheduled to be launched in about 2001, will contain a 0.85-m telescope with instruments designed to span the 3-200μm spectral range. Although other very important IR missions, such as *FIRST* and Darwin, are also being planned, *SIRTF* is imminent, and so I will focus my attention on it. Indeed, you will need to start thinking very soon about how you want to use this outstanding facility. Two unusual characteristics of *SIRTF* are its use of the "warm-launch" concept and the heliocentric orbit. The warm launch consists of cryogenically cooling only the detector systems, with the telescope eventually cooling to 6 K on orbit through a combination of radiation and the use of liquid helium boil-off gas. *SIRTF* will be launched in such a way that it will trail the Earth in its orbit around the sun. *SIRTF* will have three IR instruments: a 3 - 8μm camera, a moderate-spectral-resolution (R = 600) and low spectral resolution (R = 50) spectrograph spanning the 10 - 40μm region, and an imaging photometer for the 12 - 160μm region. To maximise the return of this roughly 3-year mission, several large-scale, systematic projects, as well as smaller-scale projects, are planned that will provide a legacy for astronomers well after the mission is over. We will now take a look at examples of these programs.

6.1. *The Search for Brown Dwarfs*

Brown dwarfs (BDs) are star-like objects which have masses below the critical value $(0.08M_\odot)$ for sustained nuclear reactions to take place. They radiate away the residual heat generated during their formation, with the more massive *BD*s having higher initial temperatures (1000-2000 K) and taking longer to cool down. (Interestingly, about half of the radiation we see from Jupiter $[0.001M_\odot]$ - obviously a planet, not a *BD* - is due to the residual heat of formation). One expects *BD*s to be intermingled with ordinary stars of all ages in clusters, in binary star systems, and as field objects. *BD*s may account for a significant fraction of the Galaxy's dark matter that is detected only by its gravitational effects, although if that is the case, the stellar IMF must have an unexpected excess of sub-stellar masses. The detection of *BD*s has important implications for our understanding not only of stellar evolution, but also of the larger scale structure and content of galaxies and the universe. The low temperatures of *BD*s make observations at

IR wavelengths critical to their discovery and exploration. No field (i.e., isolated) *BD*s have yet been detected, but there are several good candidates that are companions to other stars or are in clusters. The object GL 229B has a luminosity of $5\times10^{-6}L_\odot$ and a mass of $40M_{Jupiter}$ (Nakajima *et al.* (1995); Allard *et al.* (1996)), and *BD*s may have been identified in the Pleiades cluster (see, e.g., Martin *et al.* (1996)).

Mass [$M_\odot$]	Radius [$R_\odot$]	Temp [K]	F(2.2 μm) Jy	F(3.5 μm) Jy	F(4.5 μm) Jy	F(6.5 μm) Jy	F(8 μm) Jy
0.06	0.83	750	6.7(-5)	4.4(-4)	7.5(-4)	9.1(-4)	8.6(-4)
0.03	0.01	450	2.9(-7)	1.7(-5)	6.5(-5)	1.8(-4)	2.3(-4)
0.01	0.12	225	2.1(-13)	1.3(-9)	7(-8)	1.9(-6)	6.1(-6)
SIRTF target 85 cm 10 σ 150s				9.3(-6)	1.2(-5)	6.6(-5)	1.1(-4)

FIGURE 9. Fluxes from old Brown Dwarfs which are 10 pc from the Sun.

Figure 9, taken from the *SIRTF* Science Requirements Document (JPL D-14302), compares *SIRTF*'s target sensitivities at three wavelengths to the flux densities expected for *BD*s of several different masses that are 10^{10} yrs old and located 10 pc from the Sun. If the target sensitivities are achieved, *SIRTF* could detect, with S/N > 10, old *BD*s with masses as low as $0.03M_\odot$ in 150 s. Assuming that about 10% of the Galaxy's halo mass is composed of *BD*s with a typical mass close to $0.03M_\odot$ (corresponding to $0.0086M_\odot/\mathrm{pc}^3$), one would need to survey about 300 square degrees to the indicated sensitivities in order to detect about 50 field *BD*s in the Galactic halo. Observations of these same objects separated in time by 2 yr, during which a halo star at 65 pc would move about 2″, would help distinguish these objects as part of the true halo population based on their anticipated large proper motions. In theory, multi-wavelength IR observations will permit one to determine the *BD* temperatures, but it must be recognised that, at least for the time being, we do not know what a *BD* spectral energy distribution should look like: clearly, it will not be a simple blackbody spectrum. Once the expected spectral energy distributions are established, it may be possible to find general field *BD*s by using the type of survey approach described above for the halo objects.

Another approach in searching for *BD*s is to look for them as members of star clusters, particularly open star clusters and very young, embedded clusters. There are tremendous advantages of our finding them there. We can determine the age and distance of the *BD* because those quantities can be determined much more easily for the cluster as a whole. In addition, the *BD*s in young clusters will be younger and hotter, and therefore brighter and easier to detect, than general field *BD*s. *BD* candidates with masses as low as $0.04M_\odot$ have been identified in the Pleiades cluster (Williams *et al.* (1996)) using V, I, and K photometry. By permitting highly sensitive observations out to $10\mu m$, *SIRTF* will be able to extend the discovery of these objects down to masses of 0.01 - $0.02M_\odot$, much cooler objects emitting at correspondingly longer wavelengths. Broad spectral coverage of *BD* candidates out to these longer wavelengths will also help address the issues of cluster membership and extinction for these faint objects. Follow-up IR spectroscopy by *SIRTF* will be critical to the final phases of verification of *BD* candidates found in the larger surveys. For example, the predicted 5-15μm spectrum for the *BD* Gl 229B

(Marley *et al.* (1996)) has distinctive features of water, methane, and ammonia. Cooler objects will display features due to a host of molecules and dust grains.

6.2. *Protoplanetary and Planetary Debris Disks*

The formation of circumstellar disks appears to be a natural by-product of star formation. There is evidence that a large fraction, perhaps 50%, of all stars have disks. Because disks are likely to be the reservoirs from which planets form, the apparent ubiquity of circumstellar disks suggests that planets, too, are common. Thus, the exploration of disks around other stars bears on a broad range of profound issues relating to, for example, the formation of our solar system and the origins of life on Earth and elsewhere in the universe. *SIRTF* will play a key role in surveying and exploring disks around young stars in which planets may be forming or will form (protoplanetary disks) and around older main sequence and post-main-sequence stars in which planetary formation is probably complete (debris disks). In both cases, IR observations reveal dust particles warmed by the stellar radiation, although for the youngest stars some of the IR emission may also be due to viscous heating of the dust and gas during the accretion phase. In the protoplanetary disks these particles may be residual material from the protostellar nebula, material which may be at some intermediate stage of accumulation towards planet formation. In the debris disks, a large fraction of the IR-emitting particles may be cast off or broken off from comets or asteroids, respectively. The goals of *SIRTF* will be to determine disk properties as a function of stellar type and age and to compare them to the properties of components of our own solar system.

SIRTF will be well suited to taking a census of nearby star forming regions to determine both the frequency of occurrence of stellar disks around young stars in clusters and the disk properties as a function of the cluster age. The spectral energy distribution of a star/disk system is a strong function of the system's age (see Adams *et al.* (1987)). At the earliest (protostellar) stage, considerable material from the placental molecular cloud is still very close to, and falling onto, the star, although towards the end of this phase, some of the matter is blown back out in a bipolar outflow. The *SED* is peaked at far-IR wavelengths, indicating that much of this large amount of material is relatively cool. During the second stage, the outflow has driven much of the remaining matter away from the star, with the IR excess being much reduced. What remains of the matter at this stage is primarily the disk of gas and dust. The IR excess can continue to decrease until it is presumably at the level of the debris disks discussed below. *SIRTF* will be able to determine the *SED*s for a large number of young stars in clusters such as the Pleiades, Praesepe, and the Hyades at the telescope diffraction limit ($\lambda/D = 1.2''$ at 6μm), so that the amount and temperature distribution of the circumstellar matter can be determined. In some cases, the spatial distributions of the envelopes and, perhaps, the disks may be determined. These properties can be related to the presence of disks and envelopes as a function of cluster age and the cluster stellar content (e.g., does the presence of more massive stars in a cluster destroy the disks around nearby stars in the cluster?).

It can be argued that the discovery of debris disks around Vega, β Pic, and other almost solar type, main-sequence stars was probably *IRAS*'s greatest single achievement. *SIRTF* could greatly extend the *IRAS* survey to include many more main sequence stars in the search for debris disks. *SIRTF* will be unable to resolve many, perhaps most, of these disks (see below). Nevertheless, faint disks will be detectable, and the spectral energy distributions will be determined with broad wavelength coverage, which will permit detailed modelling of the disk structure. In **Figure 9** we show how *SIRTF*'s sensitivities compare to the brightnesses of: (1) the disk around β Pic (A5V), if that star were 100 pc away; (2) a "typical G-star disk" around 82 Eri (G5V), if that star were 20 pc away; and (3) a "typical G-star disk" around α Cen A (G2V); the "typical G-star disk" corresponds to the 100-400 AU-diameter cloud of material that Aumann & Good (1990) find around G stars in the solar neighbourhood and that is roughly consistent with the brightness expected for the Sun's Kuiper-belt cloud. The straight lines are the stellar photospheres, with the disk emission appearing as an excess at the longer wavelengths *SIRTF* Science Requirements Document). The G-star disks are faint, since they are tenuous and the stars are of relatively low luminosity, but a large number of these disks,

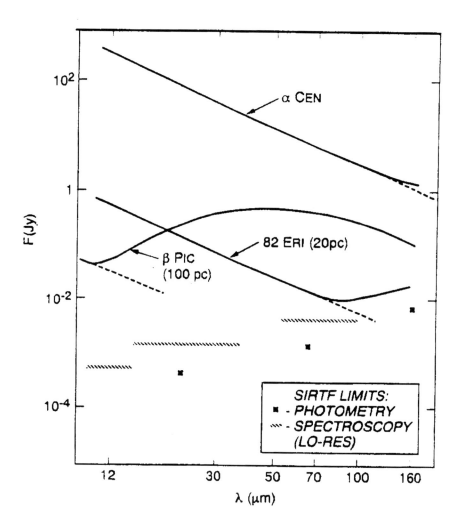

FIGURE 10. IR Spectral energy distributions of stars with debris disks. if they exist, should be detectable at 70 and 160μm out to 20 pc. β Pic's disk would be detected out to 1 kpc.

Modelling the spectral energy distribution to infer the disk structure is necessary when the disk cannot be spatially resolved, but the results are usually ambiguous. Direct imaging of the disk provides the best constraints on disk structure. Since the dust farther from a star is cooler, a disk is larger at longer wavelengths, and multi-wavelength imaging can reveal disk structure over a broad range of scale lengths, thus permitting direct determination of the disk's temperature and density structure. We have already described some of the results of follow-up imaging of the disk in β Pic, which, while nearby, optimally oriented (edge on), and perhaps unusually dense, represents the kind of information that *SIRTF* may provide about more ordinary disks. Simulations for the brightness distributions of disks like those in β Pic and for normal solar-type stars indicate that the *SIRTF* sensitivities are sufficient to detect these more ordinary disks, with several pixels spanning the detectable parts of the disks. However, resolution-enhancement techniques, which may increase the resolution by a factor of two or three, will be needed. Such techniques require a stable point spread function and reasonably high signal-to-noise ratios, which *SIRTF* will provide.

6.3. *Ultraluminous Galaxies and Active Galactic Nuclei*

We have already referred to the population of IR-luminous galaxies discovered before *IRAS*, but for which the *IRAS* survey provided the large body of observations needed to begin to examine many properties of these sources statistically. Of these sources, study of the so-called ultraluminous galaxies (ULGs) and their relationship to active galactic nuclei (*AGN*s) and quasars, which are thought to be powered by accretion onto massive black holes, has been particularly intense. The ULGs are usually defined as those with bolometric luminosities greater than about $5 \times 10^{11} L_\odot$, nearly all of which emerges in the IR. (For comparison, the luminosity of the Milky Way galaxy is about $10^{10} L_\odot$.) As in the case of the circumstellar disks, *SIRTF* will permit the discovery of many more of these objects, and the detailed determination of their broad spectral energy distributions. At $70\mu m$ *SIRTF* will reach a 5σ limit of about 1 mJy in 500s of integration, which is about 200 times fainter than the *IRAS* limit, and much fainter than *ISO*. The most luminous galaxies known, such as FSC 10214+4724, which has an intrinsic luminosity of 2×10^{13} $L_\odot$, could be detected out to z = 7, with ULGs that are five times less luminous being detected out to z = 5. It is estimated that about 10^5 extragalactic sources could be detected down to a 1 mJy flux limit, with perhaps a few percent of these being ULGs.

SIRTF spectroscopy will have a special role in determining the nature of the energy sources – starbursts or black holes – in these unusual objects. Because the central regions of these galaxies are so heavily obscured, IR emission lines, which are relatively impervious to interstellar extinction, are an especially powerful probe of the conditions there. Various line ratios provide insight into the gas temperatures, densities, excitation and the intervening extinction (Voit (1992); Spinoglio & Malkan (1992); see Section 3). The power of this technique has been recently demonstrated by observations with the *ISO* satellite Short Wavelength Spectrometer (*SWS*). Using *SWS* in a large survey of IR-luminous galaxies, including many of the ULG archetypes, Lutz *et al.* (1996) showed that the line ratios are generally consistent with excitation by radiation fields indicative of excitation temperatures of about 42 000 K. The "softness" of this radiation field suggests that enormous complexes of young, massive stars heat the IR-emitting dust. *AGN*s may contribute to the heating, but they are apparently of much less importance. *SIRTF* will use this same technique for a much larger, and generally much more distant, sample of ULGs than did *ISO*. Roughly 100 such objects at a range of redshift (some at z > 5) and luminosity would be observed and characterised in this survey program.

7. Concluding Remarks

I hope these lectures have given a reasonable idea of the exciting range of scientific problems that IR astronomical surveys can address. My goal has been to use specific examples of science programs to give an impression of how IR surveys are carried out and the kinds of information they can provide. As is always the case in such presentations, I have had to greatly restrict my coverage of the topic, with many areas, such as cosmology and the study of our solar system, being completely skipped. Likewise, there has been no discussion, or, as in the case of *ISO*, hardly any discussion, of many important IR survey programs and missions (e.g., *WIRE* and *FIRST*). It did not seem worthwhile to me to provide an exhaustive list of these programs. However, within the spirit of this Winter School, I strongly recommend that each of the attendees explore the World Wide Web to broaden his or her contact with the wonderful field of IR astronomy and the associated surveys.

REFERENCES

ADAMS, F., LADA, C. J., SHU, F. H., 1987 *Astrophys. J.* **312**, 788

ALLARD, F., HAUSCHILDT, P., BARAFFE, I., CHABRIER, G., 1996 *Astrophys. J.* **465**, L123

ANDRIESSE, C. D., 1978 *Astron. Astrophys.* **66**, 169

AUMANN, H. H., GILLETT, F. C., BEICHMAN, C. A., DE JONG, T., HOUCK, J. R., *et al.* , 1984 *Astrophys. J. Lett.* **278**, L23

AUMANN, H. H., GOOD, J. C., 1990 *Astrophys. J.* **350**, 408

BACKMAN, D. E., GILLETT, F. C., WITTEBORN, F. C., 1992 *Astrophys. J.* **385**, 670

BACKMAN, D. E., PARESCE, F., 1993 *In Protostars and Planets III*, eds. E. H. Levy & J. I. Lunine (U. Arizona Press: Tucson), P. 1253

BECKWITH, S., EVANS, N. J., GATLEY, I., GULL, G., RUSSELL, R. W., 1993 *Astrophys. J.* **264**, 152

BEICHMAN, C. A., 1987 *Ann. Rev. Astron. Astrophys.* **25**, 521

BIEGING J.H., WILNER, D., THRONSON, H.A., 1991 *Astrophys. J.* **379**, 271

BOULANGER, F., PÉRAULT, M., 1988 *Astrophys. J.* **330**, 964

BURBIDGE, G.R., STEIN, W.A., 1970 *Astrophys. J.* **160**, 573

BURKE, B. F., 1957 *Astron. J.* **62**, 90

CARLBERG, R. G., 1992 *Astrophys. J. lett.* **411**, L9

CHURCHWELL, E.B., WOLFIRE, M.G., WOOD, D.O.S., 1990 *Astrophys. J.* **354**, 247

COWIE, L. L., SONGAILA, A., 1993 *In Sky Surveys: Protostars to Protogalaxies*, ed. B. T. Soifer (ASP: San Francisco), P. 193

DE GRIJP, M. H. K., MILEY, G. K., LUB, J., 1987 *Astron. Astrophys. Supp.* **70**, 95

DE VRIES, C. P., LE POOLE, R. S., 1985 *Astron. Astrophys.* **145**, L7

DJORGOVSKI, S., SOSIN, C., 1989 *Astrophys. J.* **341**, L13

DOYON, R., JOSEPH, R. D., & WRIGHT, G. S., 1990 *In Astrophysics with Infrared Arrays"*, Ed. R. Elston (ASP: San Francisco), P. 69

DRAINE, B. T., ANDERSON, N., 1985 *Astrophys. J.* **292**, 494

EPCHTEIN, N., 1997 *In "The Impact of Large Scale Near-IR Sky Surveys"*, eds. F. Garzón, N. Epchtein, A. Omont, B. Burton, & P. Persi (Kluwer: Dordrecht), P.15

FICH, M., TEREBEY, S., 1996 *Astrophys. J.* **472**, 624

FREUDENREICH, H. T., *et al.* , 1994 *Astrophys. J.* **429**, L69

GAUTIER, T. N., 1986 *In "Light on Dark Matter"*, ed. F. P. Israel (Reidel: Dordrecht), P. 49

GENZEL, R., STUTZKI, J., 1989 *An. Rev. Astron. Astrophysics.* **27**, 41

HARPER, D.A., LOW, F.J., 1973 *Astrophys. J. Lett.* **182**, L89

HELOU, G., 1986 *Astrophys. J. Lett.* **311**, L33

JURA, M., 1993 *In "Sky Surveys: Protostars to Protogalaxies"*, ed. B. T. Soifer (ASP: San Francisco), P. 1

KLEINMANN, D.E., LOW, F. J., 1970 *Astrophys. J. Lett.* **159**, L165

KNACKE, R. F., FAJARDO-ACOSTA, S. B., TELESCO, C. M., HACKWELL, J. A., LYNCH, D. K., RUSSELL, R. W., 1993 *Astrophys. J.* **418**, 440

LADA, E. A., DePOY, D. L., EVANS, N. J., GATLEY, I., 1991 *Astrophys. J.* **371**, 171

LAGAGE, P. O., PANTIN, E., 1994 *Nature* **369**, 628

LÉGER, A., PUGET, J. L., 1984 *Astron. Astrophys.* **137**, L5

LOW, F.J., 1961 *J. Opt. Soc. Am.* **51**, 1300

LOW, F. J., BEINTEMA, D. A., GAUTIER, T. N., GILLETT, F. C., BEICHMAN, C. A., *et al.* , 1984 *Astrophys. J. Lett.* **278**, L19

LOW, F.J., KLEINMANN, D.E., 1968 *Astron. J.* **73**, 868

LUTZ, D. *et al.* , 1996 *Astron. Astrophys.* **315**, L137

MATHIS, J. S., MEZGER, P. G., PANAGIA, N., 1983 *Astron. Astrophys.* **128**, 212

MARTIN, E., REBOLO, R., ZAPATERO-OSORIO, M., 1996 *Astrophys. J.* **469**, 706

MARLEY, M. S., *et al.* , 1996 *Science* **272**, 1919

McLEAN, I., 1997 *Electronic Imaging in Astronomy (Wiley: New York).*

MILLER, E., SCALO, J. M., 1979 *Astrophys. J. Supp.* **41**, 513

MOSELEY, H., 1980 *Astrophys. J.* **238**, 892

NAKAJIMA, T. *et al.* , 1995 *Nature* **378**, 463

NEUGEBAUER, G., LEIGHTON, R. B., 1968 *"Two Micron Sky Survey A Preliminary Catalog. NASA Sp-3047"*

OLNAN, F.M., BAUD, B., HABING, H.J., DE JONG, T., HARRIS, S., POTTASCH, S.R., 1984 *Astrophys. J. Lett.* **278**, L41

PUGET, J. L., LÉGER, A., 1989 *An. Rev. Astron. Astrophys.* **27**, 161

RICKARD, L.J, HARVEY, P.M., 1984 *Astron. J.* **89**, 1520

RIEKE, G.H., LOW, F.J., 1972 *Astrophys. J. Lett.* **176**, L95

ROBIN, A. C., 1997 *In "The Impact of Large Scale Near-IR Sky Surveys", eds. F. Garzón, N. Epchtein, A.Omont, B. Burton, & P. Persi (Kluwer: Dordrecht),* P. 57

RUPHY, S., EPCHTEIN, N., 1997 *In "The Impact of Large Scale Near-IR Sky Surveys", eds. F. Garzón, N. Epchtein, A. Omont, B. Burton, & P. Persi (Kluwer: Dordrecht),* P.63

SALPETER, E. E., 1955 *Astrophys. J.* **121**, 161

SANDERS, D. B., MIRABEL, I. F., 1996 *Ann. Rev. Astron. Astrophys.* **34**, 749

SANDAGE, A., 1976 *Astron. J.* **81**, 954

SELLGREN, K. 1984, *Astrophys. J.* **277**, 623

SELLGREN, K., LUAN, L. WERNER, M. W., 1990 *Astrophys. J.* **359**, 384

SMITH, B.A. & TERRILE, R.J., 1984 *Science* **226**, 1421

SOIFER, B. T., HOUCK, J. R., NEUGEBAUER, G., 1987 *Ann. Rev. Astron. Astrophys.* **25**, 187

SKRUTSKIE, M. F., *et al.* , 1997 *In "The Impact of Large Scale Near-IR Sky Surveys", eds. F. Garzón, N. Epchtein, A. Omont, B. Burton, & P. Persi (Kluwer: Dordrecht),* P. 25

SPINOGLIO, L., MALKAN, M. A., 1992 *Astrophys. J.* **399**, 504

STACEY, G. J., 1989 *In "Infrared Spectroscopy in Astronomy", ed. B. H. Kaldeich (ESA: Noordwijk),* P.455

TELESCO, C. M., 1988 *Ann. Rev. Astron. Astrophys.* **26**, 343

TELESCO, C. M., BECKLIN, E. E., WOLSTENCROFT, R. D., DECHER, R., 1988, *Nature* **335**, 51

TELESCO, C.M., DECHER, R., JOY, M., 1989 *Astrophys. J. Lett.* **343**, L13

TELESCO, C.M., HARPER, D.A., 1977 *Astrophys. J.* **211**, 475

TELESCO, C.M., HARPER, D.A., 1980 *Astrophys. J.* **235**, 392

TELESCO, C. M., KNACKE, R. F., 1991 *Astrophys, J. Lett.* **372**, L29

THRONSON, H. A., RAPP, D., BAILEY, B., HAWARDEN, T. G., 1995 *Pub. Aston. Soc. Pac.*

107, 1099

VOIT, G. M., 1992 *Astrophys. J.* **399**, 495

WAINSCOAT, R. J., COHEN, M., VOLK, K., WALKER, H. J., SCHWARTZ, D. E., 1992 *Astrophys. J. Supp.* **83**, 111

WERNER, M.W., BECKWITH, S., GATLEY, I., SELLGREN, K., BERRIMAN, G., WHITING, D.L., 1980 *Astrophys. J.* **239**, 540

WERNER, M.W., GATLEY, I., HARPER, D.A., BECKLIN, E.E., LOEWENSTEIN, R.F., TELESCO, C.M., THRONSON, H.A., 1976 *Astrophys. J.* **204**, 420

WILLIAMS, D. M., BOYLE, R. P., MORGAN, W. T., RIEKE, G. H., STAUFFER, J. R., RIEKE, M. J., 1996 *Astrophys. J.*, **464**, 238

WINDHORST, R. A., OWEN, F. N., KRON, R. G., KOO, D. C., 1985 *Astrophys. J.* **289**, 494

Infrared Astronomy With Large Databases – The Practical Side

By DEBORAH ANNE LEVINE

Infrared Processing and Analysis Center, MS 100-22, California Institute of Technology,
Pasadena, CA 91125, USA

AND

ISO Science Operations Centre, *ESA* Villafranca Satellite Tracking Station, APTDO 50727,
28080 Madrid, Spain

The more practical side of Infrared Astronomy is presented. Topics related to dealing with the
IR sky and IR detectors are discussed, with an emphasis on space-based mid to far-IR work. The
data from the Infrared Astronomical Satellite (*IRAS*) are examined in some detail, as are the
soon-to-become public data from the four instruments aboard the Infrared Space Observatory
(*ISO*). Longer-term upcoming Infrared projects are discussed briefly.

1. Introduction

The practical side of working with the Infrared data available (or soon to be available)
on the Internet starts with knowing what is important and unique about the infrared
region of the spectrum. There are, additionally, some basic consequences of the nature
of the infrared sky and of infrared detectors and the resulting observing techniques that
should be taken into account when data are interpreted.

Astronomical data available on-line falls into three rough categories: catalogues, all-
sky surveys and observatory archives. Using uniform catalogues, such as the *IRAS* Point
Source Catalogue (*PSC*), is fairly straightforward and it is primarily important to understand issues such as the coverage, completeness and reliability of the catalogue which
are not specific to Infrared wavelengths. Uniform all-sky surveys, such as *COBE* (Cosmic Background Explorer), or the image data resulting from *IRAS*, are the next easiest
to deal with, as both the processing and the data-taking were rather uniform. To use
these data sets, one should understand the basic strategy and limitations of the instrumentation and processing techniques, but there are a limited number of such modes to
understand. Observatory archives, such as that from the Infrared Space Observatory
(*ISO*), are the most difficult to use wisely, because they contain an inhomogeneous set
of data of varying quality resulting from purpose-designed observations. In this case it is
necessary to be familiar with the types of observations possible using the observatory and
the data processing that has been applied to the data held on-line. It is not necessary
to become an expert, but it is wise to have a sufficient level of knowledge to assess if a
result taken from an on-line archive is reliable and relevant and also if further processing,
or re-reduction of the data, would be advisable before attempting to publish a result.

One of the dangers of the easy access provided by the Internet is a tendency not to
treat data with the same caution we would apply if we took the data ourselves. While
providers of on-line archives generally do their best to ensure that the data archived
and the processing applied to it is valid, the very nature of large, uniformly processed
databases is that some pieces of data will be better handled by the automatic procedures
than others. Therefore a fair amount of the content of this section is devoted to the
difficulties and peculiarities of infrared detectors and data, not because the problems are

overwhelming, or intractable, but because this is background that facilitates a realistic assessment of data found on line.

In addition to an overview of the field, the *IRAS* data set is examined in some depth, including well-known resources such as the Point Source Catalogue and Faint Source Catalogue and newer data sets such as the High Resolution Maps. *COBE* is, unfortunately, covered only briefly. *ISO*, the Infrared Space Observatory, is the first major space-based IR mission since *IRAS* and is addressed in much detail. *ISO* was collecting data until mid-April 1997 and the archive will not be fully in the public domain until ∼April 1999. However, an initial archive is expected to become available during the summer, or fall, of 1998 and the *ISO* data set will be an important database. Some software for working with *ISO* data is already available, as are some isolated data sets including the *ISO* observations of the Hubble Deep Field. The sections about *ISO* instruments contain a great deal of detail about reducing and interpreting these data, with an eye to paving the way to use of a data set which will soon be available.

2. Tricks of the Trade: Some Important Background Information

Infrared astronomy, particularly in the mid to far-IR, is complicated primarily by two things; the behaviour of commonly used infrared detectors and the complexity of foregrounds and backgrounds, due to the ubiquitousness of infrared sources. In order to use these data effectively, it is important to understand the basic sources of IR emission and the overall structure of the infrared sky at various wavelengths.

2.1. *The Nature of Infrared Emission*

Infrared emission can come from photospheres, synchrotron emission and from a large number of spectral lines, but Infrared observation is compelling, in large part because the thermal dust emission of many interesting objects peaks at infrared wavelengths. Additionally, interstellar extinction diminishes rapidly with wavelength, allowing the IR to probe quite a bit deeper into regions of heavy extinction such as the Galactic Plane than can be done in the visible region of the electromagnetic spectrum. mid to far-IR (i.e. $\lambda > \sim 10\mu m$) astronomy is done almost exclusively from space, both because it is easier to keep the telescope and surroundings cold, reducing the thermal background, and also because atmospheric windows are messy in the near-IR and all but non-existent in the mid-far IR.

2.1.1. *A bit about dust*

Dust, that is small solid particles such as graphite or silicate grains, is pretty much omnipresent in the Universe. Dust is found in such diverse places as the solar system, in dense shells around evolved stars and in the *ISM*. The most classical "dust" consists of small particles about 0.01 to 0.25μm in diameter, such as graphite and silicate grains. Some features of dust emission are usually attributed to the presence of a component of very small grains (∼4-10Å) and/or very large molecules containing >50 atoms known as *PAHs* (Polycyclic Aromatic Hydrocarbons). A number of IR spectral features are also attributed to *PAHs*. Dust becomes important because much of the energy budget of astronomical objects is re-radiated by heated dust. For the Milky Way, 30% of the energy budget goes into thermal dust emission. For distant star-forming galaxies the percentage is much higher giving rise to the so-called Ultraluminous Infrared Galaxies. Dust also provides surfaces to catalyse chemical reactions and provides a mechanism for energy transfer out of collapsing cores, facilitating star formation.

Dust emission tends to result in "excess" emission in the infrared over what would be expected from the apparent temperature of the astronomical object. It usually follows a blackbody spectrum modified by a wavelength-dependent emissivity term.

$$B(\nu) = k_\nu \epsilon(\nu) \frac{2h\nu^3}{c^2} \frac{1}{[exp(h\nu/kT) - 1]} \tag{2.1}$$

where:

$$k_\nu \sim \nu^\alpha, \alpha = 2, 1.5, 1... \tag{2.2}$$

An "infrared excess" normally arises from dust that is cooler than the source in question; however, dust emission can sometimes also be seen at much shorter wavelengths than the physical temperature of the region would imply should be the case. This can be seen in the *IRAS* data where the very cold filamentary structure of the *ISM* (cirrus), seen at 100μm, is sometimes echoed at 12μm. This is due to non-equilibrium temperature fluctuations of very small grains, or *PAHs*, in which the particle is suddenly heated e.g., by absorption of a photon, and then cools rapidly by emission in the IR.

2.1.2. A (very) little bit about lines

There are quite a lot of spectral lines of interest in the 2-200μm region, including a wealth of atomic fine-structure lines including NeII and 12.8 μm and C II at 158μm, lines due to molecular rotation and vibration, including molecular hydrogen at 28μm and a number of features attributed to *PAHs* and ices.

Several useful IR line lists can be found on the WWW. A few are at:
- `http://www.ipac.caltech.edu/iso/lws/ir_lines.html` (*IPAC*)
- `http://www.mpe-garching.mpg.de/iso/linelists/index.html` (*MPE*-Garching *ISO* Spectrometer Data Centre)
- `http://www.strw.leidenuniv.nl/` lab (Leiden database of ices)

2.2. The Nature of the Infrared Sky – Backgrounds and Foregrounds

2.2.1. Zodiacal Light

The zodiacal light (often called the "zody") is the dominant source of diffuse background over much of the near to mid IR portion of the electromagnetic spectrum. The zodiacal light arises from interplanetary dust which scatters sunlight in the optical and near-IR, but which emits thermally in the mid-IR. There is a minimum between the peaks of the scattered and thermal components at about 3μm; this is the so-called "cosmological window". The spectrum of the zodiacal light in the mid-infrared is featureless and corresponds to a temperature of 261.5±1.5K (Reach *et al.* (1996)).

The zodiacal light consists of slowly varying diffuse emission which is brightest in the plane of the ecliptic plus more concentrated bands of emission at 0° and ±9° in ecliptic latitude.

2.2.2. IR Cirrus and Cirrus Confusion Noise

The infrared cirrus is one of the most important factors affecting the success or failure of observations at wavelengths significantly longwards of 60μm. The cirrus is ubiquitous, filamentary emission from cold dust (Low *et al.* (1984)) and was named for its resemblance to terrestrial cirrus clouds. Cirrus emission is present at all spatial scales from about 4′ to 400′, with a simple power law power spectrum (Gautier *et al.* (1992)).

$$P = P_0(r/r_0)^{-\alpha} \tag{2.3}$$

where $\alpha \sim 2$ for the 1-dimensional case and $\alpha \sim 3$ for the 2-dimensional case.

The colour temperature of the cirrus is 15 to 23K (Bernard *et al.* (1994)) and the cirrus becomes very evident at about 100μm. However small grains, or *PAH*s, can show up at shorter wavelengths, around 12μm due to non-equilibrium heating as discussed in section 2.1.

Although the cirrus is most concentrated in the Galactic Plane, it is virtually impossible to find a 100% "cirrus free" piece of sky. Cirrus dramatically complicates observing at longer wavelengths and any observation at around 100μm, or so, which has not involved mapping the background should be viewed with caution. In such cases the probable cirrus environment should be investigated using whatever information is available, e.g. the *IRAS* data. Chopping, nodding or pointed on/off measurements in the presence of cirrus often result in a "bad" off position, where the cirrus background is higher than the source plus background at the target position. Cirrus can also result in spurious point source detections for scanning-mode observations – although the cirrus is extended, it is filamentary and scanning across a filament can result in detection of an apparent point source. Even when the cirrus brightness is much less than sources of astronomical interests it is still a source of apparent noise called confusion noise.

2.2.3. *Confusion Noise*

Confusion noise is the effect on detectability of astronomical sources due to undetected structure in the background at scales smaller than the resolution, or aperture of the instrument used for observations. The problem is essentially one of looking for a needle in a haystack; the "bumpy" background thwarts detection in the same way that variations in signal due to detector noise would, but do **not** integrate down with time. Confusion noise can be caused by extended structure like the infrared cirrus, or it can be caused by a multitude of undetected point sources. The latter situation can occur in very crowded regions, such as near the Galactic Centre or, at longer wavelengths, can be due to large numbers of faint galaxies. The actual amplitude of the cirrus noise depends upon a number of factors, including the observing strategy used.

Using the power law for the cirrus power spectrum (equation 2.3) the apparent noise due to cirrus structure when two symmetric off positions are used is given by:

$$\sigma_{cirrus} = E_0(d/d_0)^{1-\alpha/2}P_0 \tag{2.4}$$

d is the diameter of the beam and d_0 is the spatial scale corresponding to P_0. E_0 is a constant that takes into account the background subtraction strategy and instrumental resolution. For the case of a diffraction-limited telescope operating at 100μm, this can be simplified (Helou and Beichman (1990)).

$$\sigma_{cirrus} = 2.65E_0(\lambda/D)^{2.5}B_0^{1.5} \tag{2.5}$$

where B_0 is the brightness at 100μm in Jy/sr and E_0 is $1.15\ 10^{-3}$ for an annular reference with the same width as the beam. This allows estimation of the level of cirrus contamination in an arbitrary region of sky using, e.g., *IRAS* or *COBE* data.

For confusion due to extragalactic sources it can be shown that:

$$\sigma_{conf}^2(f_\nu) = -\Omega_{beam}\int_0^{f_\nu} f^2\frac{dN(f)}{df}df \tag{2.6}$$

(Helou and Beichman (1990)) approximating:

$$N(\nu, f_\nu) = Kf_\nu^a \tag{2.7}$$

leads to a 20% confusion limit (i.e. $\sigma_{conf}=0.2f_{conf,\nu}$) determined by:

$$[\frac{-a}{2+a}\Omega_{beam}K(f_{conf,\nu})^a]^{0.5} = 0.2 \tag{2.8}$$

Band	λ	$\Delta\lambda$	Reference Intensity	Flux Density
	μm	μm	W/m^2/μm	Jy
U	0.36	0.04	4.22E-08	1823
B	0.44	0.1	6.4E-08	4130
V	0.55	0.08	3.75E-08	3781
R	0.7	0.21	1.75E-08	2858
I	0.9	0.22	8.4E-09	2268
J	1.3	0.3	3.1E-09	1615
H	1.6	0.2	1.2E-09	1050
K	2.2	0.6	3.9E-10	629
L	3.4		7.1E-11	274
L'	3.5	0.9	6.10E-11	249
M	5.0	1.1	2E-11	167
N	10.2	6	1.1E-12	38
Q	20.0	5.5	7.3E-14	10

TABLE 1. Johnson scale zero points and equivalent flux densities

In both cases, the confusion noise is proportional to the telescope beam size. The only way to improve detectability in the presence of confusion is to improve resolution.

2.3. *Some Nuts and Bolts*

There are a few conventions that are peculiar to IR Astronomy and some others which are particularly relevant. This section lays out some of the nitty gritty details such as calculating colour corrections and confusion noise.

2.3.1. *Flux and Magnitude scales and IR units*

In the Near Infrared ($\ll 10\mu$m) the Johnson magnitude scale is typically used, with the zero points defined by Vega (an A0 V star). The exact values for the zero points tend to vary somewhat from observatory to observatory, so it is necessary to find out what values have been used. In the Mid to Far infrared it is more customary to use the Jansky as the unit of flux density, 1Jy $= 10^{-26}$ W/m^2/μm, and then Jy/sr as the unit of surface brightness. An approximate conversion between the magnitude scale (m) and Janskys (S) can be made if the zero points (S$_0$) are known:

$$m = -2.5log(S/S_0) \tag{2.9}$$

Table 1 gives one set of zero points and their equivalent in Janskys.

2.3.2. *Colour Corrections*

Infrared observations are often made with rather wide filters. For example, the *IRAS* 12μm filter has a *FWHM* of 7μm. The response of the detector is actually proportional to the flux transmitted through the filter – the in-band flux. The in-band flux is a function of both the spectral response of the filter and the spectral shape of the emitting source. The measured flux is normally quoted at a reference wavelength, typically the central wavelength of the filter, calibrated assuming a particular spectral shape for the source. If the source of interest has a spectral shape that differs dramatically from that assumed for calibration, then the actual flux at the reference wavelength may differ from the quoted flux at the reference wavelength significantly and a colour correction must be applied.

The colour correction factor "K" can be calculated if the filter response function R

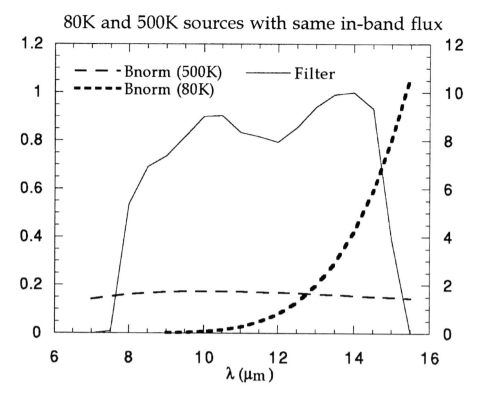

FIGURE 1. 80K and 500K sources with equivalent in-band flux in the *IRAS* 12 µm filter

(ν), the spectral shape used in calibration $f_{cal}(\nu)/f_{cal}(\nu_{ref})$, and the spectral shape of the source itself $f(\nu)/f(\nu_{ref})$ are known. Here the spectral shapes are normalised by the flux density at the reference wavelength.

The reported flux at the reference wavelength, $f_{rep}(\nu_{ref})$, is determined by the in-band flux F:

$$F = \int f(\nu)R(\nu)d\nu = \frac{1}{f(\nu_{ref})} \int \frac{f(\nu)R(\nu)}{f(\nu_{ref})} d\nu = \frac{1}{f_{cal}(\nu_{ref})} \int \frac{f_{cal}(\nu)R(\nu)}{f_{cal}(\nu_{ref})} d\nu \quad (2.10)$$

The colour correction factor K is defined by:

$$f(\nu_{ref}) = f_{rep}(\nu_{ref})/K \quad (2.11)$$

Noting that the reported flux is that which would have been attributed to the calibrator given the same value of the in-band flux it can be seen that:

$$K = \frac{\int \frac{f_{cal}(\nu)R(\nu)}{f_{cal}(\nu_{ref})} d\nu}{\int \frac{f(\nu)R(\nu)}{f(\nu_{ref})} d\nu} \quad (2.12)$$

In the example shown in the figure, if a 500K source were the basis of the calibration at a reference wavelength of 12µm, the flux density quoted for the 80K source would be over-estimated relative to the actual flux density at 12µm by a factor of about 2.

2.4. *The Nature and (Mis-) Behaviour of Infrared Detectors*

The earliest infrared detectors were bolometers. A bolometer absorbs photons and heats up; it is the resulting change in temperature which is measured measured. These devices are moderately sensitive, with Noise Equivalent Power (*NEP*) around 10^{-15} W/Hz$^{1/2}$. *NEP* is a common figure of merit for IR detectors and it is the signal power which would give signal-to-noise of one in a system with 1Hz bandpass.

The longer wavelength infrared detectors you are most likely to encounter in currently available archival data sets are photoconductors, either as single elements, or small arrays. This is the type of detector used on *IRAS* and *ISO*. With these detectors, the spectral coverage depends on the material used. The sensitivity achieved depends significantly on the amplifier used, but the *NEP* is typically 10^{-15}-10^{-17}W/Hz$^{1/2}$. As discussed below, the nature of these detectors is such that slow response and significant vulnerability to the effects of radiation are expected. These properties must be taken into account when data are evaluated.

Another category of IR detectors is the modern solid-state arrays used primarily in the near infrared, for example the *NICMOS* arrays on the Hubble Space Telescope. These devices show performance increasingly similar to CCDs, but have somewhat higher dark noise and can sometimes stabilise slowly after significant change in illumination.

2.4.1. *Intrinsic and Extrinsic Photoconductors*

Basically, an intrinsic photoconducting detector is a block of semiconductor material (commonly Si, Ge, PbS, InSb, HgCdTe or GaAs) with a bias voltage placed across it. Illuminating photons of energy exceeding the bandgap energy of the semiconductor will create electron/hole pairs, which then migrate across the potential to generate a measurable current. The signal from such a detector is generally characterised in terms of the responsivity, S. The current, I, produced in the detector by illuminating source of power P is given by I=S×P. Responsivity is normally given in amp/W. It is important to note that these detectors must have a high electrical resistance, and hence a high degree of purity, in order to generate a large signal while minimising noise – among other things, the high resistance tends to slow the response of the detector as it increases the inherent time constant τ_{RC}=RC of the circuit including the detector itself.

In theory, the responsivity of such a detector increases linearly as a function of wavelength until the cut-off wavelength, corresponding to the bandgap energy of the material, is reached. This is because any photon above the bandgap energy produces a singe electron/hole pair, so that the output of the detector depends only upon the number of such photons and not on their actual energy. In reality, the response trails off somewhat near the cut-off wavelength and also for very high energies. Because the cut-off wavelength is determined only by the bandgap energy, the spectral response of a given intrinsic photoconductor depends upon the semiconductor material used. Intrinsic detectors made of high-quality materials, in particular Si and Ge, tend to be rather well-behaved, but there usage is limited to quite short wavelengths; for Si the cut-off wavelength is 1.1μm and for Ge it is 1.8μm

Extrinsic photoconductors essentially work the same way as intrinsic photoconductors, but they use doped semiconductor materials (e.g. Ge:Ga, or gallium-doped germanium and Si:As). The added impurity (dopant) provides charge carriers that are liberated at much lower energies than the bandgap energy of the semiconductor material, hence extending the spectral response of the detector dramatically. There are Ge:Ga detectors on *ISO*, for example, which are sensitive out to 200μm even though Ge has a 1.8μm cut-off. The more heavily the material is doped, the more sensitive it is, but the increase in the number of available charge carriers reduces the electrical resistance and increases the noise, even at low temperatures. Thus they often have rather large volumes, but modest impurity concentration. The large volumes make them more vulnerable to radiation hits, which liberate charge carriers and cause spikes or "glitches" in the output. Unfortunately they can also disrupt the crystal structures causing slow-to-resolve changes in responsivity and noise characteristics of the detector. The responsivity may increase, but the noise usually increases even faster. Ge based detectors are particularly prone to these problems. It is usually possible to restore the material to its undamaged state by "curing" the detector. Curing methods include heating the detector to anneal it, flooding it with light, or boosting the bias voltage to "flush" the charge carriers. Extrinsic detectors also sometimes show a sort of overshoot response or "hook" anomaly and can sometimes generate spontaneous output spikes. These effects can complicate calibration and increase the effective noise of the detector. In addition to doping, some extrinsic detectors are also placed under pressure, which makes the charge carriers easier to liberate and extends the spectral response of the detectors. These are called stressed detectors.

BIB (Blocked Impurity Band) detectors are an attempt to compromise between the advantages and disadvantages of extrinsic detectors. They use a very heavily-doped layer, which is connected to a thin, high-purity layer that provides the high electrical resistance, to absorb the photons and create free charge carriers. The bias voltage applied across the two layers, with one contact at the blocking layer and the other at the highly-doped "active" layer. Since the highly doped layer can be made lower-volume than a less-heavily doped detector for the same sensitivity, *BIB* detectors should be somewhat more radiation-hard than an equivalent single-material detector. The high doping also extends the wavelength response, without the need to stress the detector. *BIB* detectors are relatively new technology and will figure prominently in upcoming infrared instruments.

3. Using IRAS Data

3.1. *Introduction to IRAS*

The Infrared Astronomical Satellite (*IRAS*) was launched in January of 1983 into a sun-synchronous near-polar orbit. *IRAS* was a joint project of the US, UK and the Netherlands; its mission was to perform an unbiased, sensitive all sky survey at 12, 25, 60 and 100μm. The satellite design and survey strategy were optimised for maximally reliable detection of point sources. Pointed observations, known as Additional Observations, or *AO*s, were also taken, interspersed with the survey observations. *IRAS* was not built with imaging primarily in mind, despite which, *IRAS* images have proved remarkable. However, there are some peculiarities of *IRAS* data, particularly evident in image data, of which the user should be aware. For more detail about the design and performance of *IRAS*, see the *IRAS* Explanatory Supplement (Beichman *et al.* 1988).

3.1.1. *The IRAS Spacecraft and Telescope*

IRAS carried a f/9.6 Ritchey-Chrétien telescope with a 5.5-m focal length and a 0.57-m aperture, mounted in a liquid helium cooled cryostat. The mirrors were made of beryllium and were cooled to approximately 4K.

The focal plane assembly contained the survey detectors, visible star sensors for position reconstruction, a Low Resolution Spectrometer (*LRS*) and a Chopped Photometric Channel (*CPC*). The focal plane assembly was located at the Cassegrain focus of the telescope and was cooled to about 3K.

3.1.2. *The IRAS Instruments*

The survey array consisted of 62 rectangular infrared detectors, arranged in staggered rows, such that any real point source crossing the focal plane as the satellite scanned would be seen by at least two detectors in each wavelength band, with a predictable delay in response due to scanning. See figure. Most of the detectors in each band had standard-sized apertures (see **Table 2**), with one or two being half-sized.

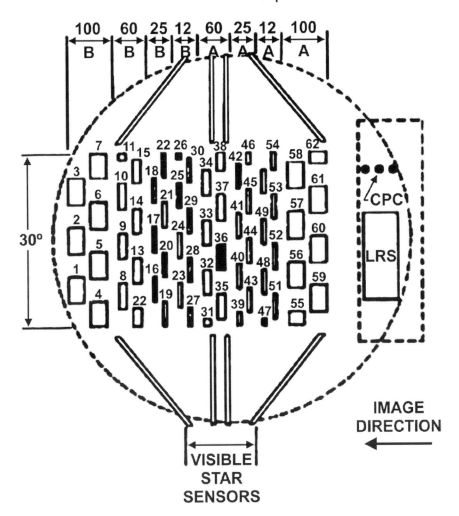

FIGURE 2. The IRAS Focal Plane Array

The *LRS* was a slitless spectrometer sensitive from 7.5 to 23 μm with a resolving power of about 20.

The *CPC* operated during some pointed observations. It mapped sources simultaneously at 50 and 100μm, and used a cold internal chopper for flux reference. However, the focal plane temperature was lower than expected, which resulted in *CPC* detector anomalies that rendered *CPC* data very difficult to use.

In **Table 2**, the centre wavelengths are in microns. For the survey array, the *FOV* is determined by the rather large detector mask size and determines the native "resolution" of the data in that band. The resolution of the *IRAS* image data is not governed by the resolution of the telescope, which was diffraction limited longwards of 12 μm, but by the size of the detectors.

Center Wavelength	No. working detectors	FOV (arcmin)	Bandpass (μm)	Detector Material	Average 10-σ Sensitivity (Jy)
		SURVEY ARRAY			
12	16	.75x4.5	8.5 – 15	Si:As	0.7
25	13	.75x4.6	19 – 30	Si:Sb	0.65
60	15	1.5x4.7	40 – 80	Ge:Ga	0.85
100	13	3.0x5.0	83 – 120	Ge:Ga	3.0
		CPC			
50	1	1.2	41 – 63	Ge:Ga	7.0
100	1	1.2	84 – 114	Ge:Ga	7.0

Slit width (arcmin)	Wavelength Range (micron)	Detector Material	Resolving Power
	LRS		
5.0	8 – 13	Si:Ga	14 – 35
7.5	11 – 23	Si:As	14 – 35

TABLE 2. Properties of the *IRAS* Detectors

3.1.3. *The IRAS Survey Strategy*

The *IRAS* orbit was a sun-synchronous near-polar (99°) orbit, which precessed by about a degree each day. The celestial sphere was divided into "lunes", bounded by ecliptic meridians 30 degrees apart. The survey covered the sky by "painting" each lune in overlapping strips as the satellite scanned through the lune during each orbit. The design of the focal plane combined with the survey strategy allowed a great deal of redundancy, which could be used to ensure highly reliable detection of point sources. The design of the survey array guaranteed that more than one detector would see each point source during a single scan. This was called seconds-confirmation. Successive scans over the same region were also checked for reproducibility of the detection and this was called hours-confirmation. Reproducibility checks on longer time scales – weeks confirmation – could also be performed. Since most (96%) of the sky was covered by at least two hours-confirming scans (*HCONs*) and 2/3 of the sky was covered by a third (which used a slightly different observing strategy). Some sources, near the ecliptic poles, had more than three hours-confirmed coverages.

One *HCON* was separated from the next by up to several months. Thus the data combined into a single image from the survey may have been taken over a period of many months, and are subject to variations in the actual observed foreground due to the changing geometry with which the zodiacal light was observed, as well as to variations in detector response.

3.2. *Features of the IRAS Data*

The *IRAS* detectors used similar materials to those used for most of the *ISO* Instruments, so many of the "features" seen in *IRAS* data will also figure in the discussion on the *ISO* instruments. Others are a function of the design of the telescope/focal plane array and are peculiar to *IRAS*. Most of these "features" are evident in the *IRAS images* only, not in the published catalogues.

3.2.1. *Point Spread Function (PSF)*

IRAS point sources do not appear circular primarily because the detector apertures were rectangular. In fact, the *PSF* of an *IRAS* source is determined both by the detector mask shape and the geometry of the scans covering the source. Thus, in a raw *IRAS* image point sources appear elongated, with the narrow dimension in the scan direction and the larger dimension determined by the cross-scan width of the detector. The *IRAS* Explanatory Supplement (Beichman *et al.* 1988) has illustrations of the focal plane indicating the detector sizes. Near the ecliptic poles, the *PSF* may even be X-shaped.

For unenhanced co-added images (e.g., *FRESCO*) the resolution is approximately $1' \times 5'$, $1' \times 5'$, $2' \times 5'$ and $4' \times 5'$ at 12, 25, 60 and 100μm, respectively. For "medium resolution" images (e.g., the *IRAS* Sky Survey Atlas [*ISSA*]) all four bands are smoothed to the 100μm resolution. For enhanced resolution *HIRES* images, the resolution varies depending upon the input data and the processing.

3.2.2. *IRAS detector transients – the "AC/DC Effect"*

The *IRAS* detectors suffered from a delayed response to flux-steps resulting in slow stabilisation. Since different detector materials were used in different bands, the effect is band-dependent. It is most significant at 12μm and negligible at 100μm. This type of delayed response on relatively short time scales is often referred to a detector "transient" in the *ISO* data; the same terminology is employed here to highlight common behaviour of extrinsic photoconductor detectors.

Because *IRAS* was continually scanning, the amount of "stabilisation" time the detectors experienced at a given flux level depended upon the size of the structure. Point sources always had a given amount of "dwell time" and the detectors stabilised to a rather well-determined fraction of what the "stabilised" response would have been. At the *IRAS* survey speed of 3.85 arcminute/s, the detectors were able to stabilise for structures on the order of $30'$ in extent, or larger. This leads to a difference in calibration for rapidly changing flux (point sources) and nearly constant flux (extended) structure. This calibration difference became known as the "*AC/DC* effect", where the calibration appropriate for point sources is known as the *AC* calibration, and the calibration appropriate to very extended structure is known as the *DC* calibration. The Point Source Catalogue V2.0 uses the *AC* calibration, but the *IRAS* Sky Survey Atlas uses the *DC* calibration. All non-point-source-filtered *IRAS* image products produced before April 1993 were *DC* calibrated. This includes early *HIRES* and *FRESCO* images and *ISSA*. After April 1993, *FRESCO* and *HIRES* images were switched to the *AC* calibration. Point source fluxes obtained by aperture photometry with appropriate background subtraction on *AC* calibrated images should be consistent with the *PSC2*. To bring point source fluxes, measured from *DC*-calibrated products, to the *AC* calibration, the fluxes must be divided by 0.78, 0.82, 0.92 and 1.0 at 12, 25, 60 and 100μm, respectively

It should be noted that neither calibration is strictly correct for structure on spatial scales intermediate between point sources and $30'$. Inspection of the plots in the *IRAS* Explanatory Supplement (Beichman *et al.* 1988) must be used to derive the intermediate-scale corrections and these are not very well determined.

3.2.3. *IRAS detector drifts and sky variation – Stripes*

Bright or dark stripes sometimes appear in *IRAS* image data in the scan direction (i.e., orthogonal to the wide dimension of the *PSF*). They are the result of detector response variations on long time-scales ("responsivity drifts") and real background/foreground variations from scan-to-scan during the survey. Much effort has gone into developing methods to minimise the effect of scan-to-scan variations. These techniques are collectively known as de-striping. Comparing the original Sky Brightness Maps (*SKYFLUX*, also sometimes called the *HCON* Images because a separate set was created for each hours-confirming coverage) with the newer *ISSA* images provides a dramatic demonstration of the effectiveness of de-striping. Nevertheless, *IRAS* images often retain striping at some low level. *HIRES* may tend to amplify any low-level striping present, particularly in relatively featureless fields. It should also be noted that de-striping improves the images cosmetically and allows detection of structure that would otherwise be obscured by the scan to scan variations, but it to some extent disguises real uncertainty in the absolute level of the background.

3.2.4. *Nonlinearities in IRAS – Flux-dependent Responsivity Changes*

The survey array detectors exhibited non-linear responsivity variations with the total flux falling on the detector. This effect can be quite large at 60 and 100μm, but is not well characterised. Because the zodiacal background observed by *IRAS* varied substantially during the survey, this means there are calibration differences between different observations of the same source at different times. This can be particularly noticeable when *HCON* 3 images are compared to or combined with *HCON* 1 and *HCON* 2.

3.2.5. *IRAS Transient effects – Shadowing*

Transient effects are closely related to the Shadows or trails are produced in the scan directions at 60 and 100μm, particularly upon crossing the galactic plane, especially near the Galactic Centre. The final calibration alleviated this effect to some extent.

3.2.6. *IRAS Transient effects – Point Source Tails*

Bright point sources may appear to be smeared out in the scan direction(s). This can happen in any band, but is most dramatic at 12 and 25μm. In extreme cases the point source "tail" may extend for a degree or more. If the source was scanned both descending and rising, it will have tails in both directions, and the tails may resemble stripes; tails can be distinguished from stripes because they diminish with distance from the bright source. This effect is not well enough modelled to be removed from the data.

3.2.7. *IRAS and Glitches*

The *IRAS* detectors were sensitive to cosmic-ray hits, which resulted in non-confirming spikes in the detector output known as "glitches". The *IRAS* data has been de-glitched, but occasional radiation hits remain in the data and appear as very transitory spikes.

3.2.8. *Optical Crosstalk and other optical effects*

Very bright (> 500Jy) point-like sources may have a spurious, very Characteristic, six-pointed star shape. This is due to reflection from the telescope secondary mirror struts. Approximately 5% of the peak flux may be contained in the star pattern. There is no known method for removing this effect. Resolution enhancement processing may exaggerate, or break up, the crosstalk pattern, if it is present.

Another unusual effect is that occasionally a bit of debris may have lit up the entire focal plane simultaneously, causing a non-physical regular pattern in the images. In such a case, all the detectors respond simultaneously, rather than in any confirming pattern. This can be easily seen in the scan data.

3.2.9. *Saturation*

The survey detectors could saturate for very strong sources (>1000Jy).

3.3. *Using the Main IRAS Catalogues*

The *IRAS* Mission produced a large number of catalogues including:

- **Point Source Catalogue (*PSC2*)**

The *PSC* was the first and remains the best known of the *IRAS* Catalogues. It contains fluxes and other information for some 245 000 point sources seen by *IRAS* . The redundant sky coverages were used in the *PSC* processing to maximise reliability and to eliminate moving (solar system) sources. The reliability of the *PSC*, i.e. the probability that a source in the catalogue is a real astronomical source, is 99.9% for sources flagged as moderate or high quality detections and the catalogue is complete to 1.5Jy for |b| >20° (except at 100µm). The fluxes for the *PSC* sources were obtained using carefully constructed point source templates designed to reproduce the actual effective detector response function.

- **Faint Source Survey Catalogue (*FSC*)**

Unlike the *PSC*, the *FSC* was derived from co-added *IRAS* data, in order to maximise sensitivity. The *FSC2* contains 173 044 point sources above about |b| >10°. It is 90% complete in that region to ~200mJy. Low galactic latitude sources were too often confused to be considered reliable. The reliability of the high quality sources in the *FSC* is 99% at 12 and 25µm and 94% at 60µm. There are no 100µm **only** sources in the *FSC*.

- **Faint Source Survey Reject File (*FSCR*)**

The *FSCR* contains 593 516 point sources derived from the co-added *IRAS* data not included in the *FSC2*. This includes faint, confused and low galactic latitude sources. However, it should be noted that the *FSCR* contains **fewer** sources in and near the Galactic Plane than does the *PSC*.

- **Low Resolution Spectrometer (*LRS*) Catalogue**

The *LRS* Catalogue contains 8-22 µm spectra of 5425 *PSC* Sources which were bright enough to have good *LRS* spectra.

- **Catalogued Galaxies & Quasars Observed in the *IRAS* Survey**

This catalogue contains fluxes and other information for 11 444 *PSC* sources that have been associated with catalogued galaxies and quasars.

- **Small Scale Structure Catalogue (*SSS*)**

The *SSS* was an attempt to apply processing similar to the *PSC* to somewhat extended sources. Instead of a point source template, square response functions of various widths up to 8′ were used. The *SSS* contains 16 740 extended sources.

- **Serendipitous Survey Catalogue (*SSC*)**

This is a catalogue of 43 866 point sources that were serendipitously observed during the pointed observations (*AOs*).

- **Catalogue of *IRAS* Observations of Galaxies**

This is a collection of observations of 85 large (> 8′) galaxies. It consists of a collection of contour plots and brightness profiles, as well as integrated fluxes. This was originally published by Rice *et al.* (1988).

- **IRAS Nearby Galaxy High Resolution Atlas**

This is a collection of observations of 30 large optical galaxies, processed with *IPAC*'s resolution-enhancement software (*HIRES*). It was originally published by Rice 1993. The galaxies are presented at ≈ 1′ resolution.

3.4. *Using IRAS Image Products*

The *IRAS* Sky Survey Atlas is available directly on the WWW in several forms as well as from data "warehouses" such as the *NSSDC*. An *IRAS* Galactic Plane atlas using a tuned *HIRES* processing is also on the WWW. *FRESCO* and *HIRES* are available from the Infrared Processing and Analysis Centre (*IPAC*) via an email-request as described below.

3.4.1. *IRAS Sky Survey Atlas (ISSA)*

IRAS conducted its survey during a ten-month period in 1983. The initial release of infrared images of the sky in 1984 and 1986, known as the Sky Brightness Images, or *SKYFLUX*, had limited sensitivity due to zodiacal emission and calibration limitations. *IPAC* reprocessed the *IRAS* survey data with improved calibration and de-striping, with the modelled zodiacal emission removed, and with the separate hours-confirming sky coverages (*HCON*s) co-added. *ISSA*, the *IRAS* Sky Survey Atlas, is the result of this reprocessing.

The *ISSA* images are used in the software tool *IRSKY*, available from *IPAC*, to characterise the infrared sky at *IRAS* wavelengths.

3.4.2. *Description of ISSA*

ISSA is a set of standard *FITS* images projected in RA and DEC with units of MJy/str. Each image is $12.5° \times 12.5°$ with $1.5'$ pixels. The spatial resolution is approximately $4'$ (in-scan) by $5'$ (cross-scan) in all bands. For each field and wavelength there are an intensity images for each hours-confirmed sky coverage (*HCON* 1, *HCON* 2 and *HCON* 3) as well as a co-added image (called *HCON* 0). Scan coverage and noise images were also generated and are available from *IPAC* on request.

ISSA images are calibrated on the "*DC*" scale and may underestimate flux for smaller structures. The lowest ecliptic latitude fields (|ecliptic latitude| > 20 deg), which have residual emission from the zodiacal bands, and because of their relatively low quality were released as *ISSA*-Reject fields.

The scientific motivation for *ISSA* was to present the infrared sky as seen by *IRAS* at spatial scales greater than $5'$. The combination of calibration improvements and the removal of most of the zodiacal emission, results in a sensitivity that is limited by detector and confusion noise across most of the sky.

Full details of the *ISSA* images and their construction are given in the *ISSA* Explanatory Supplement (Wheelock *et al.* 1993).

3.4.3. *Some notes on using ISSA data*

The *ISSA* images are designed to give *relative* photometry for extra-solar objects and *should not be used* for determining the absolute surface brightness of the sky. Some uncertainty in the absolute background level is introduced by the process of de-striping and further uncertainties arise from the modelled zodiacal background which has been subtracted from the images. Imperfections in the zodiacal emission model result in residual zodiacal light in some places and over-correction in others. The problems are most obtrusive at 12 and 25μm, at $60°$ and $240°$ in ecliptic longitude. The effect is largest at $-15°$ in ecliptic latitude and can be as large as 2.0mJy/sr at 12μm, which is about 7% of the local intensity prior to zodiacal model subtraction. It should be noted that this large scale calibration discrepancy does **NOT** affect the *IRAS point source* calibration. One can mosaic the *ISSA* images at high ecliptic latitudes, without additional flat-fielding, to an accuracy of 0.1mJy/sr. This capability is due to the global de-striping algorithm that brought all confirming coverages of the sky to a common background level.

Non-confirming objects (objects which appear in only one *HCON*), as identified by visual inspection, were removed from the co-added images. Some non-confirming objects undoubtedly remain in the final product. The individual *HCON* images may be examined to verify reproducibility of features in the co-added fields.

Likewise, solar system objects and emission from solar system material remains in the data and may be a source of confusion. Known asteroids, as of the 1986 version of the *IRAS* Asteroid and Comet Survey, were removed prior to co-addition. The zodiacal bands appear as non-confirming extended emission bands in fields at low ($< 10°$) ecliptic latitudes. Comet tails are visible in some fields (notably comet IRAS-Araki-Alcock in fields 416 and 418). Comet trails (as opposed to comet tails) are spread out along the orbit of the comet and accumulate over time. In the *ISSA* images they appear as streaks crossing the image nearly perpendicular to the scan direction because of the orbital geometry.

3.5. *The IRAS Raw Data Archive and Scan & Image Processors*

The *IRAS* "raw" scan data are known as *CRDD* (Calibrated, Reconstructed Detector Data) and the full set of survey *CRDD* is stored at *IPAC* in a Flexible Image Transport System (*FITS*) group format merged with the pointing information. The collection of the raw data in *FITS* format is known as the Level 1 Archive. The Level 1 Archive *CRDD* have the final calibration (*PASS3*) and are used for all of the *IPAC*'s by-request data products and by xscanpi, the interactive scan co-addition tool.

3.5.1. *SCANPI (aka ADDSCAN), SUPERSCANPI and xscanpi*

SCANPI performs one-dimensional (in-scan) co-addition of the *IRAS* raw survey data. It is possible to do this interactively using the tool xscanpi, available over the Internet (see section on *IPAC* Internet services, below). *SCANPI* is useful for obtaining fluxes for point sources or slightly extended sources, and may be used to get fluxes for confused or faint sources, or to estimate true local upper limits. The sensitivity gain over the Point Source Catalogue is comparable to that obtained from other co-added *IRAS* data products: about a factor of $2-5$, depending on the local noise and number of scans crossing the target position. *SCANPI* is also useful for diagnosing source extent.

SCANPI co-adds scans which all passed over a specific target position. *SUPER-SCANPI* uses exactly the same processing steps as *SCANPI* , but co-adds scans from a number of **different** target positions. *SUPERSCANPI* can thus be used to make general flux estimates for classes of faint objects.

Basically, *SCANPI* collects from the *CRDD* all scans that pass within about $1.7'$ of the target position. The scans are re-sampled, aligned and co-added in several different ways (mean, median, weighted mean, etc.) and a point source template is fit to the co-added scans. Plots of the co-added scan and a table of various flux estimators are produced.

SCANPI is the best tool for getting fluxes of slightly extended sources, however, when run in the default mode, it cannot be expected to give accurate fluxes for sources that are larger than approximately half the detector cross-scan width ($\approx 2.5'$). *SCANPI* does allow flexibility in selecting scans for co-addition, but even special *SCANPI* reprocessing will not produce accurate fluxes for sources larger than about $4'$. For sources this extended a two-dimensional data product will be more accurate. Flux estimation is best performed on target positions which are accurate to within $15''$.

3.5.2. *FRESCO (Full RESolution Survey Co-addition)*

FRESCO is the Full REsolution Survey Co-adder. It produces co-added images of Infrared Astronomical Satellite (*IRAS*) survey data. Because the *IRAS* raw data exists as individual detector scans, the data must be mapped onto an image grid. Overlapping scans are added into the same pixel with appropriate weights (co-added).

FRESCO images have a resolution about that of the size of the *IRAS* detectors; roughly 5′ in the cross-scan direction, and between 1′ and 4′ in the in-scan direction, depending on wavelength.

FRESCO employs the Maximum Correlation Method (*MCM*) (Aumann, Fowler & Melnyk (1990)) to construct co-added images. This is the same algorithm used to make enhanced-resolution *IRAS* images (*HIRES*) at *IPAC*.

FRESCO is a three-step process. First the appropriate pieces of the raw survey data are retrieved. Then a program called *LAUNDR* de-glitches and de-stripes the data. Finally, a program called *YORIC* applies the *MCM* algorithm to the *LAUNDR*'d data. The Maximum Correlation Method is an image-reconstruction algorithm. It begins by using modelled detector response functions to "observe" a flat sky, in the same way *IRAS* actually observed the target. The simulated scan data from the blank sky are compared to the actual scan data for each detector. Correction factors are then computed by dividing the actual sample detector flux by this estimated flux. The original flat field pixels are then multiplied by a weighted average of the corresponding correction factors in order to obtain a new estimated flux. In *FRESCO* processing, *YORIC* does not iterate, but stops after it has built the "iteration 1" image. This corrected image is a *FRESCO* and is equivalent to a response-weighted co-add. Iterating this process produces *HIRES* images.

FRESCO is good for studying morphology or doing aperture photometry of extended sources that easily fit within a one or two degree square field, and where resolution better than the nominal resolution is not critical. *FRESCO* images are slightly easier to use and interpret than *HIRES* images. For point sources and slightly extended sources in non-crowded regions, *SCANPI* processing will give better fluxes. The *IRAS* Sky Survey Atlas (*ISSA*) is more suitable for studying very large regions. *FRESCO* maps are *AC* calibrated.

3.5.3. *HIRES (HIgh RESolution processing)*

The resolution enhancement algorithm used for *IPAC*'s *HIRES* processor is the Maximum Correlation Method (*MCM*, Aumann, Fowler & Melnyk (1990)), as outlined above. *HIRES* images are available in the same sizes as *FRESCO*s. The *HIRES* processing follows the same 3 steps outlined for *FRESCO*, except that the process of comparing the "simulated" scan data to the actual scan data is iterated.

Many factors affect the performance of the *HIRES* algorithm, but a typical user can expect an improvement in resolution over the full resolution co-adds of a factor of four or five, with greater improvement in the cross-scan direction – i.e. the point spread function becomes more nearly circular. Among the factors which affect the performance of the *HIRES* processor are: amount of coverage, separation (in time) of coverages, flatness of background and degree of "busyness" of the field.

HIRES is suitable for studying morphology, or doing aperture photometry. *HIRES*-derived fluxes agree with the *PSC2*s to within ∼ 20%, similar to *FRESCO*. Most of the uncertainty is due to background estimation uncertainties. *HIRES* can produce good results for relatively weak sources. However, the achieved resolution and tendency to produce artefacts does depend upon source strength, coverage and background complexity. Some rules of thumb for choosing suitable targets follows:

• The source should be "sufficiently bright", about 1Jy for point sources, but *HIRES* can sometimes pull faint structure out of somewhat confused regions. Below 1Jy, results depend strongly on the background complexity.

• The target should have good detector coverage. It is desirable to have both a moderately large number of scans and as much variation as possible in scan angle. Coverage can be estimated from the all-sky coverage maps in the *IRAS* Explanatory Supplement (Beichman *et al.* 1988). More information can be gained about the coverage of a given nearly point-like target by processing it with *SCANPI*, which also produces detector scan track plots within a $4'$ window of the input position.

• Very bright sources (> 1000Jy) may suffer from hysteresis, saturation and optical crosstalk (see above).

HIRES produces a large number of diagnostic maps. The best documentation on the use of these is available from the *IPAC* WWW Pages. In summary they are:

• **The Surface Brightness Maps**

These are the basic *HIRES* maps. Intensity is in units of MJy/str.

• **The Coverage Map**

The coverage map gives the sum of the weighted detector response at each image pixel over all detector samples. The actual number of detectors contributing to a particular pixel is roughly twice the value of that pixel in the coverage map. Examination of the coverage map can indicate non-uniformities in detector coverage.

• **The Correction Factor Variance Map**

The correction factor variance map gives the statistical variance about the mean correction factor (computed for each detector) for that pixel. If all the detectors agreed exactly on what the flux of a given pixel should be, all the correction factors would be one and the correction factor variance at that pixel would be zero. For a typical default-processed source with good signal to noise, the correction factor variances at iteration 20 should be between 0.001 and 0.01. Areas with relatively large correction factor variance either indicate the presence of data which do not agree with the majority of the scans (a "bad" scan, for example) or noisy, or saturated regions. On the other hand, a high correction factor variance *may* indicate that the source is not yet as resolved as is possible and that the field would benefit from further iteration.

• **The Photometric Noise Map**

The photometric noise map indicates the internal photometric error of the detector samples resulting from the averaging of those that overlap, and is essentially the standard deviation of contributions to each pixel. It does not include absolute errors such as calibration errors. These maps can thus be used to show the relative noise across an image, but not the absolute level of the photometric noise.

• **Beam Sample Maps**

The beam sample maps are a fairly sophisticated tool for estimating the achieved resolution in the *HIRES* maps. They are the result of processing hypothetical point sources placed on a smoothed version of the actual background.

• **Detector Track Map**

This is a map that indicates which detector's centre hit each pixel. If the centre of a detector "footprint" (two dimensional modelled response function) hit a pixel, its ID is recorded in that pixel. If more than one detector hits a given pixel, the last to cross is kept. If no detector centre hit the pixel, the value zero is kept.

HIRES processing can produce some unusual results and artefacts; it is an excellent amplifier of the peculiarities of the *IRAS* data. The most common artefacts are:

• **Ringing**

A ring of low flux may appear around a point source. This ring can contain up to 10% of the flux of the point source. Ringing increases the uncertainty in the background estimation for aperture photometry. Ringing most commonly occurs on relatively high backgrounds. Changing de-striping methods or using a different flux bias may help.

- **Blank Pixels/Coverage Depletion**

By default, *YORIC* discards any negative data values passed to it. Negative values can occur with some de-striping methods if the baseline is overestimated. This is most likely in crowded fields. When these negatives are discarded, regions of low – or in extreme cases, **no** – coverage result. Regions of no or low coverage are prone to develop bright features which are clearly non-physical; they normally appear sharp-edged and smaller than any point source in the field. Photometry in low coverage regions is suspect, even in the absence of artefacts. Low coverage regions are easily identified in the coverage map.

- **Aliasing**

Aliasing occurs typically at 12 and 25μm in low coverage areas. Instead of being essentially a two dimensional gaussian, the effective beam takes on a highly complex, unreasonable shape, rendering interpretation of image structure exceedingly difficult. This can normally be diagnosed from point sources in the image and from the beam sample maps.

A few additional things to think about when using *HIRES* images are the following.

- **Dynamic Range**

The dynamic range of *HIRES* is limited on small spatial scales ($\sim 10'$) to about 1:100. At small spatial scales it is best to work at intensities within 5% of the peak value.

- **Photometry**

An important thing to consider about *HIRES* images is the flux calibration. In general, fluxes measured from the intensity maps agree with those of the Point Source Catalogue (*PSC2*s) to within 20% for point-like sources. Background estimation when doing aperture photometry with *HIRES* can be tricky, particularly when ringing is present. In addition you must consider the effect of detector transients; *HIRES* is "*AC*" calibrated and for extended sources you may need to apply a source size dependent flux correction.

- **Ratio Maps**

Making *HIRES* ratio maps is a non-trivial task! Note that the resolution in an image not only varies from band to band, but can also vary from point to point within a single map, largely due to coverage variations and differences in the rate with which the algorithm converges.

There are several possible solutions to this dilemma. The first is to smooth both maps a **lot** in order to overwhelm resolution differences. However, this approach tends to defeat the purpose of using *HIRES* in the first place. A second possibility is to examine intermediate iterations for the better-resolved band to try to find a point-spread function which nearly matches that of the less-resolved band. A more sophisticated approach is to use the *IRAS* simulator mode of *YORIC* to cross-simulate the bands.

3.6. *IRAS Data and Related Services available on the Internet*

There are 3 main *IRAS* WWW sites:

- `http://www.ipac.caltech.edu/` (*IPAC*, The Infrared Processing and Analysis Center)
- `http://www.sron.rug.nl/irasserverman.html` (*SRON*, Space Research Organization Netherlands)
- `http://ast.star.rl.ac.uk/isouk/iras/iras_overview.html` (Rutherford Appleton Laboratory, *RAL*)

Aside from these sites, the *IRAS* data can also be accessed through the Astrophysics Data Facility (*ADF*) and the Point Source Catalogue along with other *IRAS* catalogues may be found in many catalogue collections.

The site at *RAL* distributes *IRAS* data products via a system called Starlink, designed to support UK astronomers. The *SRON* site uses an email interface to register users and distributes data and the *GIPSY* software for processing the data. *FITS* images of processed data can also be requested, and the *LRS* data set is well-supported.

The most extensive *IRAS* services are offered by *IPAC*. Some data may be accessed directly from the WWW, the scan data can be manipulated from a telnet interface, and there is a by-request processing service accessible by email. Here is an overview of what is available:

- Documentation
- Data on the WWW
  - *ISSA*-**PS**

  "postage stamp" server for the *IRAS* Sky Survey Atlas (*ISSA*). *ISSA* contains medium resolution (4 arcminute) data for the entire sky as covered by *IRAS* in all four *IRAS* wavebands
  - *IRAS* **Galaxy Atlas**

  a high-resolution-processed (about 1 arcminute) atlas of the Galactic Plane and regions in Orion, Taurus-Auriga and around ρ Ophiucus at 60 and 100μm.
- Interactive Software Services accessible from the On-line Services Page
  - *IRSKY*

  a sophisticated tool originally developed for *ISO* observation planning and now extended to *SIRTF* planning, which allows display of the IR sky from the *IRAS* Sky Survey Atlas (*ISSA*) data set and overlay of sources from the *IRAS* Point Source Catalogue, the *IRAS* Faint Source Catalogue and the Hubble Guide Star Catalogue. A subset of the catalogue data for the overlaid sources can be retrieved. It is also possible to overlay the *ISO* or *SIRTF* focal plane and display raster patterns. *IRSKY* further estimates background levels and limiting noise from *IRAS* and *COBE DIRBE* data. This includes a zodiacal light model and estimates of confusion noise both from galaxies and from the infrared cirrus. *IRSKY* is easy to use and is invaluable for getting an overview of what the infrared sky looks like at a particular location.
  - *XSCANPI*

  a telnet tool to co-add and analyse raw *IRAS* scan data as described in the section on *SCANPI*, above.
  - **XCATSCAN**

  a telnet tool to search the *IRAS* catalogues in conjunction with a number of other astronomical catalogues
- By-request Data services
  - *FRESCO*

  Full Resolution Survey Co-addition as described above.
  - *HIRES*

  High resolution processing as described above.
  - *SCANPI* **and** *SUPERSCANPI*

  Scan co-addition and template-fit processing for raw *IRAS* scan data in a batch mode, as described above.

4. A Very Brief Introduction to COBE and COBE Data and Services on the Internet

The Cosmic Background Explorer (*COBE*) as launched in 1989 and took data into 1990. It carried 3 instruments. *FIRAS* (Far IR Absolute Spectrophotometer) produced moderate spectral resolution data from 105μm to 1cm. *FIRAS* analysed data sets and all-sky maps have been archived. *DIRBE* (Diffuse IR Background Experiment) produced low spatial resolution maps in 10 wavelength bands from 1.25-240μm and polarimeteric data in 3 bands. The *DMR* (Differential Microwave Radiometers) produced differential data at longer than IR wavelengths using two horns separated by 7 degrees.

The *COBE* home page is at:

- http://www.gsfc.nasa.gov/astro/cobe/cobe_home.html

The data services available from the *COBE* home page include quite extensive documentation – the explanatory supplements are on-line in postscript format as well as a number of shorter pieces of information. The data products can be retrieved by ftp from the website and some images are available to browse directly. There is also a *FIRAS* spectrum browser tool. The site also distributes *COBE* analysis Software, to be used with *IDL*.

5. MSX data on the Internet

The Midcourse Space Experiment (MSX) was launched 24 April 1996 carrying SPIRIT III, with IR instrumentation operating from 4.2 to 26μm. It included experiments designed to measure the Zodiacal background and to resolve confused regions. Some early results images of the Galactic Centre are available from the Celestial Background Team Home Page. The data will be made available to community when the analysis and data processing are completed.

MSX sites are at:

- http://gibbs1.plh.af.mil/ (Celestial Background Team Home Page)
- http://www.ipac.caltech.edu/ipac/msx/msx.html
- http://msx.mrl.navy.mil/ (MSX Home Page)

6. Using Data from the Infrared Space Observatory

6.1. *Introduction to ISO*

ISO † (Kessler *et al.* (1996)) was launched 17 November 1995, with a planned 18 month lifetime limited by the expected duration of the liquid helium; *ISO*'s second birthday occurred during this Winter School and all systems were still working nominally. The liquid helium was finally expected to last until mid-April 1998, based on a series of measurements made of the thermal mass remaining in the cryostat. This prediction was highly accurate as the helium was finally exhausted on April 8th 1998.

ISO's orbit was highly eccentric, with a 24-hour period. The satellite's perigee was at 1000 km and the apogee was at 70 000 km, so *ISO* passed through the Van Allen radiation belts twice daily. The *ISO* instruments were not operable in the radiation environment of the belts, so its useful "science window" was limited to about 16 hours daily. Full coverage of the science window was provided by two ground stations, one at Villafranca del Castillo, Spain and the other at Goldstone, California, USA.

ISO carried a 60-cm Richey-Chretien telescope with f/15, diffraction limited beyond 5μm. There has been no evidence of defocus, or aberration. The full 20$'$ field of view was split into four 3 $'$ fields which each went to one of the four science instruments (see figure): the infrared camera (*ISOCAM* or *CAM*, Cesarsky *et al.* (1996)), the

---

† *ISO* is an *ESA* project with instruments funded by *ESA* Member states (especially the PI countries: France, Germany, the Netherlands and the United Kingdom) and with the participation of *ISAS* and *NASA*

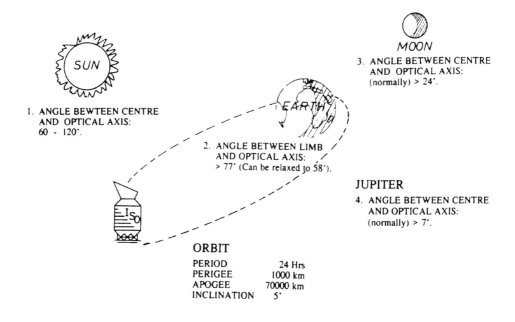

FIGURE 3. The *ISO* Orbit

long wavelength spectrometer (*LWS*, Clegg *et al.* (1996)), the imaging photopolarimeter (*ISOPHOT* or *PHT*, Lemke *et al.* (1996)) and the short wavelength spectrometer (*SWS*, DeGraauw *et al.* (1996)).

ISO's pointing performance was excellent. The specification for the absolute pointing error (for a blind pointing) was 11.7″ but the actual *APE* at the end of the mission was <1.5″! The performance in orbit was already a factor of two better than the specification. Additional work (driven by *SWS* and *ISOCAM*) which mapped field distortion on the star tracker mid-mission. (~Revolution 264; *ISO*'s chronology is often expressed in terms of the number of 24 hour orbits since launch or "Revolutions"). The stability of the pointing was also excellent, with a jitter of about 0.5″ over 30 seconds and a drift of less than 0.1″ per hour. Observations were carried out with the improved pointing from about Revolution 264 and data products in the archive will have pointing reconstruction done to the same accuracy in most cases. Most *ISO* instruments had a fairly large field of view and were little affected by the improvement in pointing performance, but it was important for *SWS* and the smallest pixel scales of *ISOCAM*.

Unlike *IRAS* and *COBE*, *ISO* was an observatory. This means that the archival legacy of *ISO* will be non-homogenous, rendering large statistical type studies difficult. *ISO*s observations were mostly made with Astronomical Observation Templates (*AOTs*) which are well-defined observing modes with a relatively small number of parameters. *ISOCAM* used 3 *AOT*s, *LWS* 4, *SWS* 4, and *ISOPHOT* 13. In addition, quite a bit of *ISO* data (including all polarisation work and a great amount of calibration observations) has been taken with non-standard commanding using the Calibration Uplink System, or *CUS*. Efforts are being made to bring all the data to as uniform a level of processing and format as is reasonable given inherent differences in instruments and observing modes.

ISO had the capability to execute staring observations, raster maps or nodding/beam-

The ISO Focal Plane Projected Onto the Sky
For Roll Angle = 0

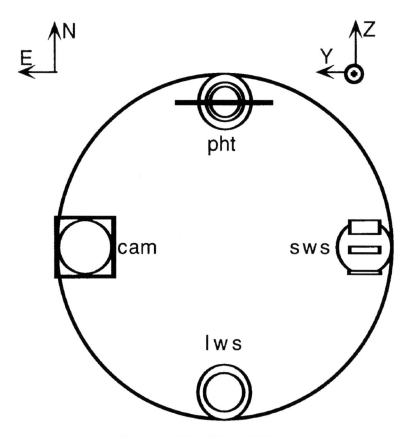

FIGURE 4. The *ISO* Focal Plane

switching observation. *ISO* could also track a number of solar system objects. The tracking is carried out by using a series of linear rasters, so that very fast-moving objects may move significantly within a very small aperture during a raster point. Depending upon the instrument and *AOT*, all the above modes were used for observing. *ISOPHOT* also contained a focal plane chopper and could execute chopped observations.

Because *ISOPHOT* was the only instrument which can chop, it was often the case that background measurements were performed as independent *AOT*s from the primary target. There were also mapping modes (*ISOPHOT* sparse maps and any map using *SWS*) in which the individual pointings are coded as separate *AOT*s. *ISO* handled this kind of "related" observation using the mechanisms of concatenation, linking or fixed-time observations. Concatenated observations have been carried out one after the other in sequence. Linked observations were rare and involved intervention for a follow-up observation. Fixed time observations were, as the name implies, carried out at specific times during the mission. The usage of these modes meant that one element of using *ISO* archival data will be identifying what other data sets are needed to reduce properly the data set of interest.

In addition to the generally available *AOT*s and the Calibration Uplink System used by the instrument teams, three of the four instruments (*SWS* being the exception) also had serendipitous modes of observation which either operated in parallel to the primary observation (at a random pointing several arcminutes from the target being observed) or operated during satellite slews, or both. The data from these modes is being analysed by consortia which hold the proprietary rights to them and is likely to enter the archive no sooner than a year after the end of the mission. These data represent random, but fairly homogenous mini-surveys. Each mode is discussed briefly with the observing modes for the instruments, below.

The philosophy of the *ISO* Project has been to process the raw telemetry using a "pipeline" or Offline Processing System (*OLP*) to several stages of reduction which are made available to the observer. The stages of reduction planned for the active mission were:

- Edited Raw Data (*ERD*), or easily-readable lightly cooked telemetry
- Standard Processed Data (*SPD*), which would have "instrumental effects" removed, but no final calibration applied
- Auto-analysis Results (*AAR*), which would be fully reduced quick-look data

The actual degree of processing at each level varies from instrument to instrument, and in the *ISO* archive attempts will be made to make the degree of reduction available more uniform from instrument to instrument.

ISO was, at the time of writing the bulk of this text, still an active observatory. The bulk of science data from *ISO* entered the public domain in May 1998, and will continue entering the archive until ~1 year after the liquid helium was depleted. Calibration observations made during the routine portion of the *ISO* Mission are in the public domain and will be part of the initial archive, which was available around 3 months after liquid helium depletion, or late summer/early fall 1998. Some data was being taken under special arrangements for the proprietary rights, including he Hubble Deep Field observation and the "solicited proposals" which begin with a proposal id of "ZZ" and have a proprietary period of only 1 month. Access to these data is discussed under WWW services, below.

General documentation about *ISO* is available from the Home Page, as well as all the technical documents. The special Astronomy and Astrophysics issue, "First *ISO* Results" (1996, A&A, 315 (2) L27-L400) is a source of basic descriptions of the instruments and examples of early results from all four instruments. Currently, the Observer's Manuals and their Addenda give the basics of instrument design and operation and the Instrument Data User's Manuals (*IDUM*s) give detailed information about the data products, processing steps, information about further data reduction and analysis and necessary caveats. The *ISO* Data Products Document (*IDPD*) gives gory details of the *FITS* formats of all files. This collection of documents, all available from the *ISO* Home Page, will eventually be superseded with a (multi-volume) *ISO* Explanatory Supplement. In addition, some information about the data and data reduction are available in the documentation for the various "Interactive Analysis" packages, or *IA*s, which now exist in some form for all four instruments.

6.1.1. *Important Common Properties of the* ISO *Data*

All *ISO* instruments were affected by cosmic particle hits or "glitches". Glitches not only caused an immediate strong detector response but could also cause long-lasting changes in the detector responsivity. De-glitching data is very important, and in most cases some smaller glitches remain in the data, as well as the long-lived glitch "tails". A further effect of glitches is that sub-critical glitches, that is small glitches which are not significantly above the detector noise and, hence, not readily visible to de-glitching algorithms, lead to increased apparent noise. Thus, the solar cycle and the resulting space weather affected observing conditions. In an extreme example, Revolution 722 was recently "rained out" by a solar flare; none of the data taken during that revolution proved usable.

Because *ISO*'s detectors were primarily extrinsic photoconductors which share the common properties of such detectors in greater or lesser degree, all *ISO* instruments were subject to responsivity changes. Responsivity changes which take place on long time scales (a substantial fraction of a revolution) are usually referred to as detector "drifts" and these are typically measured in some way through the use of internal calibration sources. Responsivity changes which result in a slow asymptotic approach to a final value when the flux on the detector changes are usually referred to as "transients". Increasing flux steps (upward transients) and decreasing flux steps (downward transients) usually behave differently and to some degree each detector element, or pixel, experiences a slightly different transient response. The effect of detector transients is most obvious in *CAM* and *PHT* data but is also present in the spectrometers. In the *ISO* data, transients are generally handled either by trying to model them, or by waiting until the detector response is nearly stabilised. It may also be possible in some cases to calibrate the fractional response as was done for *IRAS* resulting in the "*AC*" and "*DC*" calibrations.

The sections below contain general information about *ISO* and the *ISO* Instruments as well as some "cookbook" type examples of data reduction steps. Since knowledge about the instruments is still rapidly evolving, these examples should be considered as rough illustrations only, and the current documentation including the WWW pages and *IDUMS* should always be consulted.

6.1.2. *Using ISO Satellite Data Files and Cal-G Files*

The most commonly used satellite data files are the instantaneous pointing history file (*IIPH*) and the per-raster-point version (*IRPH*). All *ISO* data files follow a consistent naming convention: the first four letters identify the file type; this is followed by a 7 digit number which carries the revolution number in which the observation was carried out (3 digits); followed by a 2 digit sequence number related to scheduling; and another 2 digit number which is related to the original observer's proposal. The files end with the extension .fit. The identifying number is sometimes referred to as the *TDT* number of the observation and uniquely identifies the observation. The file types are sometimes referred to by the four letter designation, e.g. "*IIPH*" above. (*ISO* observers who received their data on the original CD-ROMs may note that part of this file name becomes a directory name in the layered structure on the CD).

Other satellite files are less frequently used and contain housekeeping information, orbital parameters etc. These files and their usage are documented in the *ISO* Satellite *DUM*.

ISO observations are also usually accompanied by a set of general calibration files called *Cal-G* files. These files are relevant to a given version of pipeline processing and may vary from instrument to instrument and from *AOT* to *AOT*. They can be used with *IA* packages or with a package like *IDL* to reduce data, or they can be examined in the case some spurious effect is suspected. These calibration files contain information such as conversions from instrumental signal levels to astronomical units, spectral response functions, libraries of darks and flats and so forth. It is generally the case that interactive analysis packages use a newer version of *Cal-G* files than is available from the pipeline processing. *Cal-G* files are often distributed along with new releases of these tools. The content, format and usage of these files are documented in the *IDUMS*.

6.1.3. *Handling FITS Binary Tables*

The vast majority of *ISO* data is in *FITS* binary table format. Binary table format (also used by *COBE*) is convenient for both true tabular data and for storing in the same file multiple images with differing header information. The disadvantage to *FITS* binary tables is that they are not directly readable by eye and are still only read by a subset of astronomical software, notably the Astrolib routines in *IDL* and *ISO*-dedicated software. *IPAC* provides a set of " first-look tools" that take *AAR* files and some satellite files from *ISO* and converts them into *ASCII* tables, or standard *FITS* images, to facilitate getting a quick look at the most-reduced form of the data.

Some useful tips for reading *FITS* binary files using *IDL* and Astrolib are in the *ISO-CAM IDUM*. Documentation of the *FITS* Standards is at
`http:///www.cv.nrao.edu/fits/documents/documents.html` and *FITS* binary tables are described in Cotton, Tody & Pence (1995).

So what is a *FITS* binary table anyway? A *FITS* binary table consists of the familiar *FITS ASCII* header, a table header and binary data organised as rows and columns, with each "column" having an independent data type and dimensionality. In this manner, the pixels that make up an image can be can be treated as a column, such that the binary table contains image header type information for each "row" or image.

6.1.4. *A FITS Binary Cookbook*

What follows is a sequence of Astrolib commands in *IDL* that can be used to read in and plot the instantaneous pointing history from the *IIPH*.

```
fxbopen,u1,'filename',1,iiph_hdr
print, iiph_hdr
fxbhelp,u1
fxbread,u1,ra,'ra'
fxbread,u1,dec,'dec'
fxbread,u1,utk,'utk'
plot,ra,dec,ynozero=1
```

Alternatively, the *IPAC* first-look tool `flsat` could be used to generate an ascii table, which could be plotted with any plotting package.

Images are slightly trickier to handle directly in *IDL* because the information about the dimensions must be extracted and used to dimension arrays to hold the images. The following example is much easier to handle using the *IPAC* first-look tool `flcam` to generate a (potentially very numerous, depending upon the observation) set of *FITS* image files and then using a standard image display utility, such as SAOimage, to display them. But here is how to read in an *ISOCAM* map file (*CMAP*) and display it using only *IDL* with Astrolib. Note that the *CMAP* file contains individual *CAM* image frames and each image is 32×32 pixels.

```
fxbopen,u1,'filename',1,cmap\_hdr
fxbread,u1,bscale,'bscale'
fxbread,u1,bzero,'bzero'
fxbread,u1,allpix,'array'
scalpix=fltarr(1024,nrows)
for n=0,rows-1 do scalpix(*,n)=float(allpix(*,n))*bscale(n) + bzero(n)
fxbread,u1,naxis1,'naxis1'
fxbread,u1,naxis2,'naxis2'
```

```
nb=naxis1*naxis2 - 1
images=fltarr(32,32,nrows)
for n=0,nrow-1 do
images(*,*n)=reform(scalpix(0:nb(n),n),naxis1(n),naxis2(n),1)
for n=0,nrows-1,3 do
tvscl,rebin(images(*,*,n),10*naxis1(n),10*naxis2(n),sample=1)
```

6.2. ISOPHOT

ISOPHOT is *ISO*'s imaging photopolarimeter; it was really three sub-instruments sometimes referred to as "detector groups": *PHT-P*, *PHT-C* and *PHT-S*.

PHT-P was a multi-band, multi-aperture photometer with apertures ranging from 5 arcseconds to 3 arcminutes. *PHT-P* had three single-element detectors:
- P1 is a Si:Ga detector using 10 filters in the 3.3-16.6μm range
- P2 is a Si:B detector using 2 filters at 20 & 25μm
- P3 is a Ge:Ga detector using 2 filters at 60 & 100μm

PHT-C was a multi-band camera operating at the longer wavelengths using two small arrays. The wavelength range of C100 overlapped with that of P3. Neither C100, nor P3, was a well-behaved detector, but the redundancy of the 9 pixels in C100 may sometimes render the data easier to interpret. The detectors were:
- C100 was a 3×3 Ge:Ga array with 43.5″ square pixels using 6 filters in the 50-105μm range
- C200 was a 2×2 stressed Ge:Ga array with 89.4″ square pixels using 5 filters in the 120-200μm range

PHT-S was a pair of low resolution ($\lambda/\Delta\lambda$ ¡ 100) grating spectrophotometers (SS & SL). Each uses a 64 element Si:Ga array, and they shared a common 24″ by 24″ aperture. The 2.5-5μm range is covered by SS and 6-12μm by SL. *PHT-S* is currently the most well understood of the *ISOPHOT* detector groups.

All of the *ISOPHOT* detector subgroups have cold readout electronics (*CRE*), bias supplies and thermal monitoring and control electronics. All the detector subgroups except *PHT-S* used internal fine calibration sources (*FCS*) to calibrate the drift response of the detectors, which can be very significant for some of the *ISOPHOT* detectors. *PHT* also has a focal plane chopper for making chopped measurements; however, the chopped measurements frequently display obvious detector transients and to date have not been well-calibrated.

6.2.1. *ISOPHOT Observing Modes*

PHT-P had 6 *AOT*s:
- *PHT*03, multi-filter photometry, stare, raster or chop
- *PHT*04, multi-aperture photometry, stare, or chop
- *PHT*05, absolute photometry, dark and *FCS*, stray light, stare only
- *PHT*17, 18, 19 sparse map sequence that had to be concatenated

PHT-C also had 6 *AOT*s, most of which are similar to those of *PHT-P*:
- *PHT*22 multi-filter photometry, stare, raster, or chop
- *PHT*25 absolute photometry, dark and *FCS* stray light, stare only
- *PHT*37, 38, 39 sparse map sequence
- *PHT*32, oversampled mapping

The *PHT*32 mapping sequence was fairly complicated, using both raster pointings and the chopper to build up oversampled coverage to a variable degree. Transient effects sometimes cause "ghosting" in *PHT*32 maps.

PHT-S only had one *AOT*:

- *PHT*40 spectrophotometry, stare, raster, or chop

ISOPHOT also had a Serendipity mode in which the C200 detector was used during slews, to make scanning observations at 175μm (Bogun *et al.* (1996)).

Two more *AOT*s, *PHT*50 and *PHT*51, were released only after the Winter School had finished. They were used to make polarization measurements with *PHT-P* and *PHT-C* respectively. These observations were actually implemented using the Calibration Uplink System.

6.2.2. *Using* ISOPHOT *data*

As is the case with *LWS* and *SWS*, *ISOPHOT* sensitivity in orbit was lower than predicted from the ground. Additionally, the calibration of *ISOPHOT* has proceeded rather slowly and will be completed during the post-operation phase of *ISO*, which will last for three years after the liquid helium was depleted. The reasons for this include the fact that the severity of the effects of space-weather are greater than anticipated, both because sub- detectable glitches increase noise and because glitches also induce short term transients. It is also the case that the drift and transient effects have proven difficult to mode, measure, or correct. Finally, the complexity of the instrument with its many subsystems and observing modes renders the calibration and understanding of the instrument a very large job. There are so many pieces to the puzzle, and a stepwise approach to calibration means that all of these many small pieces must be completed and interlinked before the final picture emerges.

A further contributing factor to the difficulty of getting a handle on *ISOPHOT* performance, and particularly that of long wavelength detectors has been that observers have frequently failed to take the structure of the 100μm sky into account when designing observations – in particular many chopped observations have been made where there is significant structure in the background.

However, this does not mean that *ISOPHOT* data is not useful. Although *ISOPHOT* suffered from some of the same limitations as the *IRAS* data and the sensitivity is not as much better as was expected, *ISOPHOT* still covered wavelength ranges that *IRAS* did not and had a higher resolution. Further, the understanding of the instrument is still making headway.

There are a number of properties of the instrument that are essential to understanding how to work with *ISOPHOT* data. The first of these is the long- time scale change in responsivity, or "detector drift", which is due to a slow increase in the responsivity as the detector is exposed to the radiation environment. Not only does the increase in responsivity cause a change in the calibration of the instrument, but it is also accompanied by an increased noise. These effects are worst after passing through the radiation belts, so the detectors were cured at the beginning of the so-called "science window" in each revolution. They were also re-cured about midway through when ground station for communication with *ISO* was changed (the "handover window"). P3, C100 and C200 were the detectors most affected by drifts. These detectors can show changes in responsivity of up to a factor 2-3.

The term "detector transients" is normally used to refer to the delayed and occasionally erratic response of a detector to a change in the flux falling on it. This is particularly problematic in chopped measurements, where the detector response may not have reached a large fraction of its final value when the flux falling on the detector changes. In some cases (e.g. *PHT-S*) the detector response curve was fairly stable and, in fact, the initial signal jump was a fairly stable fraction of the final, stabilised, signal. Under these conditions it should be possible to model the detector response and calibrate the chopped data. Thus far this has only been satisfactorily achieved for *PHT-S*, and the recommended approach is to use well-stabilised measurements. P1, P3 and C100 were the detectors most affected by transients. P2 and C200 stabilised fairly quickly.

Another factor that may complicate the *PHT* puzzle is the possible presence of non-linearities for some flux regimes in some detectors. Nonlinearities are fairly common in this type of detector system and are present in both *IRAS* data and in the *LWS*.

The *ISOPHOT* flux calibration is currently quoted as ±30% absolute and ±10% relative where relative means filter-to-filter. These values are only fully applicable to well-stabilised staring, or nodding measurements of point sources with moderate flux, where the flux lies within the calibrated range of the *FCS*.

FCS measurements were made at end of each staring/chopping *AOT* and at the beginning and end of mapping *AOT*s for all detector subgroups except *PHT-S*. *FCS*1 has been calibrated using astronomical standards. *FCS*2 has only been calibrated from *FCS*1, early in the mission, and its use has not been encouraged.

Some general points in working with *ISOPHOT* data include:
• Avoid using default responsivities, especially for the drift-prone detectors. (Defaults are unavoidable with *PHT-S*, but it was fairly stable).
• *FCS*1 is the main calibrator
• The pipeline often produces similar results to careful reduction and is useful as a quick-look, especially for maps, but it is very difficult to assess the data in its reduced state.
• In general it is best to reduce the data from the *ERD* state using the ISPHOT Interactive Analysis tool (*PIA*) so proper treatment of glitches and instrumental effects can be verified.

6.2.3. *A Word about PIA*

The *PHT* Interactive Analysis software is developed jointly by *ESA* and the *ISOPHOT* consortium. *PIA* is available from the *PIA* Home page at the *ISOPHOT* data centre at *MPIA*, in Heidelberg:
• http://isowww.estec.esa.nl/manuals/PHT/pia/pia.html

PIA works under *IDL* versions from 3.6.1 on. It is an interactive tool with a fairly easy to use graphic user interface (*GUI*) which allows the astronomer to look at the data and assess the quality of the data and effectiveness of the reduction steps in (A LOT OF) detail. It is very difficult to work with *PHT* data without using *PIA* and the basic steps in data reduction illustrated below refer to steps available in *PIA*.

6.2.4. *Basic steps in* PHT *data reduction*

The following is a crude recipe for the general steps involved in reducing *ISOPHOT* data using pia.

Start pia
Read in the data from the pipeline *ERD* file (PPER, P1ER, P2ER or PSER for *PHT-P*, C100, C200 or *PHT-S*)
Select a measurement to work on: first reduce the *FCS* measurement, then the relevant observation(s). Where otherwise unspecified, the *PIA* defaults are generally used.

Start with *ERD*. The *ERD* data consists of integration ramps which need to be fit and the slope determined in order to derive a "signal"
Linearise the ramps
De-glitch at read-out level
Do linear fit (" 1st-order polynomial") this yields *SRD*

Note that there is an option for ramp subdivision, which is useful for relatively long ramps, which would otherwise not provide enough statistics for the signal-level de-glitching to be effective.

Continue working with the *SRD*:
de-glitch at signal level
subtract dark current
process (with or without drift recognition or drift modelling) to get a signal per chopper plateau or *SCP*

Note that a reset interval correction is being tested and is believed to be an important correction in some cases. It is sometimes recommended to use "drift recognition" for long staring integrations. Drift recognition finds the well stabilised portion of the data and deletes the rest. Drift modelling can be very successfully applied to chopped P40 (*PHT-S*) observations.

Continue working with the *SCP*:
For *FCS*:
calculate responsivity
choose a responsivity to apply, normally *FCS*1.
Stop here and go back for the astronomical measurement.

For an astronomical measurement:
apply chopper vignetting correction if chopper is used
do background subtraction (within measurement if chopped)
calibrate to in-band power using selected responsivity – this yields *SPD*

Continue working with *SPD*:
extract flux density
assign astronomical coordinates

For single-pointing data, the data are now essentially fully reduced. For maps go on to "Astrophysical Work".
Steps done in Astrophysical Work:
For single-pointing data, combine filters for SED plots or apertures for curve- of-growth, etc.
For maps, it is necessary to go to the "Mapping" section and produce the map.
For spectra apply spectral response function and wavelength calibration.

6.3. ISOCAM

ISOCAM is an infrared camera operating in the 2.5-18μm range, consisting of two independent 32x32 pixel arrays:
- *CAM-SW* uses InSb CIDs in the 2.5-5.5μm range
- *CAM-LW* uses a Si:Ga photoconductor array in the 4-18μm range

Each has discrete filters and Circular Variable Filter(s) (*CVF*) ($\lambda/\Delta\lambda \sim 35$). Each has a lens wheel that allows selection of a 1.5,3,6 or 12$''$ per pixel field of view. In addition, the entrance wheel contains polarisers.

Unlike the other *ISO* instruments *CAM* is always on, even during perigee passage. *ISOCAM* observes in "parallel mode" whenever it is not the " prime" instrument.

6.3.1. *ISOCAM Observing Modes*

All of the following modes (except for polarisation which is only used for the LW detector) can use either *LW* or *SW*. All modes allow optional dark, calibration and "clean" operations, but these are rarely used. In fact, the use of clean can *induce* severe transients. The dark flags are commonly used in *CAM* observations for a purpose other than making dark current measurements. They have a series of special values that are used in order to control the configuration of the camera when it arrives on target, in order to prevent saturation in the default, rather sensitive, parallel mode. There are three standard *AOT*s and a polarisation mode, which uses the Calibration Uplink System.

• *CAM*01 is the basic mode using single pointings, or rasters, in one or more filters including *CVF* settings or "filters". A common strategy is "microscanning", i.e. executing a raster with step size of order several pixels to provide redundancy which facilitates de-glitching.

• *CAM*03 is used to make beam-switching observations.

• *CAM*04 is used to make *CVF* scans over a wavelength range.

• "*CAM05*" is the pseudo- *AOT* for polarisation observations

• the *CAM* Parallel Mode continuously takes images, usually at 6.75μm, when *SWS*, *LWS* or *PHT* is being used (Siebenmorgan *et al.* (1996)).

6.3.2. *Using* ISOCAM *Data*

ISOCAM performance in orbit is as predicted from ground to a remarkable degree. The sensitivity of the *LW* detector is excellent: it is possible to detect a 300mJy point source at 10σ significance in 2000 seconds at 15μm using the 6″ per pixel field of view. Detection of sources as faint as several tens of *micro*Janskys have been reported. *CAM*'s sensitivity is often limited by flatfield noise/glitch remnants. The *ISOCAM* photometric calibration is about 20% in practice and the repeatability is a few %. Transient effects, and the effect of transient removal are somewhat difficult to assess.

There are a few points to be aware of in the *CAM* data:

• *LW*'s column 24 is "dead". The detector itself is believed to be functioning nominally, but the data from it is missing from the telemetry.

• The *CVF* has significant optical ghosting.

• "lens wheel jitter", an irreproducibility of the position of the lens wheel, causes an overall astrometric uncertainty of one to two pixels due to an unpredictable shift of the image on the array whenever the lens wheel is repositioned.

• Saturation of *LW* is serious – very bad cases have been known to leave artefacts in the flatfield for DAYS.

• Transient behaviour is a critical limitation, but progress has been made in fitting transients.

• Each pixel had its own transient behaviour and flux history and would stabilise at its own rate. Therefore transient behaviour is implicated in flat-field noise.

• It is currently believed that detector flux history over times much longer than an observation will have to be taken into account to do the best possible job of transient handling.

• The glitch rate was typically about as anticipated; de-glitching is fairly successful but (very) severe glitches may leave behind (very) long transients after the glitch itself has been removed.

• there are a number of "illegal" configurations which observers have used. These did not always cause the observations to fail in an obvious way, so some study of the observing configuration is recommended.

• The *AAR* gives very representative results, especially for moderate flux levels but does not include transient correction and uses library flat fields, so quality is limited in many cases.

- It is generally best to reduce the data from the *ERD* or *SPD* level using *CIA* and/or *IDL* directly, and to carefully assess each correction made. The most important corrections occur when processing the *CAM* data from *SPD* to *AAR* and it is usually adequate to start reduction from the *SPD* level.

6.3.3. *A Word about CIA*

The *CAM* Interactive Analysis software is developed by *ESA* and the *ISOCAM* consortium. *CIA* works with *IDL*, version 3.6 or 5.0. *CIA* is currently available only on approval, but is expected to be freely distributed by the *ISO* Data Centre at Vilspa by the time the *ISO* Archive is available.

CIA has a basic *GUI* but is more often used in command-line mode, using a number of pop-up *GUI* tools. Data-handling in *CIA* is somewhat complex and learning the system well takes some study! *CIA* offers a number of alternative routines for de-glitching, transient fitting and other data reduction steps, so several methods can be tried and compared.

6.3.4. *Basic steps in CAM data reduction using CIA*

The following is a cookbook recipe for reducing and *ISOCAM* raster (*CAM*01) in *CIA*.

- Start *CIA* and load the *SPD* (CISP) data. (Loading the data takes several steps.)
  - " Slice" the *SPD* into a set of *SCDs*
    — spdtoscd, 'filename', sscd, spd_dat='path'
    — sscd_info, sscd

This creates many *SCDs*. An *SCD* is an *IDL* data structure holding *SPD* data for a *CAM* "state", that is, a given configuration of lens wheel, filter wheel, etc. of the camera. *SCDs* exist for intermediate configurations as well as "useful" states. The procedure above also creates an entity called an *SSCD* which keeps track of the *SCDs*, as can be seen when the sscd_info command is used.

- "Clean" the *SSCD*, i.e., discard the "useless" states and correlate like configurations.
  - cleaned_sscd=sscd_clean(sscd)
  - sscd_info, cleaned_sscd(0)

Note that cleaned_sscd is an array with an element per configuration, and carries information that identifies the data belonging to like configurations.

- For a raster observation (*CAM*01) load a configuration into a raster 'prepared data structure'
  - raster_pds=get_sscdraster(cleaned_sscd(0))
  - help,raster_pds,/str
  - x3d,raster_pds.cube

Note that x3d is a *GUI* tool for browsing the data. Now the data are ready to perform the actual data reduction steps, for the first configuration now stored in raster_pds.

- Subtract the dark

The routine called calib_raster searches the library of darks (which in the pipeline are stored as *Cal-G* files) to find a close match, scales it as necessary for gain and subtracts it. Currently the time within a revolution is not taken into account in calib_raster, but is a procedure under development.

- old_raster=raster_pds
- calib_raster,raster_pds,/dark
- x3d,raster_pds.cube

Note that old_raster is useful for making comparisons. It is a good idea to save the old data at every step, to facilitate experimentation and evaluation of the results.

- De-glitch the data

By default, calib_raster uses the Multi-resolution Median Transform method also called 'mm' (Starck *et al.* (1996)). This method essentially discards structures which persist for less than the dwell time at a given position.

- calib_raster,raster_pds,/de-glitch
- x3d,raster_pds.cube
    o Deal with the transients

Handling the slow response to flux steps ("transients") is the most controversial step. The most conservative approach is to wait for all the responses to stabilise, but complete stabilisation rarely occurs. There are 5 transient **correction** methods available in the current *CIA* and these are described in the *CAM IDUM* in some detail, as well as in the *CIA* documentation.

(*a*) the *IAS* method "fit1"
(*b*) the *IPAC* method "fit2" It is considered most useful for downward transients
(*c*) the Saclay method "fit3"
(*d*) the inversion method "inv"
(*e*) the Vision method "vision"

There are varied opinions about what version of transient removal to use. Some are more successful in some circumstances than others. The routine calib_raster's default is the conservative S90 method, which finds the first 90% stabilised frame and discards the preceding data. The *CAM* Team are currently recommending the inversion method when sufficient frames are present in the data (50 to 100) and the Saclay method otherwise. A good rule of thumb is to try several methods and make sure any conclusions you draw don't change!

- calib_raster,raster_pds,stab='inv'
- x3d,raster_pds.cube
    o Flat-field the data

The default in calib_raster is to build an "automatic" flat from the data cube. This is a normalised median image with the "bad" column filled.

- calib_raster,raster_pds,/flat
- x3d,raster_pds.cube
    o Build the mosaic from the raster pointings

The default is to build the raster without (re)projection.

- calib_raster,raster_pds,/flat

The mosaic is a new element in the structure: raster_pds.raster.

- View the result!
    o tviso,raster_pds.raster

Note that this full suite of data reduction can be done in one step...

- calib_raster,raster_pds,/dark,/deglitch,stab='fit2',/flat,/raster

...but this is not encouraged! It is better to proceed methodically and evaluate the result of each step.

Some additional corrections are available and are important in some cases. There is a second order dark correction, which basically filters the data in Fourier space, which can be effective for deep measurements of point sources. There is also a correction for optical distortion, which is relevant for, large (i.e. 6-12″ per pixel) fields of view.

- Finally: write the data out to disk as a *FITS* image
    o raster2fits,raster_pds,name="something.fits",dir='path'

Note that slightly different *CIA* routines are used for *CVF* scans, staring mode observations and beam-switching observations.

6.4. *Overview of ISO Spectroscopy*

ISO's spectroscopic capability is unique; the two dedicated spectrometers have excellent resolution. In addition to *SWS* and *LWS*, both *ISOCAM* and *ISOPHOT* have some spectroscopic capability. To summarise:

- Short Wavelength Spectrometer (*SWS*)
  - 2.38-45.2μm, $\lambda/\Delta\lambda \sim$1500 (grating)
  - 11.4-44.5μm, $\lambda/\Delta\lambda \sim$ 30 000 (*FP*)
- Long Wavelength Spectrometer (*LWS*)
  - 43-196.9μm, $\lambda/\Delta\lambda \sim$ 200 (grating)
  - 47-196.6μm, $\lambda/\Delta\lambda \sim$ 8000 (*FP*)
- Spectrophotometer (*PHT-S*)
  - 2.5-12μm, $\lambda/\Delta\lambda <$ 100
- *CAM- CVF*
  - 2.5-18μm, $\lambda/\Delta\lambda \sim$ 35

The two spectrometers have a number of characteristics in common. The format of the end result of the pipeline data reduction, the Auto-Analysis Result, has similar structure and information content. In both cases, the *AAR* is not a "simple" spectrum with wavelength, flux and uncertainty, but is still broken into many subspectra per detector, scan direction, scan number, target line, etc. This means that the data structure can be somewhat confusing, with the same wavelength appearing in several detectors, multiple scans, etc, and, likewise, with data from several (widely-separated) wavelengths appearing associated with the scan of a single spectral line. For both spectrometers, data are read from all detectors all the time and serendipitous data are generally present.

For both spectrometers the spectral resolution and sampling vary with wavelength and calibration differences and responsivity changes lead to detector-detector continuum mismatch which is not compensated for perfectly in the pipeline processing. In both cases, some degree of residual instrumental signatures remain in the *AAR* and imperfections in the spectral response function calibration files can result in spurious features in the spectra.

6.5. *A brief word about ISAP*

The *ISO* Spectral Analysis Package (*ISAP*) is developed by a large collaboration which include members of the *LWS* and *SWS* Instrument teams, the *IAS* and *IPAC*. *ISAP* can be obtained by public ftp from sites at *IPAC* and the *ISO* Spectrometer Data Centre at the Max Planck Institute of Astronomy in Garching.

- ftp.ipac.caltech.edu in /pub/software/isap/ (UNIX)
- mpeia2.mpe-garching.mpg.de in isap/ (UNIX & VMS)

ISAP is used to further reduce the *AAR* data from *LWS*, *SWS* and *PHT-S*. It is an *IDL* package, which runs under *IDL* version 4.01c, or later. *ISAP* has both *GUI* and command- line interfaces.

ISAP can be used to view and subset the data according to scan, detector number, scan direction, etc. This allows identification of the useful data and determination of the existence of any problems with fringes, or mis-matched scans, due to transient effects. For very large data sets, containing many line measurements, the *AAR* file may need to be split at the command-line level into separate data files for each line. Once the data have been inspected, *ISAP* can be used to process the data to get a simple spectrum. Its capabilities include masking bad data, averaging the multiple scans and overlapping detectors, incidentally de-glitching with some form of median filter. *ISAP* can also be used to combine the data from all detectors for a full-range scan, if desired. *ISAP* further contains general analysis functions relevant to spectral data including unit conversions, line and continuum fitting, de-reddening, synthetic photometry and some modelling (e.g. a zodiacal light model).

ISAP is generally easy to use from the *GUI*. The data can be read and special functions applied using menu buttons on the main panel. A range of buttons for subsetting the data by detector, scan number, etc. are also fairly intuitive to use. The plot display can be zoomed by drawing a box with the left mouse button. One very important feature of *ISAP* is that data are *selected* by drawing a box with the right mouse button, which pops up an extensive menu of operations that can be performed on the data, such as averaging, re-binning, etc.

6.6. *The LWS*

The Long Wavelength Spectrometer (*LWS*) covers 43-196.9μm at moderate and high resolution. The *LWS* has a circular aperture of 1.65$'$, which is the diffraction. limit at 118μm. The *LWS* grating is ruled at 7.9 lines per mm and mounted on an up- down scanning mechanism. The grating operates in second order from 43 to 94.6μm and in first order from 94.6 to 196.9μm. The Fabry-Perots can be introduced into beam path using selection wheel; the *FPS* is used from 47 to 70μm and the *FPL* is used from 70 to 196.6μm.

LWS has 10 detectors, the 5 *SW* detectors have a sensitivity range of about 10μm around the peak and the 5 *LW* detectors have a range of about twice that. The detectors are laid out in a staggered arrangement: *SW*1 *LW*1 *SW*2 *LW*2 *SW*3 *LW*3 *SW*4 *LW*4 *SW*5 *LW*5 The *LWS* detectors are extrinsic photoconductors:

- *SW*1 Ge:Be
- *SW*2-5, *LW*1 Ge:Ga
- *LW*2-5 stressed Ge:Ga

The *LWS* uses 5 Reference Illuminators to calibrate the detector response.

6.6.1. *LWS Observing Modes*

LWS has 4 *AOT*s. Rasters can be performed in any of them and the 10 detectors are always read out and present in the data, although for the modes using the Fabry-Perot only one detector is expected to produce usable data.

The *LWS AOT*s are:

- Grating range scan (*LWS*01)
  - up to the full wavelength range
  - option to sample at 1, 1/2 ,1/4 or 1/8 resolution element
- Grating measurement of lines (*LWS*02)
  - up to 10 lines, specified number of resolution elements around the line
  - an alternative narrow-band photometry mode also exists, in which the grating is not scanned, resulting in 10 data points
- *FP* range scan (*LWS*03)
  - both grating and *FP* are scanned
- *FP* measurement of lines (*LWS*04)

LWS also has a Serendipitously Parallel Mode in which limited data from the narrow-band photometry mode is extracted from housekeeping telemetry when other instruments are in use.

6.6.2. *Working with LWS data*

The *LWS* wavelength calibration is very good: ± 0.10 resolution elements for the grating and ± 40 km/s for the *FP*s. Improvements are still being made and ± 20 km/s accuracy is expected for *FP*s in the archive. The flux calibration is accurate to 10-30% for the grating and between 30% and a factor of two for the *FP*s. The calibration is less accurate for extended sources and is best for bright sources (i.e. $\sim 10^4$ Jy).

The effect of glitches on the *LWS* is more severe than was predicted and the achieved sensitivity is lower than expected. *LWS* is particularly badly affected because the detectors are large. Data which has been taken in a mode with little redundancy (i.e. with only a single up/down scan) cannot be adequately de-glitched.

There are fringes present in the spectra of extended or off-centre sources. The fringes appear at a fixed wavenumber of ~ 3.6-cm^{-1}. There are also spurious lines which sometimes appear in the spectra of sources which have very strong emission in the near-IR. These features are due to a leak in the blocking filters and result in broad emission features which appear in different wavelengths for different detectors and which are seen in physically adjacent detectors.

There is an uncertainty in the positioning of the grating of up to 0.25 resolution element. This is an uncertainty in the ability to predict the position of the grating, its actual position can be measured to somewhat higher accuracy. This uncertainty does not affect the data in grating modes, but does affect *FP* data, especially in the mixed-mode (L03).

The *LWS* beam profile is somewhat narrower than expected.

As with the other *ISO* instruments, the *LWS* detectors show drift effects which are tracked using the internal reference sources. They may also show persistent responsivity changes after severe glitches. They show transient effects, also called flux-history dependent responsivity changes, which are usually manifested as up and down scans which do not match.

When working with *LWS* data, it is important to remember that the pipeline products are calibrated assuming a point source. For extended sources, aperture corrections must be applied, and this is not perfectly straightforward because the point spread function is larger than the aperture at the long end of the *LWS* range. It is also the case that background subtraction is normally done using a separate "off" position, so users of *LWS* archival data will need to identify and retrieve the off source measurement, as well as the target spectrum.

The pipeline for *LWS* produces two main *AAR* files, the *LSAN* file and the *LSNR* file. Normally *ISAP* reduction proceeds from the *LSAN* file which has the following properties:

- responsivity and drift corrections have been applied
- dark current subtraction has been performed
- the units are W/cm^2/μm for the grating and W/cm^2 for the *FP*
- no averaging of scans has been done
- no "stitching" of adjacent detectors (offsets can be at the 10-20% level) has been done
  - no generation of maps is done for rasters
  - bad data points are flagged and assigned the value 0, but are not deleted

For faint sources, the *LSNR* may be needed. The *LSNR* file is the same as the *LSAN* except that illuminator corrections and dark current subtraction have not been performed. For very faint continuum sources, the dark current may be overestimated, causing negative values, which are problematic for the analysis software.

6.7. *A Word about LNIA and LIA*

Until recently *LWS* had no *IA* and the *LWS* Team at RAL re-reduced data on request using pipeline versions in development. This process was dubbed (somewhat tongue-in-cheek) the *LWS* Non-Interactive Analysis system, or LNIA. The standard procedure was to have the data re-processed and then proceed from the *AAR* in *ISAP*.

There is now an *LIA* package available with *ISAP* which can be used in the *ISAP* command-line environment to perform pipeline-like steps with some changes of parameter. It is available from the same places that *ISAP* is available. *LIA* uses the same version of the pipeline software as the official pipeline uses, so it is not exactly equivalent to reprocessing the data.

6.8. *Basic Steps in LWS data reduction with ISAP*

What follows is a summary of the steps typically performed when reducing *LWS* data from the *LSAN* file using *ISAP*. The shifting and averaging of adjacent detectors is a matter of personal preference – it can be argued that this process "hides" the uncertainty in the continuum level due to the detector drifts.

- Start the *ISAP GUI*
  - isap (For the command line version the command is normally "isap com")
- Read in the *AAR* data (*LSAN*)
  - buttons for all detectors, scans, scan directions and lines will appear
- Take a look at the data
  - start by turning on all the buttons and choosing "plot" to get an overview
  - to get a feel for the data structure it is useful to use the colour coding available in "plot options" and to plot all lines for one detector, then all detectors for one line, etc.
  - For L02, L04 you may want to determine the active detector (in "Special Functions" if the spectral lines are not obvious
- Discard the bad data points
  - in "Special Functions"
- Discard any remaining obvious outliers by selecting them with the right mouse button and choosing "zap"
  - Filter out residual glitches and average scans for each detector
  - plot some data to work with (1 detector or all)
  - check that up and down scans are consistent and can be combined
  - select the data to average by drawing a box with the right mouse button.
    — using the popup menus Average the scans
    — usually median, or medmin clipping is recommended. Medmin alternates between discarding the maximum and minimum outliers and is useful if there are some negative spikes in the data.
    — use "do not average across detectors" if more than 1 detector was selected
  - repeat for each detector unless you selected them all the first time
    — note that only the "detector" selection button remains. This is OK, *ISAP* will still be able to recognise multiple scans when averaging subsequent detectors.
- If working with a range scan, remove the detector-detector offsets if desired
  - plot all the detectors
  - select all the data
  - using the popup menu choose "Shift" then select a reference detector
    — *SW*3 and 4 are usually good choices (detector numbers 2 or 3)
- De-fringe if needed (for *LWS*01)
  - plotting all detectors for a given line should clarify if obvious fringes are present

- o "Special functions"
- Re-bin the spectrum if desired and average across detectors
- Now you have a "single" spectrum
- save the data

6.9. *The SWS*

The Short Wavelength Spectrometer (*SWS*) covers the wavelength range 2.38- 45.2μm, with moderate to high resolution: $\lambda/\Delta\lambda$ 1000-2000 for the grating and 20$\times$ higher for the Fabry-Perot which operates from 11.4-44.5μm.

SWS actually is comprised of two nearly-independent grating spectrometers:
- *SW*
  - o grating 100 lines/mm, first 4 orders, 2.4-13μm
- *LW*
  - o grating 30 lines/mm, first 2 orders, 11-45μm
  - o both *FPs* use *LW*, first 3 orders of grating are used
- *LW* and *SW* can be scanned independently and can take simultaneous data through a common aperture

SWS uses 3 apertures (14″×20″, 14″×27″ and 20″×33″) to cover the full wavelength range of the grating. The orientation of each aperture is fixed within the focal plane so its orientation on the sky is a function of spacecraft attitude. Both grating sections use all three apertures for some wavelengths, but a change of aperture requires re-pointing the satellite, so only one aperture can be used at a time. This restricts the combinations of lines that can be done with *LW* and *SW* simultaneously.

SWS had 6 detector arrays: there are two 1×12 linear arrays for each grating and 1 detector pair for each Fabry-Perot. When all the combinations of aperture, grating order and detector that are needed to cover the full range are taken into account, there are 12 combinations, known as *AOT* bands, for the grating and 5 for the Fabry-Perot. A summary of the *AOT* bands follows:
- *SW* Grating:
  - o Band 1 (ABDE) covering 2.38-4.08μm uses InSb
  - o Band 2 (ABC) covering 4.08-12μm uses Si:Ga
- *LW* Grating
  - o Band 3 (ACDE) covering 12-29μm uses Si:As *BIBIBs* (Back Illuminated Blocked Impurity Band Detectors)
  - o Band 4 covering 29-45.2μm uses Ge:Be
- Fabry-Perot
  - o Band 5 (ABCD) covering 11.4-26μm uses Si:Sb
  - o Band 6 covering 26-44.5μm uses Ge:Be

SWS makes use of internal calibration sources to calibrate drift effects for all bands.

6.9.1. SWS *Observing Modes*

SWS has four basic observing modes. Each of them incorporates dark current measurements and photometric checks. *SWS AOTs* can be used in staring mode only, and mapping can only be done by concatenating a series of fixed pointings. *SWS* is the only *ISO* instrument which has no parallel, or serendipitous mode.

The four *SWS AOTs* are:
- Grating Full Range Scan *SWS*01
  - o scans full *SWS* range
  - o choice of 4 "speeds" at 1/8,1/8,1/4,1/2 full resolution and corresponding to 13, 25, 50, 100 minutes integration time.

- o 1 up-down scan
- Grating Line Measurement *SWS02*
  - o up to 64 lines in each *AOT*
  - o multiple up-down scans
- Grating Partial Range Scan *SWS06*
  - o up to 64 wavelength ranges at full resolution
  - o multiple up-down scans
  - o possible reference scans of continuum
    - — this feature was removed in June 1997
- *FP* Line Measurement *SWS07*
  - o up to 64 *FP* lines
  - o 1 up scan
  - o *SW* behaves as for *SWS06*

6.9.2. *Working with* SWS *data*

In general *SWS* performance is good but, as with the other *ISO* instruments, there are a number of things to bear in mind when reducing the data. Glitches are significant and can leave long term transients. Detector signal jumps are occasionally seen, where an entire detector block experiences a responsivity increase that decays over a scan.

Detector transients are generally seen in *SWS* specifically in Bands 2 and 4. The detector ramps take some time to stabilise and up and down scans do not always match. In some cases, the profile of bright lines may be distorted and will be different in different scan directions. The reference scans which were an option in *SWS06* were discontinued because they often induced significant transients.

Band 3 with its *BIBIB* (Back Illuminated Blocked Impurity Band) detectors tends to be somewhat noisy, and the noise in Band 3 is sensitive to the temperature environment. In fact, it was noticed in mid-mission that Band 3 was often unusually noisy following *ISOCAM CVF* scans, which can cause some warming, and the scheduling policy was changed to avoid putting *SWS* observations soon after *CAM CVF* observations. Because *SWS* has relatively small apertures it is particularly sensitive to pointing errors. The improved pointing affected *SWS* significantly, and the work to make the improvements was partially triggered by the *SWS* team. A pointing error of about 6″ in the cross-dispersion direction corresponds to up to 40% flux loss and errors in the dispersion direction result in flux loss and some wavelength uncertainty, although an error of about 4″ is still within the accuracy of the calibration. *SWS* is also somewhat sensitive to jitter in the pointing. The beam profile is quite steep in the dispersion direction and jitter may increase the apparent noise if the source is already near the edge of the beam. Jitter in crowded fields can also increase the apparent noise.

SWS has fringing due to interference in the blocking filters. Band 3 is the most severely affected by the fringing. The fringes shift somewhat with the position of the source in the aperture, and although the fringes are taken into account in the relative spectral response function, they may be out of alignment with the actual fringes, leading to residual fringes in the reduced data.

There are also some spurious features in the relative spectral response function which was measured in the laboratory before launch. The features appear at 9.35, 10.05 and 11.05μm and are probably due to transients during the measurements. There is additionally a small (10%) 13μm leak appearing at 27μm,

The flux calibration for *SWS* is good, about 5-30% for the grating in the absence of significant transient effects. The wavelength calibration is very good, $\lambda/5000$- $\lambda/12000$ for the *SW* section and $\lambda/8000$- $\lambda/16000$ for the *LW* section. In fact, *SWS* measurements have been used to improve the literature accuracy of the determination of rest wavelength of some lines (Feuchtgruber *et al.* (1997)).

SWS makes use of significant redundancy in the sampling. Each of the 12 detectors in each array covers about 1/3 of a resolution element. and the grating steps are about 1 resolution element each, so that typically 2-3 detectors see each wavelength during each scan.

When reducing *SWS* data, the *AOT* Bands should in general be processed independently as they use different apertures, grating orders, etc. They can be joined near the end of the reduction, if desired.

6.9.3. *A Word About SIA*

The *SWS* Interactive Analysis Package (*SIA*) is currently available only at *ISO* Data Centres. It is distributed on approval, to experienced users. This policy is not expected to change when the *ISO* Archive is available.

SIA can perform the same functions as the pipeline, but some parameters can be tweaked. The data quality obtained by working from *AAR* in *ISAP* compared to working from *ERD* or *SRD* in *SIA* is not quite as good, but comparable.

6.9.4. *Basic Steps in Reducing SWS Data with ISAP*

The following is a cookbook similar to the cookbook for *LWS*, which outlines the typical steps used in reducing and combining *SWS* data in *ISAP*.

- Start the *ISAP GUI*
- Read in the *AAR* data (*SWAA*)
  - buttons for all detectors, scans, scan directions and lines appear
- For *SWS*01 convert the line tags to *AOT* Band tags using the option found in "Special Functions"
- Take a look at the data, using the plot options to investigate various combinations of *AOT* band, detector, scan, line etc.
- For each *AOT* Band or Line:
  - plot the band minus scan 0, scan direction 0 if present.
  - select the data by drawing a box with the right mouse button, which will cause a new window of options to pop up.
  - From the new window choose "Make_*AAR*" to turn the selected subset of data into a new file which has valid detector tags etc.
  - Edit bad data points by hand by selecting them with the right mouse button and using "zap".
    — note that real lines should appear in more than 1 detector and in both scan directions
  - average the up-down scans and re-bin the data
    — first check for transient effects
    — select the data with the right mouse button
    — choose "Average"
    — on the first pass just average the scans – don't average the detectors
    — median clipping or min-med clipping are the suggested filtering modes.
    — choose a bin size near the resolution
- Still working within a single *AOT* Band, align the 12 detectors
  - select data
  - use "shift" to bring detectors to a common level.
- average the 12 detectors within the *AOT* band, using "Average" with the average across detectors option.

- repeat the process for each AOT band and each line in the data.
- If desired, merge the AOT-band data sets and combine them using "shift" and "average".
- Smooth data if desired

6.10. *ISO Data and Services on the Internet*

In addition to the distribution of Ias, as detailed in the relevant sections above, a great deal of information and some *ISO* data are available on the WWW. Some sites of interest and the services provided are:

- The *ISO* Home page: `http://isowww.estec.esa.nl/`
  - News
  - Manuals and FAQs
  - Pointers to ftp sites and www sites where *ISO* data are available or will soon be available from the solicited proposals
  - Eventual site of the *ISO* Post-Mission Archive
  - Other documents pertaining to *ISO* and its instruments
  - Information about meetings relevant to *ISO*
  - Log of observations and information about quality control and shipment of data
  - Information about discretionary time proposals
  - Press releases and published papers
- *ISOCAM*: `http://iscam1.ias.france`
  - Documentation and information about the Data Centre (in French)
- *CIA* home page: `http://sapwww.saclay.cea.fr/www/isoms/cia_home.html`
- *ISOPHOT* Data Centre: `/http://www.mia-hd.mpg.de/M` *PIA*/`Projects/`*ISO*
  - *ISOPHOT* news
  - *ISOPHOT* documentation
  - *PIA* Information
  - information about the Serendipity Survey and other projects
- `http://ast.star.rl.ac.uk/isouk/isouk.html`: The United Kingdom *ISO-IRAS* Support Homepage
- *SWS*: `http://www.sron.rug.nl/iso/`
- The *ISO* Spectrometer Data Centre:
`http://www.mpe-garching.mpg.de/iso/isosdc.html`
  - Information about the Data Centre
  - Preprints and Press releases
  - *ISAP* access and documentation
- US *IPAC ISO* Science Support Center: `http://www.ipac.caltech.edu/iso/iso.html`
  - FirstLook Tools used to convert the *ISO FITS* binary files for basic satellite data files and the pipeline end-products into easier formats for a quick-look.
  - *ISAP* access and documentation
  - Information about the Science Support Center
  - Notes and Tips on using *ISO* Data
  - List of *ISO* papers published in the major astronomical journals
  - *IRSKY* observation planning/sky visualization tool
- LIRA, *ISO* Support in Japan: `http://koala.astro.isas.ac.jp/iso/iso.html`
  - Data Reduction guides for all instruments in Japanese
- The European Large Area Survey (*ELIAS*): `http://artemis.ph.ic.ac.uk`
  - Detailed information about one large *ISO* project
- *ISO* Hubble Deep Field: `http://artemis.ph.ic.ac.uk`
  - Raw and Reduced *ISOCAM* data of the Hubble Deep Field

o Information about data reduction

o documentation and publications

7. Coming Attractions

The following is a short summary of some infrared projects in progress or on the horizon, all of which are expected to result in significant archival data sets. They are presented roughly in chronological order.

7.1. *2MASS*

2MASS (The Two Micron All Sky Survey) is a new all-sky survey using two matched 1.3-m telescopes; one at Mount Hopkins and the other at *CTIO*. The survey cameras use 256×256 infrared arrays with 2″ pixels sensitive at the J, H and K bands. *2MASS* is expected to be ∼50 000 times more sensitive than earlier surveys. The telescope at *CTIO* saw first light 6 December 1997. Data products should begin to be available around 1999 and the survey is expected to be complete in 2002.

2MASS information can be found at:

- http://pegasus.phast.umass.edu/GradProg/2mass.html
- http://www.ipa.caltech.edu/2masss/

7.2. DENIS

DENIS (the Deep Near Infrared Survey of the Southern Sky) is a deep survey in the southern hemisphere at I, J andK. *DENIS* is on-going.

DENIS information is available from:

- The Leiden Data Analysis Centre: http://www.strw.LeidenUniv.nl/denis/
- The Paris Data Analysis Centre: http://denisexg.obspm.fr/

7.3. WIRE

WIRE (The Wide-field Infrared Explorer) is a *NASA* small explorer mission, due to launch in March 1999†. *WIRE* is designed to answer questions about the moderate to high redshift universe through a deep 12 and 25μm survey. *WIRE* will be a 30-cm telescope with 128×128 Si:As IR arrays with 15.5″ pixels.

More information about *WIRE* can be found at:

- The *WIRE* Home Page: http://www.ipac.caltech.edu/wire/

7.4. SOFIA

SOFIA is the Stratospheric Observatory for Infrared Astronomy. *SOFIA* will be a modified Boeing 747 carrying at 2.5-m telescope and is the successor to the Kuiper Airborne Observatory. *SOFIA* is being developed jointly by *NASA* and the German Space Agency (*DLR*). *SOFIA* operations are expected to begin in 2001 and to continue for ∼20 years.

SOFIA information is available at:

- The *SOFIA* Homepage: http://sofia.arc.nasa.gov/

7.5. SIRTF

SIRTF (The Space Infrared Telescope Facility) is a *NASA* Great Observatory expected to launch in 2001 with a 2.5 year lifetime. *SIRTF* will be placed into solar orbit trailing the earth. The telescope will not be significantly larger than *IRAS* or *ISO* – 85-cm –

† Noted added in proof: The launch ultimately failed due to a severe hydrogen leak which drained the satellite's coolant before the leak could be brought under control. The satellite's mission was completely lost.

but *SIRTF* will carry large-format infrared arrays. *SIRTF* will perform imaging and spectroscopy in the 3-180μm wavelength range using three instruments:

   (*a*) The Infrared Array Camera (*IRAC*)

   (*b*) The Infrared Spectrograph (*IRS*)

   (*c*) The Multi-band Imaging Photometer for *SIRTF* (MIPS)

   IRAC uses 256×256 pixel arrays to image a 5′ field of view at 3.5, 4.5, 6.3 and 8μm. *IRS* will perform low-moderate resolution spectroscopy in the 5-40 μm regime using 128×128 pixel arrays. MIPS will image at 12,30,70 and 160μm and perform low resolution spectroscopy from 50-100μm.

   More *SIRTF* information is available from:

- The *SIRTF* Home Page: `http://sirtf.jpl.nasa.gov/sirtf/home.html`
- The *IRAC* Home Page:
`http://cfa-www.harvard.edu/cfa/oir/Research/irac/firstpage.html`
- The *IRS* Home Page: `http://astrosun.tn.cornell.edu/SIRTF/irshome.htm`

7.6. *FIRST*

FIRST (Far InfraRed and Submillimetre Telescope is a European Space Agency (*ESA*) Cornerstone mission. *FIRST* will carry a 3m telescope and instrumentation to perform photometry and spectroscopy in the 85-600μm range. *FIRST* is expected to launch ∼2006 with at 4.5 year lifetime.

   More information about *FIRST* can be found at:

- `http://astro.estec.esa.nl/SA-general/Projects/First/first.html`

8. Summary

   Although only a limited amount of infrared data is currently available on the Internet, primarily the *IRAS* and *COBE* data sets, plus a small amount of *ISO* data, increasing amounts of data will be going on-line. In particular, the *ISO* archive will first become available in the summer-fall of 1998 and will be the first observatory style infrared archive. Because infrared backgrounds are complex, particularly at ∼100μm, and infrared detectors are particularly vulnerable to radiation hits and also tend to exhibit transient responses and long term drifts, using infrared data sets wisely requires a modicum of knowledge about the techniques and pitfalls of infrared astronomy. These have been examined in some detail for the *IRAS* and *ISO* data sets.

REFERENCES

H.H. Aumann, J.W. Fowler and M. Melnyk, 1990, AJ, 99, 1674

"*IRAS* Catalogues and Atlases: Explanatory Supplement", 1988, ed. C.A. Beichman, G. Neugebauer, H.J. Habing, P.E. Clegg, and T.J. Chester (Washington, DC: GPO).

J.P. Bernard, F. Boulanger, F.X. Desert, M. Giard, G. Helou & J.L. Puget, 1994, A&A, 291, L5

S.Bogun *et al.* , 1996, A&A, 315, L71

C.J. Cesarsky *et al.* , 1996, A&A, 315, L33

P.E.Clegg *et al.* , 1996, A&A,315,L39

W.D. Cotton, D. Tody & W.D. Pence, 1995, A&AS, 113,159

Th. de Graauw *et al.* , 1996, A&A, 315, L49

Feuchtgruber *et al.* ., 1997, ApJ, 487,962

T. N. Gautier, F. Boulanger, M. Perault & J. L. Puget, 1992, AJ, 103, 1313

G. Helou and C. Beichman, 1990, Proc 29, Liege Intn. Astr. Coll., *ESA* SP 314, also available from ftp.ipac.caltech.edu in /pub/irsky/doc

M.F.Kessler *et al.* , 1996, A&A, 315, L27

D Lemke *et al.* , 1996, A&A, 315,L64

F.J. Low *et al.* , 1984, ApJL, 278,L19

W. Reach *et al.* , 1996, A&A,315,L381

W. Rice, et. al, 1988, ApJ Suppl,68,91

W. Rice, 1993, AJ,105,69

G. H. Rieke, 1994, "Detection of Light from the Ultraviolet to the Submillimeter", (Cambridge University Press)

R. Siebenmorgan *et al.* , 1996, A&A, 315, L169

J-L Starck, F. Murtagh, B. Pirenne & M. Albrecht, 1996, PASP, 108. 446

S. Wheelock, et. al., (1993), "Explanatory Supplement to the *IRAS* Sky Survey Atlas", (Pasadena: JPL)

Astronomical Archives In The Optical And In The Ultraviolet

By RUDOLF ALBRECHT & PIERO BENVENUTI[1]

Space Telescope European Coordinating Facility, European Southern Observatory, D-86748
Garching, Germany†

This paper explores the requirements and the implementation considerations for astronomical science data archives in the optical and in the *UV* wavelength range. Examples of space and ground astronomy facilities are described. An attempt is made to project the current developments into the future.

1. Introduction

The science of astronomy has a long tradition of keeping archives of observations. This has been very fortunate, as we have been able to use the description of astronomical observations, like, for instance, the solar eclipses described by Chinese astronomers, in the dating of historical events. Kepler's laws of the motions of the planets are an early example of archival research: Kepler used observations performed by Tycho de Brahe to derive the laws.

At the beginning of the twentieth century, astronomical archives consisted mainly of plate archives. Photographic plates have the advantage that the detector and the storage medium is one and the same. The advantage is that storage is very efficient and compact, the disadvantage is that the act of storing the data will alter them. Astronomical plates are susceptible to all kinds of environmental influences which will ultimately lead to the destruction of the data.

Electronic archives were first produced by the radio astronomers. In radio astronomy there is a wide gap between the output of the detector and the scientifically meaningful rendition of the data. This gap has to be filled by software, and it has always been clear that keeping the original data was an advantage because it was possible to re-analyse them with improved calibration software.

The International Ultraviolet Explorer (*IUE*) satellite pioneered large scale archiving of space astronomy data. Originally conceived as a safeguard against accidental loss of the data, the *IUE* archive quickly became a research tool in its own right. Based on this example, other space astronomy projects such as the Hubble Space Telescope (*HST*) were conceived with the science data archive as an integral part of the project.

In the case of the *HST*, there was another important consideration which mandated an archive. The total cost of the project was considerable, the incremental cost of the archive is negligible in comparison, but the efficiency of the facility is increased enormously, thus making the project more cost-effective. Similar considerations apply to big-science ground based facilities like the European Southern Observatory (*ESO*) Very Large Telescope (*VLT*). As a consequence the *VLT* science data archive is being developed.

There are also very practical aspects which mandate the establishment of a science data archive. Archive research can do more than the Telescope Time Allocation Committee will allow: more orbits, longer time lines, more data, synoptic work, and the

† Affiliated to the Astrophysics Division, Space Science Department, European Space Agency

production of catalogues. Moreover, some events simply cannot be observed again, for example SN1987A. Archives are essential to monitor performance, trends, and the calibration history of the science instruments. A very practical reason is that operational experience with the instruments leads to re-calibration and re-processing of the data. Finally, some instruments, like the Faint Object Spectrograph (*FOS*) and the Goddard High Resolution Spectrograph (*GHRS*) of the *HST* simply do not exist any longer. However, data obtained with those instruments are still accessible through the *HST* science data archive.

While the data are being stored in the science data archives in near real time, they are not made available to the community right away: the original Principal Investigators (*PIs*) of the observing programs get exclusive rights to the data for a certain period of time, usually one year. After this proprietary period the data are made available and other scientists can use them for their research.

Archive data are also being used to monitor the performance of the instruments and to develop or improve calibration procedures. For this reason access is granted to proprietary data to the instrument scientists. However, it is being understood that the data will not be used for independent research.

2. Archive Evolution

There was a subtle evolution from the "plate vault" to the electronic "active" archive. When digital data (digicon, reticon, *PCS*, *CCD*, ..) began to be collected, they were kept only by the observer. This may still be true today in many observatories, but the situation is changing. A side effect was that gradually observations were not recorded any more and even the "Log Book" (which served to catalogue the plates taken) was abandoned.

A change was triggered by space experiments (such as *TD1* and *Copernicus*), in which the recording of observations was "imposed" by the strict procedures of the spacecraft operations. But the concept of "science archive" was not initially there: when *IUE* was first launched the "archive" was thought of as a collection of magnetic tapes, essentially the output of the daily operation.

The significant difference which turns a collection of data, into an archive, is the uniformity of the data processing which allows to compare or combine data sets with each other and with data obtained with other instruments. In the case of *IUE* this has led to products like the Uniform Low Dispersion Archive (*ULDA*), which will be described later.

3. Digital Data Volumes

Before 1970, photomultiplier tubes produced several readings per second. Storing these data was no problem.

Beginning circa 1970, 1-D scanning techniques (area scanner, spectrum scanner) produced kilohertz data rates. Storage media at the time were punched paper tape, which was later replaced by magnetic tape. At this point, storage problems started. At the same time plate scanning machines like the *PDS*- 1000 produced high data rates and high data volumes, but the data were generally not kept. Plates were actually re-scanned, if required.

In the mid-seventies, vidicons, the *IPCS*, and early *CCD*s typically generated 256×256 integer pixels and produced several K words every few seconds. Ancillary data became important (darks, flats). Radio astronomy produced even higher rates.

In the early eighties, space missions produced high quality data at a rate faster than can be properly utilised, or even properly calibrated. Detectors grew to 1K pixels square, the data volumes grew to Mbytes per day. Archives became essential at that point. The policy of proprietary versus public data was widely implemented, driven by *HST*.

Circa 1990: Hubble Space Telescope launch. 1-2 Gbytes per day were generated initially and routinely processed and archived. Archive research programs are evaluated by the *HST* Time Allocation Committee (*TAC*) and funded just like observing proposals.

At the present time, detectors are at 4K by 4K, and growing. Data rates are at 5 Gbytes per day per facility. Soon panoramic detectors will reach the physical size of photographic plates. *VLT* type facilities will produce of the order of 100 Gbytes of data per night.

4. What is an Archive: Archive vs. Catalogue

4.1. *The past*

Before the introduction of *CCD*s, an archive in the optical domain was an ordered collection of photographic plates (images, or spectra) which was safely kept and retrievable with reasonable effort. The normal way to keep order in such an Archive was to enforce a (hand-written) Log of the observations, which registered the most important details, such as object name, *RA* and *DEC*, filter, instrument setting, type of emulsion, sky condition, etc., plus a unique ID number of the plate.

A catalogue was (and still is) a free-standing database, normally concerning a class of object, which is the result of a (generally) uniform data extraction, calibration, or analysis process. Usually, a catalogue is the result of a lengthy research programme (typically a publication for the A&A Supplement). It is clear that, under these concepts, no one would have tried to re-do the work, i.e. to access the same observations (extracting them from the archive) and to re-extract the data (see below for the difference).

The Centre des Donnees Stellaires de Strasbourg (*CDS*) has played a key role in the early controlling, handling and distributing a wealth of these classical catalogues, first limited to stellar data, later including extended objects. There was a clear, even physical, separation between the observations (which themselves might have been kept in an archive) and the catalogue: only the latter was intended to be accessible and distributed.

A notable exception is represented by the photographic surveys (Palomar, *ESO*, *SRC*), where the archive itself was carefully duplicated by photographic means and distributed to the subscribers.

4.2. *The present*

Today, both archives (i.e. the raw data) and catalogues (i.e. the processed data) are intended to be easily accessible by the astronomers. In this new context, a catalogue can be either the classical catalogue, or the database which describes the content of an archive. The introduction of the concept of the relational database has further extended the scope of this new concept of a catalogue, by joining several databases.

Because the data are now as easily accessible as the catalogues, it is quite possible that different users analyse the same set of observations with different methods and produce different catalogues (e.g. extracting and characterising objects from the Hubble Deep Field using different algorithms, like *DAOPHOT*, *SExtractor*, etc.).

In order to make the transition from the past, to the present situation, several steps had to be undertaken: we believe that *IUE* played a key historical role in this transition, as will be seen.

5. Archive Requirements

5.1. *User interface*

When users talk about the features of a science data archive, they usually refer to the part of the system which is visible to them: the user interface. Indeed, the user interface is probably the most important aspect of an archive. Users are more willing to work with a somewhat slower system as long as the look-and-feel is right and they get the impression that the system works for them. On the contrary, the most powerful system will not be accepted by the users if the user interface is counterintuitive, or requires too much learning before meaningful results can be achieved, such as using *SQL* (Standard Query Language, a system independent database access and programming language) as a user interface.

The current state of the art for user interfaces are Web-based interfaces, implemented as either as server-side scripts, or as client-side applets, or as a combination of both (Rasmussen (1995)). Several years ago there was an intensive discussion raging within astronomy, with different groups engaged in "religious wars" about which user interface to standardise around. This has been rendered obsolete by the Web, which offers a standard paradigm, programmability, graphics, platform independence and community-wide availability, without the need for off-site support. There is a lesson in this, and we should heed it: move with the market and invent your own solution only if absolutely necessary.

5.2. *System requirements*

Science data archives are usually implemented using hierarchical database management systems. For large archives like the *HST*, or *VLT* archives, we consider it cost-effective to go with commercial products, even though their price is considerable. However, given the savings in terms of development time, debugging, maintenance, upgrades, etc., we come out ahead. We want to build a science data archive and we do not want to develop database management software. And a price of even 10^5 is low when seen in the context of a 10^9 project.

This is not to say that there is no need to develop additional system components. Data ingest, quality control, backups, user request handling, operator interface, status reporting, exception condition processing, etc. require a significant amount of software development.

An important aspect is the control of proprietary data. It goes without saying that no mistakes must be committed and strict control and security measures must be put in place. At the same time easy access must be possible for the *PI*s and for the instrument scientists. To handle all these functions in a smooth and transparent manner requires powerful software and well-developed operating procedures.

5.3. *Data formats*

An example of a successful one-of-a-kind effort in astronomy is the *FITS* format. Originally intended as an interchange format for multi-dimensional data sets on half-inch magnetic tapes, *FITS* has successfully transitioned to other media and to ambitious applications. It is evident that the data distribution format of any science data archive has to be, or at least has to include as one of the options, the *FITS* format.

However, this does not necessarily mean that *FITS* has to be used as the internal data format of the archive. For a variety of reasons an alternative format might be more suitable, or more cost effective. If multiple output formats are being supported, the time and resource requirements to re-format the data have to be taken into account.

An exciting possibility is to store the data in a very compact, even compressed form, even to distribute them to the user in this form, and to us client-side software (such as applets) to expand the data.

5.4. *Bulk storage*

Bulk storage is probably the most problematic aspect of science data archives. The oldest bulk storage media for astronomical archives, stone tablets, papyrus scrolls and, much more recently, books, might not have the storage capacity of modern media, but they are certainly durable, with demonstrated shelf lives in the range of millennia. They can be read without special hardware, albeit requiring some knowledge of the sometimes anachronistic encryption languages.

Even photographic plates last for a very long time. Some plates in the archives of the major observatories date back to the middle of the last century. With a minimum of care it should be possible to store them without loss of information for another century.

This situation reversed when archive data were stored on magnetic tape. As a minimum, magnetic tapes had to be re-written periodically because of the fact that the magnetically encoded information on the tape deteriorates because of stochastic processes on the magnetised layer of the medium. However, the relative stability which was present during the sixties, when 800 bpi magnetic tapes were the industry standard, vanished when, in ever faster progression, the standard moved to 1600 bpi, to 6250, to helically encoded tapes, to optical disks, to CD-ROMS, and to magneto-optical media. Some of these bulk storage media have projected shelf lives of decades and maybe even centuries, but it turns out that their useful life time is limited by the availability, or, more correctly, the maintainability, of the reader hardware: first generation optical disks are still good, but it is not possible to read them any longer because technology has moved on and the original hardware is no longer available. Keeping such hardware alive is impossible because of the unavailability of qualified technicians and the lack of spare parts. It also does not make much sense because new generation instruments and detectors require the most recent generation of bulk storage media.

We thus find ourselves in the strange situation in which the life time of the archive bulk storage media is not a multiple of the life time of the project, but in which, on the contrary, the life time of the bulk storage media is an ever smaller fraction of the intended life time of the archive. Coming to grips with this is a considerable challenge.

6. Software Engineering

Scientists often assume that, just because they write programs in support of their personal research, they know all about software engineering. This is about as correct as assuming that somebody can run a restaurant just because that person is able to fix scrambled eggs. The situation is similar to what used to be the case in electronics: about 25 years ago astronomers designed and built their own photoelectric photometers, amplifiers, counters, etc. This is not the case any longer. It has become obvious that electronics design is best be done by professional electronics engineers, and that the most cost-effective solution is to buy off-the-shelf products whenever possible.

Most research software is one-of-a-kind, poorly documented, works only in severely limited conditions, for a limited variety of data, and only with the original author around. Such software is usually not portable between different hardware platforms, or different operating system. It fails in uncontrolled manners, and maintenance and upgrades are impossible.

It is clear that this cannot be tolerated for production software of the type required to operate a science data archive, which has to guarantee a high degree of service and which has to be maintained even in the absence of the original author. Production software has to be of a clean design and a consistent implementation, which follows clearly established standards. The documentation has to be thorough and uniform, and of sufficient depth to allow the analysis of exception conditions. If the software fails, it has to do so gracefully and not catastrophically. To the maximum extent possible, the software has to be transportable: it is the very nature of an archive to have to survive for a long time; this implies that changes in technology which occur on a time scale of between 2 and five years have to be expected. In order not to have to re-develop basic archive capabilities on the same time scale, it is important that the software survives several technology upgrades, without requiring fundamental changes.

An important element which is quite often ignored is configuration control. This is the process of proposing, authorising, implementing, documenting and testing changes to the software, plus the timely notification of the users of what has been changed and how. Countless person-hours have been wasted on improperly implemented changes, when software developers modified the software, quite often with the noble intent to make improvements and to correct problems, but failed to communicate this to the users properly. It is most irritating when software performs differently today from the way it performed yesterday, and occurrences like this have contributed to many misunderstandings.

Problems of the kind described above multiply if the software has to be distributed to other installations, in different geographic locations. Off- site maintenance of software requires dedicated and sustained effort, which in turn requires the appropriate resources. Even more problematic is a situation in which some software elements are supplied by an industry software contractor. Scientists are often insufficiently aware of the fact that both the goals and the methods of industry are different from the goals and methods of science. It is absolutely essential that industry contracts be handled by persons with a sound professional software engineering background.

It is one of the biggest challenges for a project manager to produce operations, or production software, in a research and development environment. Done properly and skilfully, it is possible to use the two approaches for mutual benefit and to develop reliable software in a cost effective manner; done carelessly it leads to unstable systems, which either fail, or which have to be retrofitted at large expense.

7. The Pioneering role of IUE

The International Ultraviolet Explorer (*IUE*) spacecraft was the first real observatory-type mission. It allowed a large number of astronomers to access the ultraviolet spectral range in an easy and efficient manner. It goes without saying that the mission taught us many lessons on how to operate such facilities, and also on what has to be avoided.

IUE was a collaboration between *NASA* (US), *ESA* (Europe) and *SRC* (later *SERC*, now *PPARC*, UK). *NASA* provided the launch vehicle (a Delta launcher), *ESA* built the spacecraft and *SRC* contributed the science instruments. This 3-agency co-operation resulted in two-thirds of the observing time for *NASA/GSFC*, one sixth for *ESA*, and one sixth for the UK. Ground stations were established at *GSFC*, near Washington, in the USA, and at the *ESA* satellite tracking station in Villafranca del Castillo, near Madrid, in Spain.

IUE was designed to do spectroscopy from an oscillating geostationary orbit (i.e. minor interference from the Earth, continuous telemetry link, flexible scheduling). The telescope had an aperture of 45 cm, and was a f/15 Ritchey-Chretien. The primary mirror was made of beryllium, the secondary of fused silicirca. In all, 80% of the light was collected within 1 arcsec. The field of view was 16 arcmin. The spacecraft featured a 43 degree cutaway hood for Sun avoidance.

The science instruments consisted of spectrographs with two similar paths, one (short wave, *SW*) for 1150 - 2000Å, the other (*LW*) for 1800 - 3300Å. Each path had two modes: high dispersion (1Å/mm)/high resolution (0.1Å) and low dispersion (60Å/mm)/low resolution (6Å). Each mode had two slits: one circular (3 arcsec diameter), one long (10×20 arcsec). The detectors were a *UV* converter coupled by fibre optics to an SEC TV tube. There were two cameras per path, hence *SWP* (P for primary), *SWR* (R for redundant), and *LWP*, *LWR*. *SWR* failed during initial in-flight checkout. *LWP* had initial problems with its scan control logic, therefore *LWR* was used as prime. The scan problem was solved by a flight software fix and when *LWR* developed a bright extended spot, *LWP* was used as prime up to the end of the mission. Field acquisition and guiding was done with an image dissector (*FES*) with multimode operation (imaging and guiding).

When *IUE* was first launched in February of 1978 the projected life time was 3 years with a possible extension to 5 years. It was actually operated for 18.5 years, ending on Sep 30, 1996, at 18:44 UT. It had accumulated 104,470 spectra (the historic last images are *SWP* 58388 and *LWP* 32696).

Initially, the archive was intended for the safekeeping of the data. An important change, however, was in place: the routine processing of the data. The output was not only the "raw data" but also the different steps of the processing (photo-geometric corrected image, rotated sections of the image, "gross", "background", "net" and "calibrated" spectra). The "header" was supposed to contain all details, both on the observation and on the processing.

This intention was good, but the implementation was poor: the header assumed the existence of an "electronic" input file (*RA*, *DEC*, etc.) for each observation, which was supposed to be the output of a sophisticated "automatic scheduling programme", which, although started was never completed or used as such due to a variety of circumstances.

The observations were scheduled almost in real time. A new, simpler but essential, observing log was developed during the early months of operation and manually filled in by the Telescope Operator in at the time of the observations. It was around that time that the "*IUE* astrophysical classification two-digit code" was invented.

Not surprisingly, the calibration S/W (and even the calibration concept) was not working as expected and underwent numerous changes (minor and major, see the history of *IUESIPS* in the *IUE* Newsletters). After about three years of operations it became clear that:

(*a*) *IUE* could continue to operate for much longer,

(*b*) the calibration was reasonably understood and stable, i.e. the output was very uniform.

On that basis, the concept of the *IUE ULDA* (Uniform Low Dispersion Archive) was developed: re-process all the data with the latest pipeline software and offer the product to the community. It was very fortunate that the size of the extracted Low Dispersion database matched the storage capability of the time.

The concept of an archive of accessible data as a science tool was developed, the first Atlases of stellar spectra were produced, followed later by others (Planetary Nebulae, SNe, etc.). After few years, variability studies became important: Seyferts, cataclysmic variables, etc. Again, the availability of a uniform, easy to use, archive made these studies possible.

After 18.5 years of operations, the individual observation programmes have lost their identity (and their primary scientific goal, perhaps...): almost the totality of the papers presented in Seville, at the Conference on "Ultraviolet Astrophysics beyond the *IUE*

Final Archive" (Nov. 1997), were about classes of objects, or phenomena, as derived by a global analysis of the archive data.

However, the high dispersion equivalent of the *ULDA* has yet to be produced. In addition to the resource requirements, there are other obstacles like the fact that the knowledge of the instruments and their development is rapidly vanishing.

In spite of these shortcomings, the *IUE* science data archive has been quite successful. Indeed, it might be the case that the most ambitious investigations, based on *IUE* data, will be done years after *IUE* was decommissioned. It must be stressed that this is possible because:

(*a*) calibration was controlled,
(*b*) the pipeline was controlled,
(*c*) information was attached to data,
(*d*) data are easily accessible.

This is the legacy of *IUE* to future projects: if any of the above elements is missing, one cannot build an Archive.

8. Data Analysis Systems

One of the prerequisites for the efficient calibration and analysis of astronomical data is a good data analysis host system. The system must be interactive, but also support the data calibration pipeline.

Data analysis host systems allow the astronomer to interact with the data in a convenient manner, and to be able to access all components of the system in a consistent and coherent way. Ideally, the data analysis host system should run on different platforms, allow interactive and batch (script) processing, and have a large user community which contributes analysis software to the system. Examples of such systems are *AIPS* (Astronomical Image Processing System produced by *NRAO*), *MIDAS* (Munich Image Data Analysis System produced by *ESO*) and *IRAF* (Interactive Reduction and Analysis Facility produced by *NOAO*).

Some of the problems which *IUE* encountered had to do with the fact that such systems were either not available or in their infancy, and by the fact that astronomers were insufficiently aware of the operational importance of such systems.

The first image processing "system" was *VICAR* (Video Image Capture And Reduction) produced by *JPL*. It consisted of a collection of "main programs" which followed consistent rules and conventions, and which were able to efficiently access image data. The main motivation, however, was the idiosyncratic job control language of the IBM 360 series of computers, which made it necessary to use software to convert commands which an astronomer could understand to IBM JCL. When minicomputers like the PDP-8 and the PDP-11 came along an interactive version of *VICAR* was developed, but most of the application software could not be ported because it was written specifically for the IBM 360.

IUESIPS (*IUE* Spectral Image Processing System) was initially based on interactive *VICAR*, but it never became widely used due to porting limitations. Instead, other systems, like *IHAP* (Image Handling And Processing, an early product of *ESO*) were employed. *IHAP* was totally non- portable and required a clone of the hardware it ran on at *ESO*. It was, however, quite successful because of the fact that it was available on La Silla, and it was used at *ESO* Garching by almost all European astronomers who obtained spectral data with *ESO* telescopes.

On the American side *IUE* gave rise to *IDL* (Interactive Data Language). *IDL* has since become a successful commercial product. It is still being used for spectral data analysis, utilising the large analysis software libraries which have been collected at *GSFC* and *IPAC*. *IDL* has been further developed to handle *HST* data and an enormous quantity of analysis software has been produced. More recently *IDL* has had considerable commercial success in non-astronomical applications, such as medical imaging. This will ensure the survival of *IDL*.

Before the advent of personal work stations and networks, image processing systems were run on large mainframe computers. Image processing stations required special display hardware. Because of the need to tune system parameters to support a particular system, it was usually not possible to support more than one such system. Another reason why the attempt to support multiple systems quite often failed was the fact that astronomers were insufficiently aware of what is described in the section on software engineering. The notable exception to this was *AIPS*, which as the first system used professional software engineering procedures to produce and distribute the software.

It was during that time (the early '80s) that holy wars were fought over which system should be the standard. The group which had an early advantage, because of central management and the realisation that computer networks were required for efficient operations, was *STARLINK* in the UK. However, they gave up their advantage when they failed to deliver one product and then improving it, but rather started the improvement cycle before anything useful was produced. *STARLINK* has been quite successful in providing data analysis infrastructure for UK astronomy, but it did not succeed to the maximum extent possible.

In 1985, the *STScI* made the decision to use *IRAF* for *HST* data analysis. *ScI* had been funded by *NASA* to produce *SDAS* (Science Data Analysis Software), a collection of data analysis routines which were supposed to run on a data analysis host system provided by an industry contractor. This never happened, so this decision to go *IRAF* was dictated by circumstances, but it was painful for Europe: the *ST-ECF* (Space Telescope European Coordinating Facility), which had been established by *ESA* and *ESO*, and which is using the *ESO* infrastructure, had worked with *ESO* to accommodate the original *SDAS* in *MIDAS*. This suddenly became impossible and we had considerable problems explaining this to our users.

Today, it is technically easy to support more than one data analysis system. The limitations have to do with the level of resources which is required to do this properly. It is good policy to support a maximum of three systems. A good combination is *IRAF*, which today offers the largest body of astronomical data analysis software, and which allows *HST* data to be analysed and calibrated; *IDL*, which is a very mature commercial product, with a significant amount of astronomical data analysis software, and which is very powerful and easy to use; and a local system which supports local instrumentation, or the local wavelength range (radio, or high energy).

There is no one best system. In selecting a system you must define your requirements. If you work with *IUE* data go for *IDL*. If you work with *HST* data go for *IRAF*. **DO NOT** succumb to the temptation to develop your own system. Don't be afraid of commercial products. They are cost effective: the cost for an educational *IDL* license is a fraction of the annual salary of a software engineer, but there are much larger potential savings.

In addition to the technical requirements for data analysis host systems, there is another aspect which is quite often overlooked, but which is very important. To develop software within the framework of a widely used and supported data analysis host system, will make sure that the software will survive changes of hardware and software technology. One example is the move towards Linux which is happening right now. With the introduction of the Pentium chip PCs have become powerful enough to be useful for astronomical data analysis. Given their market penetration, they are of course significantly cheaper than Unix stations. The Linux operating system offers the possibility of porting software from Unix workstations to PCs. The *IRAF* group at *NOAO* has ported *IRAF* to Linux, so all software developed for *IRAF* will execute on PCs. This has considerable cost advantages for the astronomical community.

9. Case Study 1: Hubble Space Telescope (HST)

9.1. *Considerations*

The *HST* Science Data Archive has been in the *HST* science operations concept from the very beginning (Woodhole Report, circa 1975) Even without the cost increase of the project due to the Fine Guidance System and Shuttle problems it was clear that *HST* was going to be a very expensive mission.

Several policy decisions were made (Giacconi (1982)) to maximise the return on investments:

(*a*) - *HST* data were to become public after a proprietary period of one year.

(*b*) - No re-observations were to be done unless justified on grounds of science.

(*c*) - "Institutional" calibration, i.e. all calibration observations and the routine calibrations were to be performed by *ScI* personnel, rather than the *PI*.

(*d*) - Archive research proposals were to be accepted. They had to go through the *TAC*. In the USA they are funded just like observing proposals.

In the case of the *HST*, the Science Data Archive is an integral part of the science operations.

The *HST* and its science instruments impose severe requirements on the ground support system. The orbit is a low earth orbit at 600 km, with about 90 min orbital period. This implies severe observing (scheduling) constraints and is reflected in the fragmentation of the data.

The initial science instruments (*SI*s) were:

- The Wide Field/Planetary Camera (*WF/PC*) with 4 times 800×800 *CCD*s per mode, exposure times as short as seconds.

- The Faint Object Camera (*FOC*) which produces up to 512×512 pixels per mode. The shortest exposure time is about 10 minutes.

- The Faint Object Spectrograph (*FOS*) and the Goddard High Resolution Spectrograph (*GHRS*): a few K per data set, depending on the operational mode

- The High Speed Photometer (*HSP*): one "pixel" per mode (PMT), but millisecond resolution

- The Fine Guidance System (*FGS*), which can be used as an "astrometric" instrument. It produces 200 samples per second per *FGS*.

The current *SI*'s (after the 1997 Servicing Mission) include:

- *WF/PC2*: essentially identical to *WF/PC*, better in the *UV*.

- *FOC*: unchanged. Hardly used.

- The Space Telescope Imaging Spectrograph (*STIS*): frame sizes are $1K \times 1K$ (*CCD*s), $2K \times 2K$ (*MAMA*s)

- The Near Infrared Camera and Multi-Object Spectrometer (*NICMOS*): 256×256 images, many readouts.

- Corrective Optics Space Telescope Axial Replacement (*COSTAR*): replaced the *HSP* after the 1993 Servicing Mission and corrected the spherical aberration problem for *FOS*, *GHRS*, and *FOC*. It produces no science data.

- *FGS*: unchanged

During SM 99 the Advanced Camera for Surveys (*ACS*) will be installed: 2K×4K times 2 (mosaiced) times 2 (dithered) frames.

The other *SI*s will not change. *FOC* will be removed and *NICMOS* will have exhausted the cryogen.

HST has severe operational requirements. Due to the orbital geometry multiple readouts are required because of background build-up during Earth occultations. South Atlantic Anomaly passages interrupt the observations and sometimes cause problems with the electronics. Instruments like the *FOC* have severe dynamic range limitations, which have to be compensated by operational procedures.

The *WFPC* is a very efficient cosmic ray detector. To correct for cosmic rays CR-split observations are performed, consisting of two frames each.

Dithering is a technique to offset two frames by fractions of a pixel in both directions. This allows the undersampling of the *WFPC* to be solved and improves the resolution of the images. Dithering will become an operational procedure on *ACS*.

While the science data are already quite voluminous, there is a wealth of engineering data, some of which are important for the calibration of the instruments. More than 6000 parameters are being monitored at widely different sample rates The engineering stream is extracted from the telemetry and stored at *GSFC*. An engineering data subset is included in the science data: *FGS* data, *SI* and *OTA* data.

The problem was to specify which engineering data were going to be needed to understand properly the science data. For example: magnetometer readings to describe accurately the magnetic field of the Earth are required in order to properly calibrate the *FOS*. This was not known at the time of launch and therefore the data were not included in the science data stream. The problem now is that access to the engineering data is exceedingly difficult.

There are implied requirements:

• Not only the raw data had to be stored but also the calibrated data.

• Raw data are generally 12 to 16 bit unsigned integers. Calibrated data are floating point and thus require more storage.

• Ancillary data: error maps, bad pixel masks, etc., etc.

Thus the baseline for handling *HST* data was initially 1 - 2 Gbytes per day. With the new generation of instruments this has increased to up to 5 Gbytes per day. In the future we expect that this will at least double.

This has to be supported reliably and correctly for 365 days per year, 24 hours per day, throughout the expected 15 year life time of the *HST*. This is a heavy commitment.

9.2. *Implementation*

Early concepts consisted of an archive using 6250 bpi magnetic tapes. On the basis of the expected data rate and the expected life time this would have required a tape storage area of some 30×10×10 meters. Such a room exists in the basement of the *ScI*.

The science data archive was considered a part of the Science Operations Ground Systems. *SOGS* was to be produced by an industry contractor (*TRW*), not the *ScI*. When *TRW* had cost and schedule problems *ScI* was instructed to build an interim Data Management Facility (*DMF*) and the Archive contract was re-opened for bids. The *DMF* was based on optical disks, thus the tape room became superfluous.

DMF at *ScI* and the Science Data Archive at *ECF* (and at the Canadian Astronomical Data Center (*CADC*) with which both *ScI* and *ECF* collaborated) were developed concurrently. There were valiant efforts at co-ordination. Some aspects were successful, others were not: *DBMS* compatibility was successful; the attempt at media compatibility was generally successful; the attempt at software compatibility was not; and the attempt to have a compatible user interface oscillated.

The current implementation of the *HST* science data archive at the *ScI* (the Data Archive and Distribution System, *DADS*) was produced by an industry contractor. At the *ECF* the Archive has become part of the *ESO (VLT)* archive. *VLT* will drive the archive requirements in the future.

Media compatibility was enforced. This turned out to be very expensive. Data compatibility was achieved through interface agreements, not through s/w compatibility.

Because of the high media costs *ECF* pushed for CD-ROMS. While their storage capacity is one order of magnitude smaller, they are two orders of magnitude cheaper, making CD-ROMS more cost-effective. Because of their penetration of the entertainment and home computer market they have a longer operational life. All CD-ROM hardware (readers, writers, robotic disk changers, etc.) can be procured at a fraction of the cost of equivalent hardware for large format optical disks. The capacity problem is being solved for us by the market: Digital Versatile Disks (*DVDs*) are being introduced right now.

The user interfaces at *ScI*, *CADC* and *ECF* are different, but similar. They are Web interfaces, making use of the universal availability of powerful Web browsers and the functionality offered by Java and by applets.

An important development was the cross link to other databases (e.g. to *CDS* for name resolving). Today this is easily possible, but it was a pioneering achievement when it was introduced in 1986.

9.3. *Recent developments at* ST-ECF

Because of the substantial increase in data volume after SM '97, we had to find ways to economise, i.e. to decrease the total effective volume of data (Pirenne *et al.* (1995)). As a first step we identified those data which nobody had ever requested, which included all engineering data.

Beginning in 1995 *CADC* and *ECF* developed the concept of on-the-fly- calibration: whenever users request a data set, they do not get the calibrated data, but a re-calibration is performed using the best calibration reference files and the most suitable calibration software. This was initially offered as an option, but it is now becoming the standard procedure (Crabtree et al. (1997)). This will allow us not to archive any longer the calibrated data and to rely on routine on-the-fly- calibration.

We also investigated data compression. This works especially well on the raw (i.e. integer) data. The smaller total amount of data can be stored on CD-ROMs and kept on line using low cost robotic devices. Data get expanded and calibrated during retrieval.

It is conceivable that we will need data which we do not store any longer. The fall-back option for this contingency is to use the *ScI* archive.

9.4. *Current operations (at* ECF *as of November 97)*

HST data are processed by the *OPUS* pipeline at the *ScI* and archived there. Copies of the optical disks are shipped to *ECF* and data subsets are burned onto CD-ROMS. Copies are produced for the *ScI*.

Preview for non-proprietary data is possible. Limited quick-look data analysis capabilities have been implemented as Java applets. Data distribution is (will be) from CD-ROMs, robotic systems minimise human intervention. On-the-fly calibration is used routinely. There is a high degree of commonality with the *CADC*, but almost none with *ScI (DADS)*

The current data volume is in excess of 3 Terabytes (even with the reduced data volume during the last half year) We receive and process about one *DADS* OD disk per working day (4 Gbytes), there are typically two deliveries per month (implying a maximum waiting time of about 2 weeks). This volume is condensed into about 10 CD's per month (i.e. insignificant in terms of volume and media cost)

We distribute about 1 Gbyte per working day. There are, on the average, more than 30 different users per month, typically issuing several requests each. Data distribution is on magnetic tape, on CD-ROMs, and, as bandwidth increases, through the Internet.

In the near future we will archive raw data only and maintain the archive data in compressed form. De-compression will be done during data retrieval and the data will be calibrated on-the-fly before being sent to the user.

A possible further development is to do the de-compression and calibration steps in the client (user) machine. Applets might be employed to achieve this. At the current time there are still severe limitations in terms of speed and operational reliability, but on the long term the data will be shipped out through the Internet, and they will include codelets which decide what the target machine is to do with the data.

9.5. *Lessons learned*

Small is beautiful. The *ECF/CADC* were able to move and adjust much more rapidly than the *ScI*. On the other hand, the tail cannot wag the dog. The larger institution calls the shots in the long run. It is important not to diverge too much.

One-of-a-kind solutions must be avoided. They are unsupportable in the long term because of cost. The same is true for software distribution and off-site maintenance and support. Possible single point failures must be taken care of. This includes key personnel.

Don't try to push technology. Astronomy is too small a market and you might end up with a high-cost unsupportable implementation. Let technology push you and move with it.

Re-processing of the data will always be necessary as the understanding of the instrument and the calibration procedures improves. This can be combined with the required change of the bulk storage medium every 5 years, which must include the re-writing of the old data on the new media.

Commonality between different development groups is a noble goal, but it cannot be enforced between groups who operate under completely different boundary conditions.

9.6. *Added value archive products*

Being free of the data production requirements at the *ECF* we have identified areas in which we can improve the quality and the usefulness of the archive output products. The first such step was the pre-view facility, followed by the on-the-fly calibration.

During 1997, we developed software to build associations. Associations are sets of several observations which are logically linked (Bristow *et al.* (1997)). Examples include CR-split observations for *WFPC2*, which is really only one observation which, for operational reasons, consists of two individual data frames. Most users want to retrieve the result, not the input frames.

Another example are dithered observations. These are sub-stepped frames which can be combined to increase the resolution (Hook (1995)). The most massive such association is the Hubble Deep Field (*HDF*), which consists of several hundred individual frames.

It turns out that there are many unplanned associations: individual repeated observations of the same field, even by different *PI*'s. To combine these observations it is necessary to have access to the *HST* jitter data to obtain the correct sub-pixel stepping.

9.7. FOS *Final archive*

The original calibration of the *FOS* was based on empirical correction factors, implemented in the conventional way via polynomial fits. Soon after the beginning of the operations unexpected stray light was discovered and various explanations were proposed. However, analysis using a software model of the *FOS* showed that this was caused by the physics of the instrument (Rosa (1997a), Rosa (1997b)) As a result software to calibrate this effect was retrofitted into the *OPUS* pipeline.

FOS also had a problem with magnetic shielding, caused by improper assembly during *SI* integration. The effect can be calibrated using a good model of the earth's magnetic field and the magnetometer data in the *HST* engineering data stream.

A project has been started at the *ST-ECF* to produce a "best and final" calibration for all *FOS* data and make it available on CD-ROMs, together with raw data, calibration files and calibration software.

9.8. *Long term possibilities*

A promising strategy is to process the *HST* science archive into a catalogue, i.e. to extract all objects from the data (images), to define an unambiguous naming and identification scheme (as it was done for the Guide Stars Catalogue), to collect ancillary data for these objects from other sources (if available, or as they become available), and to add derived (i.e. research) results as they become available. Such an archive would be a powerful research tool, enormously increasing the value of *HST* data. A small example of how this might work is the Hubble Deep Field.

9.9. *Archive research*

As already mentioned, archive research has always been in the operational concept of the *HST*. "Classical" archive research makes multiple use of individual *HST* observations and allows the combination of data from different *PI*'s, *SI*'s. A recent example is the work on blue stragglers in 47 Tuc by M. Shara.

IUE has shown us that there comes a time for synoptic investigations. This typically involves searching the full archive for data with certain characteristics. The main obstacle is that the data are insufficiently classified (the classification depends on *PI*). It also involves processing all data of a certain kind, according to a definable procedure. The problem there is the on-line availability of the data and the interface to an appropriate data analysis system. One might also consider cross-correlation between different science data archives. The non-uniformity of the different archives will have to be overcome.

10. Case Study 2: The ESO Very Large Telescope (VLT)

Although *ESO* has been carrying out large projects before, the *VLT* was a quantum jump in several respects:

• The size of the project. The "worth" of the *VLT* is a multiple of the total "worth" of the company. This is considered impossible in industry.

• Burst resource requirements. It is evident that both cash flow and staff size are large during the implementation phase.

• Staff size and staff skill mix. With the need to have dedicated staff for specialised activities the skill mix of the staff has to change with time .

• The need to outsource. Both for reasons of permissible maximum size of staff and because of the fact that specialised skills are required only during certain periods, some of the *VLT*-related work is being performed by contractors.

• Implementation of work package accounting. As *ESO* moved from the comparatively small projects of the past to the *VLT* the *ESO* accounting system had to be overhauled.

• A computer based Management Information System had to be implemented.

• Maybe the biggest change was that in a project of the size of the *VLT* no one person is capable of monitoring the whole project.

The *VLT* shares many of these characteristics with other projects. These have become known as Big Science projects. Such state-of-the-art research facilities have become quite expensive (in the range of 10^9). In order to maximise the science return from these facilities it is necessary to develop end-to-end-models: proposal submission, evaluation, planning and scheduling, implementation, acquisition, calibration, analysis, archiving. All the s/w elements in this chain must work seamlessly together; all the s/w elements must be maintainable, portable, robust; the development methodology must support multiple developers; all user interfaces must be consistent; configuration management is imperative; documentation must be adequate; etc.

All this sounds reasonable, but it is it difficult because *ESO* is a Research and Development environment and all the problems identified in the section on software engineering have to be solved.

10.1. *Space vs. ground*

The users of space experiments and large ground-based observatories are the same group of scientists. With ground-based facilities becoming as complex as space projects it is useful to employ the same methods and operational procedures. For both types of facilities, observing time is oversubscribed and the desirability of an archive of ground data (like for *IUE/HST*) is obvious. The technical means to do all that are available.

The most challenging issue is that in the past the operational procedures for ground-based facilities were dramatically different: the observers were give the telescope for a certain period of time, and it was up to the individual astronomer how the telescope was used and whether and how calibration observations were performed.

Another difference is instrument stability and calibration: space instruments are not modified, sensitivity changes are monitored, while ground instruments are left to the astronomers and undocumented configuration changes are possible. If a standard, configuration controlled mode of operation is not enforced, the archive will only contain non-uniform data which cannot be used by others than the original observer. Funnily enough, users accept that space telescopes are operated in service mode (e.g. *HST*), but want to be the owner of ground telescopes.

The classical operational mode of a ground-based telescope will not really be possible for the *VLT*: the system is too complex. In addition, flexible scheduling is desirable in response to changing observing conditions and in order to maximise the efficiency of the telescope. Service observations require breaking of the observing programs into schedulable pieces and the dynamic queuing of observations. Finally, the interferometry mode is exceedingly complex and needs careful preplanning.

The *VLT* will be operated in a way which is more similar to the way space facilities are being operated, than to the operations of a conventional telescope. Of course, some lack of uniformity in ground-based data is to be expected because a ground-based instrument will not be operated unchanged for 18.5 years. The easy accessibility of the ground-based instruments is, at the same time, an advantage. However, it has to be carefully controlled.

To make the process described above possible an end-to-end data flow model was developed for the *VLT* (Grøsbol & Peron (1997)). This spans observation planning, observation scheduling, implementation of the observations and interaction with the telescope control system. The calibration pipeline and archiving steps are integral parts of this process.

10.2. *Archive operations*

The *VLT* archive at *ESO* was developed from the *HST* archive. This ensures a high degree of continuity for the *HST* users, as well as practical operational experience with representative data. The common archive facility was first extended to include data from other *ESO* telescopes such as the refurbished NTT, which in more than one way serves as a prototype for the *VLT*. Our experience with the production and the distribution of output data will make sure that *VLT* users receive their data in a timely and reliable manner.

There are several major differences between the *ESO/ECF HST* archive and the *VLT* archive. The *VLT*, consisting of four 8-m unit telescopes and a number of smaller telescopes, will produce much more data than the *HST*. In addition, we never had the near real-time acquisition requirements for *HST* data.

Another difference is the geographic distribution. It is clear that part of the *VLT* archive has to be located close to the telescope to act as a data acquisition facility and to safeguard against data loss. On the other hand, the research oriented part of the archive will have to be located in Europe, where the research will be done.

10.3. *Research with the* VLT *archive*

It is evident that the *VLT* archive will utilise many of the developments of the last few years, such as CD-ROM/*DVD* technology, data compression, on-the-fly calibration, etc. Beyond that the archive should also be usable as a research facility in its own right, combining access to *VLT* data with access to data from other facilities and the data from databases located elsewhere in the world. This will require a special computational infrastructure. Because of the large data volume, the need to access heterogeneous databases, and to execute different software packages, we need an environment tailored to these requirements.

The Research Station is the next step after the personal workstation. It consists of a powerful local processor which is networked to other machines and to databases and knowledge bases. It might have a configurable personalised interface which allows access to all services and functions in a consistent and efficient manner. The emphasis should be on visualisation and conceptualisation (model building). This can be realised through multiple screens, big screen projection ("flight simulator"/planetarium), or through virtual reality (*VR*).

10.4. *Data mining*

An important development was started recently between *ESO/ECF* and *CDS*, which also involves collaborators from the Computer Science Department of the Technical University, in Munich.

Data mining as described is a step in a process called "Knowledge Discovery in Databases" (*KDD*) and includes the application of specific algorithms to produce a particular enumeration of patterns over the database. Knowledge Discovery in Databases is the extraction of implicit, previously unknown, and potentially useful knowledge from databases.

In view of the very large amounts of data generated by state-of-the art observing facilities, the selection of data for a particular archive research project quickly becomes an unmanageable task. Even though the catalogue of observations gives a precise description of the conditions under which the observations were made, it does not contain any information about the scientific contents of the data. Hence, archive researchers have first to do a pre-selection of the possibly interesting data sets on the basis of the catalogue, then assess each observation by visually examining it (preview) and/or execute an automated task to determine its suitability. Such procedures are currently used for archive research with the *HST* Science Archive. This is only acceptable if the data volume is limited.

However, already after the first year of Unit Telescope 1 operations, the *VLT* will be delivering data in quantities which make it not feasible to follow the same procedures. The *ESO/CDS* Data Mining Project aims at closing the gap and developing methods and techniques that will allow a thorough exploitation of the *VLT* Science Archive.

The basic concept is not to have to ask for individual data sets, but instead to be able to ask for all information pertaining to a set of search criteria. In addition to parameters contained in the information catalogue, the search criteria should include parameters which pertain to the science content of the observations. This implies that parameters which describe the science content have to be generated after the observations. The proper time to do this is during the ingest of the data into the archive. These parameters can then be correlated with other information

The concept is to create an environment that contains both extracted parametric information from the data, plus references to existing databases and catalogues The environment then establishes a link between the raw data and the published knowledge, with the immediate result of having the possibility of deriving classification and other statistical samples

10.5. *Determination of science related parameters*

The aim of parameterisation must be the enumeration of statistically relevant and physically meaningful parameters. Examples: integrated energy fluxes of objects, colours, morphology, distribution, identification. This will lead to data archives which are organised by objects, rather than by data sets (Albrecht *et al.* (1994)).

A promising beginning are tools like *SExtractor*. This package allows the extraction and parameterisation of objects on large image frames (Bertin & Arnouts (1995)). Electronic links will be used to collect parameters on the objects which can be extracted from other sources, for instance databases from other wavelength regions. These parameters will either be physically imported and added, or they will be attached to the objects through hyperlinks.

10.6. *Processing*

An important initial step is the classification of the objects. The challenge of classification is to select the minimum number of classes such that objects in the same cluster are as similar as possible and objects in different classes are as dissimilar as possible. However, this has to be done in such a way that membership of an object in a particular class is meaningful in terms of the physical processes which are responsible for the condition of the object. This is not always the case for traditional classification systems in astronomy: the binning criteria were determined by the characteristics of the detector and the physiology of the human classifier.

This is also not necessarily the case for statistical approaches (clustering, pattern recognition, neural network), because no physics is involved in establishing the classes.

The current emphasis in on automatic classification and on the database access mechanisms to mine terabyte-sized databases.

11. The Future

It is difficult to extrapolate into the future, especially in the area of computer development. Twenty years ago the CEO of Digital Equipment Corporation stated that nobody will want a computer in their living room. At the time of this review DEC is being bought up by a company which has become rich producing computers for people's living rooms. Dramatic developments like the Internet were not even the subject of speculations by science fiction writers, and the World Wide Web was not foreseen, even in 1990, three years before it started its explosive growth. However, some things are certain to happen. The risk is to err on the conservative side.

Technology is still proceeding exponentially. Technology drivers are entertainment and consumer electronics, not science and engineering. Multimedia applications require enormous processing power from which science computing will profit. Network bandwidth will become a non-issue as interactive TV will define bandwidth requirements. Networks will be ubiquitous and indispensable and they will be used in different and unexpected ways: in addition to information, networks will offer computing services through applets.

There will be a quantum jump in storage technology, based on nanotechnology and holographic storage devices. Low cost supercomputers will become available implemented in the form of "piles of PCs": massive parallel processing using mass produced hardware. Scaling will be done by adding machines rather than designing new ones. All computer resources will be transparently networked, so the differences between local and remote will vanish. Computers will become as unobtrusive as electric motors are today.

Programming will be done on a very high level, using demonstrational Interfaces, which let the user perform actions on concrete example objects, while at the same time constructing abstract "programs" (better: operational sequences). This involves "guessing" by the computer, i.e. the dynamic change of default values based on context. Context can be derived from a model of the operations (e.g. *CCD* calibration), or from the occurrence of previous instances. Models of operations can be represented as metacode, or as rule bases, or a combination of both. "Learning" from previous instances can be done through a trainable neural net.

The combination of demonstrational and active notebook interface (e.g. the interface to the Mathematica package on the Macintosh computer) is probably the ideal research oriented interface.

Virtual reality (*VR*) will be made available by the entertainment industry. *VR* techniques hold the key to the ultimate user interface. Current applications include military, aerospace, education, interactive video/TV.

VR removes the limitations of the 2-D screen and provides 3-D display capability. With improved natural language interfaces, data analysis/data base exploration/"discussions" become possible. In fact, most astronomers' offices already constitute a mild non-immersive *VR* environment.

11.1. *An alternative approach: Improving the astronomical research process*

The shortcoming of the developments described above is that they are, or will be, the result of technological progress (more powerful hardware) and the application of concepts which already exist, but have so far not been possible to apply. In other words, this approach can be characterised as: more of the same. However, what we have to aim for is a paradigm change, in order to achieve results which are not just more accurate, but which are of a different nature, and we have to get the computer to help us do this. A past example is the introduction of numerical integration in stellar evolution models, after analytical models had reached their limits.

In other words, we have to step back and examine the process of doing astronomical research and investigate where and how state of the art computing technology can be employed. The process of doing scientific research has only recently been defined in epistemological terms: the model of the research process as developed by Sir Karl Popper (1972) comes closest to what most natural scientists do when they "do science".

The research process starts with the input of signals, either through sensory perception, or through measuring devices, which register signals which are either too faint, or not suited for our senses. We know this step as data acquisition.

The next step is the transformation of the input data into meaningful values, quite often literally the "data reduction", from a jumble of instrument dependent individual measurements, to a much smaller, coherent and consistent set of parameters.

By injecting concepts into the collection of parameters, we construct models. Concepts range from very simple, such as a linear correlation, to the very complex, like evaporating black holes. The injection of concepts happens spontaneously and associatively, it is a result of the evolution of our brain.

Models come in two flavours, hypotheses and theories, the difference being that a hypothesis is an as-of-yet unsubstantiated and incomplete theory. Given the fact that no theory is ever complete, it is more correct to say that all models are hypotheses. This is in agreement with the historical observation that even "wrong" models served well as good hypotheses in a heuristic sense.

Good models allow us to make predictions as to future observations. They also allow us to add to our pool of concepts by abstraction and generalisation. If a model conflicts with observations, we have to discard it. Since we can never be certain that any model will forever withstand the test of future observations Popper concludes that, in science, we can never demonstrably attain the "truth".

Asking the question where in this process the most progress has been made historically, we tend to think that it has been in the first step: the introduction of ever more powerful telescopes and detectors, and the opening of more spectral windows has allowed to include, quite literally, observations of the whole Universe into the building of models.

One could contend, however, that the most progress has been made in the application of concepts: the scientific revolution (i.e. the paradigm change) during the period of enlightenment removed concepts like that of the supernatural, of magic and of the subjective, from our model building tools, which indeed provided us with the very basis of what we today call scientific thinking.

11.2. *The scientific library*

Models found through the above process are described by the scientists using a combination of natural language (with exactly defined semantic content of crucial elements, usually called technical terms) and mathematical representation. In other words, a scientific publication, and, more generally, the scientific library, constitute a knowledge base, right now encoded in the idiosyncratic literary style of different authors with different cultural and language backgrounds.

In astronomy we have converged on one main representation language which we call scientific English, the quality of which, however, differs considerably between scientists, limiting their ability to convey, as an author, or to internalise, as a reader, a scientific model.

It is thus desirable to define a meta language for conveying scientific information, which is both human readable and computer processable.

(Having said this we must stress that this only applies to the formal representation of scientific models. For all other purposes we must use English. For the past 20 years essentially all important astronomical publications have been published in English. While this is a disadvantage for the non-native English speakers, it is an enormous advantage for the science of astronomy: in no other science are all active scientists able to communicate with each other so easily. Thus, all efforts which have the potential of a deviation from this situation must therefore be forcefully resisted by the community.)

Even with all-English publications human-to-human knowledge transfer is sub-optimal and computer-assisted processing of published knowledge is impossible. First step towards a meta language are a data dictionary and a thesaurus. VISION: "publishing" will not be done in the form of papers, but as additions, or modifications to a global knowledge base containing hypotheses (models). Consistency checking, novelty evaluation, truth maintenance, etc., is then immediately and easily possible. This could eliminate refereeing, or at least make it much easier.

As a by-product, the knowledge base, or segments of it, can be mapped into different natural languages, (even languages which the contributors do not speak) and at different levels (such as textbook, or popular description).

11.3. *Electronic publishing*

In the beginning "electronic publishing" was little more than preparing a publication on a word processor and sending it to another computer. The American Astronomical Society (*AAS*) started to accept abstracts produced in this manner for *AAS* meetings in 1981. Because of early problems with standards and conventions related to formatting and special characters there was a pause of several years and a re-start with the introduction of Tex/Latex. Tex/Latex provided a very important service to our science during the past decade. It is now anachronistic and should be discontinued.

The *AAS* has pioneered electronic publishing: ApJ and ApJ Letters have been electronically available for years. More recently, "electronic-only" journals have started. Advantages are: they are quick, no shipping required, searchable, potentially processable. There are also problems: non-mature information might be distributed. What constitutes the "paper": what's on my disk, or what's on the disk of the publisher? Copyright, referencing, quoting need to be solved, and will be solved.

In addition to improved access and timely availability electronic publications have the advantage of being searchable. There are organisations like the *NASA* Astrophysics Data System (*ADS*) which specialise in such services. The aim is to free the user from having to read an increasingly enormous amount of material in order to find the desired information. Advanced search services are becoming available for a medium in which reading through all available material is totally prohibitive: The World Wide Web.

The above search services are convenient and useful. However, as of today, they are still mainly text-string oriented, and not context oriented.

11.4. *Long term goals*

The long term goal has to be to consider the body of electronically available publications as a database much like a science data archive, or an astronomical catalogue.

In analogy to knowledge discovery in a numerical database, we can then do knowledge discovery in this text database, which contains concepts, models, and hypotheses: discovery of implied, previously unknown, and potentially useful knowledge from such a database should be possible. Alternatively, candidate models can be injected into such a database with the aim of either supporting, or disproving the model.

Having the contents of this database represented in a meta language would facilitate this process enormously. However, some advances should be possible even on the basis of just text in scientific English.

It is obvious that the capability to do this would immediately lead to enormous advances in scientific productivity.

REFERENCES

ALBRECHT, R., ALBRECHT, M.A., ADORF, H.M., HOOK, R., JENKNER, H., MURTAGH, F., PIRENNE, P., RASMUSSEN, B.F., 1994, *In: Proceedings of the Workshop on Astronomical Archives, Trieste, Albrecht M.A. and Pasian, F. (Eds.), ESO Workshop and Conference Proceedings Series* **50**, 133–141

BERTIN, E., ARNOUTS, S., 1995, *Astronomy & Astrophysics Supplement Series* **117**, 393

BRISTOW, P. BENOIT, P., & MICOL, A., 1997, *ST-ECF Newsletter* **24**, 21

CRABTREE, D., DURAND, D., HILL, N., GAUDET, S., & PIRENNE, B., 1997 *ST-ECF Newsletter* **24**, 19

GIACCONI, R., 1982 *Proceedings of the Special Session of IAU Comm. 24 (Hall, D.N.B., Ed.), NASA CP-2244*, 1

GRØSBOL, P., PERON, M., 1997 *ASP Conference Series* **125**, 23

HOOK, R., 1995 *ST-ECF Newsletter* **23**, 31

PIRENNE, B., BENVENUTI, P., ALBRECHT, R., 1995 *ASP Conference Series* **125**, 290

RASMUSSEN, B.F., 1995 *ST-ECF Newsletter* **22**, 9

ROSA, M.R., 1997a *ASP Conference Series* **125**, 411

ROSA, M. R., 1997b *ST-ECF Newsletter* **24**, 14

Optical and UV Astronomy in the Internet Age

By DANIEL GOLOMBEK

Space Telescope Science Institute, Baltimore, Maryland, USA

This series of lectures is devoted to optical and UV data, with particular emphasis on observations obtained by the Hubble Space Telescope (*HST*). We explore the instruments on board the *HST* which have taken data since the telescope was launched and the instruments which will continue to observe in the future. The contents of the *HST* data archive are discussed, along with information on how to retrieve and understand *HST* observations which can be found in the archives. Other optical and UV archives are also briefly discussed.

1. Introduction

Due to the nature of *HST* data, a brief revision of data formats is presented, followed by a description of the major sites that contain information and astronomical data.

The topics discussed are:

(*a*) Astronomical Data
- *FITS* format description
- *HST* data
- Group format extension
- Reading *HST* science data
- Other *HST* files
- Associations
- Additional programs that read *FITS* files
- Other file formats

(*b*) Optical and UV information
- Pre-prints and papers
- *NED*
- *CDS* (*SIMBAD*, *VizieR*, *Aladin*, Dictionary of Astronomical Nomenclature, Information & On-line Data in Astronomy, Star's pages, Other services)
- *NSSDC*
- *Leda*
- Data
- Miscellaneous images and catalogues
- Scanned Data (*DSS*, *APS*, Plate Database)
- Optical and UV surveys
- Multiwavelength maps
- Astrometry
- Artificial Data
- Meta-databases
- Planetary information
- Organisations
- Summary

(*c*) Optical and UV archives
- Optical (*ESO*, *CFHT*, *ING*, *NOAO*)
- UV (*IUE*, *EUVE*, Astro missions, *FUSE*)

(*d*) The Hubble Space Telescope
  • Brief description
  • On-board instruments (Faint Object Camera (*FOC*), Fine Guidance Sensors (*FGS*), Near Infrared Camera and Multi-Object Spectrograph (*NICMOS*), Space Telescope Imaging Spectrograph (*STIS*), Wide Field Planetary Camera 2 (*WFPC2*))
  • Archival instruments (Faint Object Spectrograph (*FOS*), Goddard High Resolution Spectrograph (*GHRS*), High Speed Photometer (*HSP*))
  • Future instruments (Advanced Camera for Surveys (*ACS*), Cosmic Origins Spectrograph (*COS*))
(*e*) The Hubble Data Archive (*HDA*)
  • Description
  • Contents
  • Browsing and extracting
(*f*) Surfing for multiwavelength data

2. Astronomical Data

This section is devoted to the description of the standard format to store astronomical data and how to work with the files. A brief description of other data formats is also given at the end. The instructions for working and manipulating data will be given for *IRAF*, although other systems like *MIDAS* and Figaro can also used. The selection of the system depends on the support given to each one at the site at which the reader will be working.

2.1. *FITS*

All galactic and extragalactic astronomical data are stored and distributed in the *Flexible Image Transport System FITS*). As the standard, all astronomical reduction and analysis packages need to be able to read and write *FITS* files. The details of the *FITS* convention can be found in "Implementation of the Flexible Image Transport System (*FITS*)", published by *NASA*'s Office of Standards and Technology and in Wells *et al.* (1981), Grøsbol *et al.* (1988), Harten *et al.* (1988), and Ponz *et al.* (1994). These and other documents can be downloaded from `ftp://nssdca.gsfc.nasa.gov/pub/fits`.

FITS format data consists of
  • an ASCII header
  • binary data
The header contains all the information about the observation and the file in a sequence of keywords. These keywords are written in a fix 80-column format of the form:
$$keyword = value \ / \ comment$$

2.1.1. HST *data: group format extension*

HST observations use an extension of the *FITS* convention, the so-called *group format extension*. This means that a group of data is stored within the same filename. For example, all four images from each chip of the *WFPC2* camera are stored with the same name; each one constitutes a group. Spectra obtained with the *FOS* and *GHRS* might have hundreds of groups.

These *FITS* extension data needs a special way to be written to tape; i.e., the group information needs to be stored and recovered upon reading. The tapes distributed by the *HDA* will then have
  • a file with nine alphanumeric characters ending with _cvt and an extension
  • a binary table (with the `.tab` extension)

Instrument	Filename First Character
ACS	J
FGS	F
FOC	X
FOS	Y
GHRS	Z
HSP	V
NICMOS	N
STIS	O
WF/PC	W
WFPC2	U

TABLE 1. Data archive file naming conventions for *HST* instruments.

Data downloaded directly from the *HDA* consists of just one file with the .fit extension.

HST data sets in the archive have filenames composed of nine alphanumeric characters. The first one identifies the instrument used to obtain the data. The convention is shown in **Table 1**.

The other characters are related to the proposal number under which the data were obtained, the order in which they executed, and the way that they were transmitted to the ground.

2.2. *Reading HST FITS files*

To analyse the data of the first generation *HST* instruments as well as from *WFPC2*, it is strongly recommended that it be read from its original *FITS* format and stored in *GEIS* format on disk. *NICMOS* and *STIS* data should remain in *FITS* format.

The task stsdas.fitsio.catifits can be used to list the contents of a tape. An example of such listing is:

```
FITSNAME       FILENAME                    DIMENS     BITPI BSCALE   BZERO     O

1              null_image                             8
    TABLE      u27l1y01t.trl               1Fx58R       1Fx58
2              null_image                             8
    TABLE      u27l7z01p.trl               1Fx58R       1Fx58
3              u27l1y01t_cvt.c0h           800x800x4  -32F  1.       0.        U
    TABLE      u27l1y01t_cvt.c0h.tab 49Fx4R           49Fx4
4              u27l1y01t_cvt.c1h           800x800x4  16S   1.       0.        U
    TABLE      u27l1y01t_cvt.c1h.tab 49Fx4R           49Fx4
5              u27l1y01t_cvt.d0h           800x800x4  16S   1.       0.        U
    TABLE      u27l1y01t_cvt.d0h.tab 49Fx4R           49Fx4
6              u27l1y01t_cvt.q0h           800x800x4  16S   1.       0.        U
    TABLE      u27l1y01t_cvt.q0h.tab 49Fx4R           49Fx4
7              u27l1y01t_cvt.q1h           14x800x4   16S   1.       0.        U
    TABLE      u27l1y01t_cvt.q1h.tab 3Fx4R            3Fx4R
8              u27l1y01t_cvt.shh           965        16U   1.       0.        U
    TABLE      u27l1y01t_cvt.shh.tab 3Fx1R            3Fx1R
9              u27l1y01t_cvt.x0h           14x800x4   16S   1.       0.        U
    TABLE      u27l1y01t_cvt.x0h.tab 3Fx4R            3Fx4R
EOF encountered
```

The suggested parameters to read a *FITS* tape with the `stsdas.fitsio.strfits` task are:

```
   fits_files = ''flora!mtka"    FITS data source
    file_list = ''1-999"         File list
    iraf_files = ''gm"           IRAF filename
     template = ''none")        template filename
 (long_header = no)              Print FITS header cards?
(short_header = yes)             Print short header?
     datatype = ''default")     IRAF data type
        blank = 0.)             Blank value
        scale = yes)            Scale the data?
     xdimtogf = yes)            Transform xdim FITS to multigroup?
  oldirafname = yes)            Use old IRAF name in place of iraf_file?
       offset = 0)              Tape file offset
         mode = ''h")
```

Note that

```
    xdimtogf    = yes
    oldirafname = yes
```

As mentioned above, the header contains all the information about the information. Important parameters can be listed using the `stsdas.toolbox.headers.imininfo` task; resulting in, for example for a *WFPC2* image:

```
Rootname                Instrument              Target Name
U27L7502T                 WFPC2                    3C402.0


Program         = 27L           Obs Date        = 30/03/95
Observation set = 75            Proposal ID     = 05476
Observation     = 02            Exposure ID     = 310.0000000\#002
Source          = Tape Recorded Right ascension =  19:41:42.9
File Type       = SCI           Declination     =  50:36:36
                                Equinox         = J2000


First filtername  = F702W                Number of groups = 4
Second filtername =                       Data type       = real


Image type        = EXT         Exposure time (sec)    = 140.
Mode              = FULL        Dark time (sec)        = 140.
Serials           = OFF
Shutter           = A           Calibration steps done:
Kelsall spot lamp = OFF           MASK ATOD BLEV BIAS DARK FLAT
```

And for an *FOS* spectrum:

Rootname	Instrument	Target Name
YOPE0402T	FOS	3C186

Program	= OPE	Obs Date	= 3/12/91
Observation set	= 04	Proposal ID	= 2578
Observation	= 02	Exposure ID	= 5.1200000
Source	= Tape Recorded	Right ascension	= 7:44:17.5
File Type	= SCI	Declination	= 37:53:16
		Equinox	= J2000

Aperture = A-1		Detector	= AMBER	Number of groups	= 9
Polarizer = C		Disperser = H27		Parameters/group	= 19
Obs mode = SPECTROSCOPY				Naxis1	= 2064
Position angle = 339.8297		Patterns	= 12	No channels	= 512
Obs time	= 04:28:42	Readouts	= 9	Comb	= 5
Exposure/pixel = 270.		Memory clears = 1		Sub step	= 4
Live time	= 0.5	X base	= 0.	Bins	= 1
Dead time	= 0.01	Y base	= 346.	Y size	= 1
Integrations	= 1	First channel = 0		Y space	= 0.
Total exposure = 1080.					

2.2.1. Other HST files

Besides the science headers described above, there are others that contain further information about the observation:

• The *Standard Header Packet* (extension shh) contains information about the ephemerids of the observation, the positions of the Sun and the Moon during the observation, engineering data and other proposal information.

• The *Unique Data Log* (extension ulh) contains data from the satellite like temperatures in different positions, voltages, gains, etc.

• The *Observatory Logfiles* are binary tables (extension cmi and cmj) with information about the name and positions of the guide stars used in the observation, any loss of lock time if any and the jitter information.

2.2.2. Association Data

The most recent instruments installed on board of *HST*, *NICMOS* and *STIS*, have the characteristics that to create a useful science observation several exposures are combined. These individual exposures constitute an association than can then be worked on (read, calibrated, archived, and analysed) together rather than as separated observations. In the case of *NICMOS* an association is the collection of all the observations of the source and the sky that are executed as part of an observation. **NICMOS and STIS data should always remain in *FITS* format and not be read to other formats**. The syntax to operate with these files is:

filename.fits [extension number] [keyword option] [image section]

Please refer to the *HST* **Data Handbook** for more details about these files.

2.2.3. Other programs that read FITS files

Table 2 lists other programs, besides the *STSDAS* task mentioned, *IDL*, *MIDAS*, *Figaro* that can read *FITS* files (although most do not read group-formatted data)

Platform	Program	archived at
UNIX, VMS	*FITSIO*	ftp://legacy.gsfc.nasa.gov/software/
UNIX	xv	ftp://ftp.cis.upenn.edu/pub/xv
Macintosh OS	GraphicConverter	ftp://sumex-aim.stanford.edu/info-mac
Macintosh OS	PhotoshopFITS	gopher://gopher.archive.merit.edu/mac/graphics/graphicutils/
Macintosh OS	MAIA	gopher://gopher.archive.merit.edu/mac/graphics/graphicutils/
Macintosh OS	NIHImage	http://rsb.info.nih.gov/nih-image/
Macintosh OS	*FITS*view	ftp://fits.cv.nrao.edu/fits/os_support/mac_os/fitsview/
Windows	FITSview	ftp://fits.cv.nrao.edu/fits/os_support/mswindows/fitsview/

TABLE 2. *FITS* files readers

2.3. *Other file formats*

2.3.1. *Science data*

(ftp://rtfm.mit.edu/pub/usenet/news.answers/sci-data-format)

- *BDF* this is the format of the *MIDAS* and Starlink files. Only one file contains the header and binary data
- *VICAR* is one of the formats used by *JPL* to store planetary data
- *PDS* is the format adopted by *JPL* to store planetary images on CD-ROM
- *HDF* was created by *NCSA* to store and transfer data
- *GRIB* is used for meteorological data

2.3.2. *General*

- *GIF* common format to store and display images and movies
- *JPEG* compression format to store and display images
- *MPEG/QuickTime* common format to store and display movies
- *TIFF* common format to store and display high resolution images

3. Optical and UV information

Historically, any newly discovered object needed to be "confirmed" by detecting it in the optical band. Catalogues were published with stamp-sized images obtained at the largest available telescope, or from sky survey plates (e.g. Parkes catalogue identification papers in several issues of the Australian Journal of Physics and Astrophysics Supplements (e.g. Bolton *et al.* (1977)); Einstein sources (Maccacaro *et al.* (1994); QSO identification (Schneider *et al.* (1992)). This is probably still true, although not as general, as we try to "see" what we have just found.

We are probably reaching an information overload with more efficient instruments, larger detectors and an increasing number of facilities. At the same time, the technology has been catching up and has allowed us to be able to reach most of these data, regardless of where they reside. These notes are just a roadmap to guide the reader through this maze, highlighting some of the more important starting points, or stops.

Probably the most difficult part of a project is staring at a blank page and needing to fill it up, or getting some brief description and try to complete the information. The following section will have a few pointers to information, or data useful to elaborate, or to check new ideas.

3.1. *Preprints and papers*

Currently one of the best ways to stay up to date on what is being submitted is to consult the ApJ's yellow pages. This is a reactive activity; it is easier to receive a daily message from the preprint database at the Los Alamos National Laboratory (`http://xxx.lanl.gov/archive/astro-ph`). It is important to remember though that the documents in the database were not yet refereed and that they are just preprint which will be subjected (sometimes) to major review. The archive has several mirrorsand it is best to select the one which is closest for document retrieval.

Actual papers can be obtained from *NASA*'s Astrophysical Data Service, *ADS* (at `http://adswww.harvard.edu/`) and its mirror sites in France (`http://cdsads.u-strasbg.fr`) and Japan (`http://ads.noa.ac.jp/`).

Personal and institutional subscribers to the ApJ can read on-line, or down-load accepted papers (`http://www.journals.uchicago.edu/ApJ/journal`). The list of papers submitted to several journals, what used to be called the astronomical yellow pages, can be read on-line (`http://www.noao.edu/apj/ypages/yp.html`). To complete the electronic library, Astronomy and Astrophysics main journal papers (`http://link.springer.de`) and in the supplement series (`http://www.ed-phys.fr/docinfos/OnlineAetA.html`) are also on-line.

Planetary science papers in Icarus (`http://astrosun.tn.cornell.edu/Icarus/`) are also available on-line.

Pages with references and documents in Spanish can be found at the Universidad Complutense de Madrid node (`http://www.ucm.es`), or reading the electronic journal *Astronomía Latino Americana* (`http://www.inaoep.mx/~ala`). Articles of interest to amateurs are available in the on-line version of Sky and Telescope (`http://www.skypub.com/`), Astronomy magazine (http://www.astronomy.com), or the Astronomical Society of the Pacific (`http://www.aspsky.org`).

Nature also publishes an electronic edition (`http://www.nature.com`).

3.2. NASA *Extragalactic Database (*NED*)*

NED (`http://nedwww.ipac.caltech.edu/`; telnet `ned.ipac.caltech.edu`, *username*: ned) contains information about 800 000 extragalactic objects, which includes 1.3 million names and almost a million bibliographic references (since 1983), abstract (since 1988) and notes on the sources. *NED* is **the** place to start on a project that includes extragalactic sources and is freely available to anyone with Internet access.

NED includes an object-based database, it organises all information around extragalactic objects, not by catalogues (Helou *et al.* (1995)). The master list was culled from around 40 major astronomical catalogues, plus several lists from the literature (for example, PG quasars, or compact groups of galaxies). *NED* team members continuously produce the cross-identification of each source in the list.

The information for each source (if available) includes
- name(s)
- object type (from the discovery method, or original catalogue)
- position
- photometric measurements in as many bands as available and information to assess its pedigree
  - radial velocity, or redshift
  - cross-identification
  - references
  - catalogue notes

Each entry includes the reference where the information was published.

NED also includes abstracts of dissertations devoted to extragalactic studies presented since 1980. Dissertation titles and authors only are available for theses presented between 1909 and 1980.

It is very easy to search for information in *NED*, as you do not need to know much about the object. The query can be made by name, "near a name" (in equatorial, ecliptic, galactic, or super-galactic co-ordinates), an IAU formatted name, for an object near a position, or by object type. The list can also be created in such a way that object types are excluded for those retrieved inside a search radius.

If the list is long, a batch job request can be submitted. The user will send an e-mail message to `nedbatch@ipac.caltech.edu` and *NED* will reply with the desired information in one or more messages. *NED* can also be run in a client mode, making it possible to access the database from within user programs.

To complete the search, an image of the object (or the region) can be obtained. *NED* can produce images from the Digitized Sky Survey, infrared images from the *IRAS* survey and radio maps.

3.3. *A Thousand and One Catalogues in Strasborg: CDS*

The first attempt at a systematic collection of information about astronomical objects and query them in an easy way was done at the Centre des Donneés Stellaires (*CDS*) in Strasborg (`http://cdsweb.u-strasbg.fr/CDS.html`). Since 1983 the *CDS* has been embarking on a project to "collect, homologise, distribute and preserve [on-line] astronomical information" (F. Genova, personal communication). The information is not raw data, but metadatabases, created from observing logs, catalogues, papers, and images.

The services available at the *CDS* include:
- *SIMBAD*
- *VizieR*
- *Aladin*
- Dictonary of Astronomical Nomenclature
- "Information & On-line data in Astronomy" (Egret & Albrecht (1995)) complementary on-line version.
- Star*s pages
- Abstracts
- *ADS* mirror
- Opacity project atomic data
- *CFHT* documentation
- *IUE ULDA* (French host)

3.4. *SIMBAD*

SIMBAD (Set of Identifications, Measurements, and Biography for Astronomical Data) can be accessed from the *WWW* (at the *CDS* – `http://cdsweb.ustrasbg.fr/simbad.html` or at the CfA `http://www.cfa.harvard.edu/simbad`), via telnet, or e-mail. *SIMBAD* has 5 000 individual user accounts in 55 countries. An IP control password is available which allows institutional access to the service. As it is necessary to have an account, prospective users in the US (charges are assumed by *NASA*) should send a request to `simbad@cfa.harvard.edu`. Others (European charges are covered by *ESO* and *ESA*) should enquire in their libraries, or at `questions@simbad.u-strasbg.fr`.

SIMBAD contains more than 1.5 million objects, more than 4.3 million identifiers and more than 95 000 references from 90 journals (since 1950, for stars and since 1983, for extragalactic objects). The data included in *SIMBAD* evolves constantly as the measurements are updated and more references are included daily (in 1994 there were 500 new entries per month).

Its most important characteristic is the cross-identification of objects. *SIMBAD* contains a cross-index to several hundred astronomical catalogues and logs of space observatories.

It can be queried by name, co-ordinates, or reference code. *SIMBAD* is the result of a cooperative effort among several institutions. The bibliography is compiled by the Institute d'Astrophysique, in Paris, and the Bordeaux Observatory; the catalogues by the Observatoire du Midi-Pyrenées, the Paris and Besançon Observatories, Astronomisches Recehen-Institut (Heidelberg) and the Institute for Astronomy of the Russian Academy of Sciences. Corrections to the databases are made by the Montpellier Observatory.

As was the case with *NED*, *SIMBAD* can also be used in a client/server mode to retrieve essential information on objects to be included in other programs.

What started with *SIMBAD* was continued with an impressive collection of browsers and catalogues, a true astronomer's Bazaar (`http://cdsweb.u-strasbg.fr/Cats.html`).

3.5. *VizieR*

VizieR (`http://cdsweb.u.strasbg.fr/cgi-bin/VizieR`) is a collection of around 2400 catalogues (1800 already available on-line) that can be queried either by the specific entry on it, by author name, a set of keywords that specify the object, the wavelength of interest, and the mission, or through a map based on a neural network analysis of the keywords associated with the catalogues. Since having Hipparcos on-line in May 1997, the use of *VizieR* has increased exponentially.

VizieR is a cooperative endevour between the *CDS*, the Astronomical Data Center in the US, the Russian Academy of Sciences's Institute for Astronomy (*INASAN*), the National Astronomical Observatory in Japan, the Beijing Observatory in China, and the Indian Inter-University Centre for Astronomy and Astrophysics (*IUCAA*). The *CDS* created a standard which is followed by these data centres and (more importantly!) by the journals in the publication, presentation and distribution of data.

Future enhancements include links to *SIMBAD* and other external archives.

3.6. *Aladin*

Aladin is an interactive sky atlas that includes images from the Digitized Sky Survey and, for some regions (the *LMC*, *SMC* and Galactic plane, for example) special images with 1″ resolution. It also includes a cross-identification tool of the objects in the image. Aladin is only available via an X-term version, a *WWW* port is being developed at the moment.

3.7. *Dictionary of Nomenclature of Astronomical Objects*

A by-product of *SIMBAD*, the Dictionary of Nomenclature of Astronomical Objects, which can be found at `http://simbad.u-strasbg.fr/cgi-bin/Dic`, is an up-to-date electronic version of the "Second reference dictionary of the nomenclature of celestial objects" (Lortet *et al.* (1994)).

3.8. *Information & On-line Data in Astronomy*

At the *CDS* there is an on-line version of the book "Information & On-line Data in Astronomy" (`http://cdsweb.u-strasbg.fr/data-online.html`, Egret & Albrecht (1995)). Each section of the book has links to most of the archives and centers of interest.

3.9. *StarPages*

StarPages (`http://vizier.u-strasbg.fr/starpages.html`) is a collection of directories, dictionaries and astronomical databases:

• Starheads (`http://vizier.u.strasbg.fr/starheads.html`) is a list of around 4000 astronomers world-wide.

• StarWorlds (`http://vizier.u.strasbg.fr/starworlds.html`) lists more than 6000 organizations world-wide.

• StarBits (`http://vizier.u.strasbg.fr/starbits.html`) lists around 115 000 acronyms and abreviations

3.10. *NSSDC*

The *National Space Science Data Center* (*NSSDC*) provides access to and distributes a wide variety of data obtained with several *NASA* missions. Not only astronomical images and spectra are included, but also space plasma data and Earth and space photographs. It is one of the largest astronomical archives.

3.11. *Leda*

The *Lyon-Meudon Extragalactic Data Base, Leda* (telnet `leda.univ-lyon1.fr`; **login:** leda) includes on-line versions of the Principal Galaxies Catalog (more than 73 000 objects), the Third Reference Catalog of Bright Galaxies (more than 23 000 objects), HI catalogues and the Catalog of individual optical velocity measurements (*OVC*). Around 70 parameters are included for each of more than 95 000 objects. Among them, the position, flux at different wavelengths and colours, redshift or radial velocity, and surface brightness. The references for all listed measurements are also included. All the parameters are listed in a homogeneous way, calculations are made only from accepted models.

Images of many of the galaxies can be retrieved, as well as a map of the region, listing all the objects, with their PG number. The output figures can be drawn with, or without the stars in the field.

A query can be made by e-mail to: `ledamail@lmc.univ-lyon1.fr`, if the list of objects is long.

3.12. *Data*

The very handy "Handbook of Space Astronomy and Astrophysics", by M.V. Zombeck (Zombeck (1990)), is available on-line (`http://adswww.harvard.edu/books/hsaa/idx.html`) thanks to Cambridge University Press.

Other useful on-line information on astrophysical objects is stored at the *CDS* Web site (`http://cdsweb.u-strasbg.fr/data-online.html`).

3.13. *Other images, lists and catalogues*

Among the billions and billions of bits in the Web, some other useful lists and archives are worth a mention:

• Catalog of non-stellar objects (`http://adac.mtk.nao.ac.jp/prt7.html`).

• Catalogue of Messier objects (`http://www.seds.org/messier`).

• NGC images (`http://maxwell.sfsu.edu/asp/ngc.html`).

• `http://www.obspm.fr/department/darc/planets/encycl.html` for information on extra-solar planets.

Survey/ telescope	Epoch	Band	Dec. range	Limit mag.	Pixel size μm	comments
Pal QV	1983-85	V	+90:+06	19.5	25	*GSC*
SERC J	1975-87	J	−20:−90	23.0	25	*GSC*
SERC-EJ	1979-88	J	−00:−15	23.0	25	
POSS-I	1950-58	E	+90:−30	20.0	25	up to dec −18
POSS-I	1950-58	O	+90:−30	21.0	25	selected plates
POSS-II	1987-96	J	+90:+00	22.5	15	
POSS-II	1987-97	R	+90:+00	20.8	15	
POSS-II	1987-97	N	+90:+00	19.5	15	
SES	1990-96	R	−20:−90	22.0	15	
SERC ER	1990-95	R	−00:−15	22.0	15	
SERC	1986-88	V		14.0	25	Galactic plane; *GSC*
SERC	1979	V		14.0	25	LMC; *GSC*
SERC	1985	V		14.0	25	SMC; *GSC*
SERC	1985	V		14.0	25	M31; *GSC*

TABLE 3. Scanned plates at *STScI* available on-line.

3.14. *Scanned data*

3.14.1. *The Digital Sky Survey*

As part of its activities leading to the Guide Star Catalog (*GSC*), photographic plates covering the whole sky were scanned. For the northern hemisphere, a new set of shorter exposure plates (the Quick-V plates) were obtained. For the south, the *ESO/SERC* Southern J plates were used. In addition, some regions were covered by special plates. In parallel with the scanning process, the original Palomar Sky Survey (*POSS-I*) XE plates were scanned. The last and the southern plates constitute the Digitized Sky Survey (*DSS*). A detailed description of the *GSC*, its contents and the plates can be found in Lasker *et al.* (1990), Russell *et al.* (1990) and Jenker *et al.* (1990).

Once these scans were finished, the two *PDS* machines at *STScI* were completely refurbished. After an agreement with the Palomar Observatory, the plates that constitute the *POSS-I* are also being scanned. All these scans are available on-line at the following *URL*: http://archive.stsci.edu/dss/.

The original scans were made with a pixel size of 25μm, resulting in a 14 000×14 000 pixel image, with a scale of 1.67″ per pixel. A 512×512 image thus covers an area of 14′.3×14′.3. The new scans are made with a pixel size of just 15μm, resulting in a raster of 23 000×23 000 pixels.

Please note that to provide co-ordinates for targets to be observed with the Hubble Space Telescope it is necessary to use only the plates used to create the *GSC* and not the plates from the *POSS-I*.

The Digitized Sky Survey is a set of CD-ROMs containing the *POSS-I* XE and *SERC-J* scanned plates, compressed by 10×. It covers the whole sky. A 100× compressed version, called RealSky, is presently available only for the northern sky, with the southern part to be released soon. Both sets are sold by the Astronomical Society of the Pacific.

Table 3, which is based on Lasker *et al.* (1996), describes some of the characteristics of these scans.

These new scans will be used to construct the *GSC*-II. This updated version will include 2×10^9 objects (100 times more than *GSC*-I), at at least two epochs and two bands; proper motions will be calculated with an error of less than 4 mas/yr and a magnitude error between 0.1 and 0.2 mags (*GSC*-I did not contain any proper motion information and the magnitude error is 0.4 mags).

3.14.2. *Automatic plate scanner*

The University of Minnesota created the Automatic Plate Scanner Catalogue (*APS*) from the scans of *POSS-I* O and E plates with $|b| > 20°$. The catalogue consists of a list of classified objects found on both plates using a neural network algorithm to separate the stellar and non-stellar objects. The two catalogues (one for each band) are available on-line (`http://aps.umn.edu`), they contain co-ordinates, magnitudes, colours and other parameters. It provides information on 10^8 stars and around 10^7 galaxies to a limiting magnitude of better than 21 in the blue.

3.14.3. *Plate database*

The Institute of Astronomy of the Bulgarian Academy of Sciences maintains a database of wide-field plates (`http://www.wfpa.acad.bg`). The insformation is standardised and lists the object name, emulsion and filter, J2000 co-ordinates and the UT time of the exposure.

3.15. *Optical and UV surveys*

A list of optical and UV surveys was generated after IAU Symposium 179. These can be accessed at `http://www-gsss.stsci.edu/iauwg/survey_url.home`. Among them are:

3.15.1. *Optical*

• The Sloan Digital Sky Survey (*SDSS*) (`http://www.sdss.org`) will be a complete survey of π steradians, using 30, 2048×2048 CCDs in five bands and two fibre-optic spectrographs. It will consist of a photometric catalogue of 10^8 galaxies, 10^8 stars and 10^6 quasars; a spectroscopic catalogue including emission and absorption lines, images in five bands and 1-d spectra of 10^6 galaxies, 10^5 stars and 10^5 quasars and 10^4 clusters. As the size of the *SDSS* will probably exceed 10^{10} bytes, maintaing, distributing and querying it will be quite a challenge.

• *ESO* slice project (`http://bcas5.bo.astro.it/ cappi/esokp.html`) is a galaxy redshift covering 30 square degrees near the South Galactic Pole, complete to $b_J = 19.4$. It includes the redshifts for 3000 galaxies.

• Two degree Field (*2dF*) QSO redshift survey: will determine the large-scale tridimensional structure of the Universe up to $z \sim 0.1$ by measuring the redshift of 250 000 galaxies. In addition, spectra of 30 000 quasars will be obtained. This survey may be located at either `http://www.aao.gov.au/local/www/rs/qso_surv.html` or, alternatively, at `http://www.ast.cam.ac.uk/AAO/www/rs/qso_surv.html`.

3.15.2. *UV*

• *FOCA* was a small airborne telescope which surveyed 70 square degrees at 2000Å, with an angular resolution of $12''$ and $20''$, the catalogue which was generated (found at `http://www.astrsp-mrs.fr/~xbig/foca/node2.html`) includes galaxy counts and colour distributions for sources between magnitudes 15 and 18.5.

• Alexis (`http://alexis-www.lanl.gov`) is an *EUV*-X-ray satellite surveying the sky to map the diffuse soft X-ray background, to study flaring stars and to make observations of *EUV* sources. Any object detected is immediately compared with lists of known objects to determine its possible nature.

3.16. *Multiwavelength maps*

Several institutions provide on-line facilities that create maps of the sky at one or more wavelengths. Among them are:

- SKYVIEW (http://skyview.gsfc.nasa.gov/skyview.html) provides an interface to obtain images from radio through to gamma-ray wavelengths.
- SkyCat (http://archive.eso.org/skycat) accesses the *HST*, *ESO*'s NTT and the *CFHT* archives to retrieve images and displays them with a variety of options to identify objects on them.
- Astronomical Digital Image Library contains "books" with *FITS* images taken at different wavelengths (http://imagelib.ncsa.uiuc.edu/imagelib.html). It provides a convenient way to archive and distribute the data

3.17. *Astrometry*

The Hipparcos catalogue contains 118 218 astrometric entries (milliarcsecond position, parallax and proper motion) with a median error of 1mas. The *Tycho* catalogue on the other hand, includes more than a million objects with a median error of 7mas, for the brighter sources, and 25mas, for the fainter ones. More than 100 epochs were acquired for the photometry of the objects in the Hipparcos catalogue. The photometric errors range from 0.0004-0.0007 magnitudes for the brighter stars.

These catalogues are now the standards for astrometry. The catalogue epoch is J1991.25; the equinox, J2000.0. The European Space Agency distributes these catalogues as a series of 16 books and several CD-ROMs. The catalogues can be queried in the *WWW* (http://astro.estec.esa.nl/SA-general/Macc/hip.html).

3.18. *Artificial data*

Simulated data is useful, among other things, to, for example, test algorithms, check photometry results, predict exposure times, or optimise an observation using deconvolution techniques.

- Tiny Tim (http://scivax.stsci.edu/~krist/tinytim.html) creates *HST* point spread functions. A *PSF* can be generated for any instrument with any configuration, any wavelength and position. Tiny Tim runs on VAX/VMS, Unix and Linux.
- The *STSDAS* (http://ra.stsci.edu/*STSDAS*.html) synphot package simulates photometric data and spectra as they are observed with the Hubble Space Telescope. Its output is used also in the exposure time calculators that users query to determine the number of orbits needed for their projects. synphot can be used to estimate count rates obtained by other telescopes by changing the components lookup tables and instrument graph. A manual is available on-line.

3.19. *Other Meta Databases*

- Astrobrowse allows the user to retrieve information from several sites using one single query form. The user only needs to specify a position (or a name that will be resolved into a position), a search radius and a list of the services (as available from different organisations) that can be accessed. At the moment only a prototype exists on-line (http://www.clark.net/pub/warnock/ASTROprofile.html).
- *AMASE* (http://amase.gsfc.nasa.gov) is an on-line multi-mission and multi-spectral range catalogue with pointers to locate data stored in different *NASA* archives. Besides the information on each object (name, co-ordinates, colour, radial velocity or redshift, proper motion, etc.), *AMASE* provides very useful information about the observation: the mission; instrument; instrument configuration and detector used; the

pointing; time of observation and roll angle; the target; proposal and Principal Investigator.

• The Astronomical Data Center (`http://adc.gsfc.nasa.gov/`) contains more than 800 catalogues and 750 tables culled from the literature. The *ADC* has produced 3 CD-ROMs with astronomical catalogues (a fourth will have been made available at the end of 1997). A list of the recently incorporated catalogues is included in the monthly *ADC* Newsletter, distributed electronically and also stored at the *ADC* site.

3.20. *Software*

3.20.1. *IRAF (STSDAS and PROS)*

The most widely used astronomical reduction and analysis package is *IRAF*. It is available at *NOAO* (`http://www.noao.edu/iraf/`). Other packages, like *STSDAS* or *PROS*, with tasks tailored for specific reductions and analysis run under the *IRAF* kernel. Ports of *IRAF* exist, among others for Unix and Linux.

3.20.2. *ASDS*

The Astronomical Software Directory Service (`http://asds.stsci.edu/asds/`) is a collection of software-related documents and the means to perform full-text searches on them. This collection is made up of
• high-level package descriptions
• on-line documentation
Users can browse the contents, or list their software in the *ASDS*.

Another useful listing of software packages and utilities is provided by *SEDS* and can be found at `http://www.seds.org/billa/astrosoftware.html`.

3.20.3. *Starlink*

Selected packages are available on-line at the Starlink Software Store and can be accessed at `http://star-www.rl.ac.uk/cgi-store/storestop`. The distribution method is changing as the primary method in the future will be by CD-ROM instead.

3.20.4. *Software-related publications*

The latest software developments are usually presented at the Astronomical Data Analysis Software and Systems (*ADASS*) conferences. Their proceedings are published by the *ASP*. Another series were the *ESO-ST-ECF* Data Analysis conferences, whose proceedings were published by *ESO*.

3.20.5. *IDL*

The Interactive Data Language (*IDL*), developed by *RSI* (`http://www.rsinc.com`) is widely used for reduction and analysis. It is a commercial package with ports to many systems, from laptops to workstations.

3.20.6. *Linux for astronomy*

The Random Factory (`http://home.earthlink.net/~rfactory/lfa.html`) publishes a CD-ROM with a collection of the most popular reduction and analysis astronomical software that run under the Linux operating system. The packages include:
• *AIPS* and *MIRIAD* for radioastronomy.
• *IRAF* and *MIDAS*.
• *GYPSY* and *Karma* for image processing.
• *PGPLOT/PGPERL* for plotting and visualisation.
• *NEMO* for stellar dynamics simulations.

- *SAOImage* and *SAOtng* for image display.
- *Starbase* – an astronomy database.

3.20.7. *Statistics*

Some basic statistics computations are available in the `ttools` and `statistics` packages in *STSDAS*. More serious techniques, like multivariate analysis can be found in the *Statistical Consulting Center for Astronomy* (`http://www.stat.psu.edu/scca`). This site, besides being a portal to a service for statistical queries, includes also links to code (`http://www.stat.psu.edu/statcodes`).

3.21. *Planetary Information*

There are several "tours" of the solar system. The most popular ones are:
- `http://nssdc.gsfc.nasa.gov/imgcat/`
- `http://www.seds.org/nineplanets/nineplanets/nineplanets.html`
- `http://pds.jpl.nasa.gov/planets`
- `http://photojournal.jpl.nasa.gov`

3.22. *Organisations*

- American Astronomical Society (`http://www.aas.org/`).
- *CADC* (`http://cadcwww.dao.nrc.ca/`).
- European Space Agency (`http://www.esrin.esa.it/`).
- *ESO* (`http://www.hq.eso.org/eso-homepage.html`).
- *ESA* Astrophysics Division (`http://astro.estec.esa.nl/index.html`).
- *IAC* (`http://www.iac.es`).
- *NASA* (`http://www.gsfc.nasa.gov/NASA_homepage.html`).
- *ST-ECF* (`http://ecf.hq.eso.org/ST-ECF-homepage.html`).
- *STScI* (`http://www.stsci.edu`).
- UN Office for Outer Space Affairs (`ftp://www.un.or.at/oosa/homepage.html`) which organises workshops on Basic Space Science (`http://www.seas.columbia.edu/~ah297/un-esa/index.html`).

3.23. *How am I going to remember this?*

Fortunately there are on-line compilations of most of the information presented in this section. In particular, the AstroWeb consortium is probably one of the best places to start surfing. There are several versions of AstroWeb at:
- *STScI* (`http://www.stsci.edu/science/net-resources.html`).
- *NRAO* (`http://fits.cv.nrao.edu/www/astronomy.html`).
- *CDS* (`http://cdsweb.u-strasbg.fr/astroweb.html`).
- *MSSSO* (`http://msowww.anu.edu/anton/astronomy.html`).
- *VILSPA* (`http://www.vilspa.esa.es/astroweb/astronomy.html`).

All of them validate the links daily and report on those which are not accessible.

Finally, the web crawlers and cyber-robots that mine the information superhighway are very helpful to find obscure links. Some are:
- Altavista (`http://www.altavista.com`).
- Yahoo (`http://www.yahoo.com`).
- Lycos (`http://www.lycos.com`).

	Short Wavelength	Long Wavelength
Wavelength range (Å)	1150 - 1950	1900 - 3200
Resolving power	1.2×10^4	1.3×10^4
Low dispersion resolution (Å)	6	8

TABLE 4. *IUE* Spectrographs

4. Optical and UV Archives

4.1. *Optical archives*

Several observatories archive routinely their data and make it available to the community. Most of these catalogues have a name resolver data query.

4.1.1. *ESO*

The *ESO* archive (http://archive.eso.org) contains all the *NTT* observations since April 1, 1991. Data can be requested after one year from the observation date. The data is stored uncalibrated, so remember to request the calibration files as well.

4.1.2. *CFHT*

All infrared and optical images obtained with *CFHT* since late September 1992 are archived (http://cadcwww.dao.nrc.ca/cfht/menu.html). After the proprietary period of 2 years they are widely available through the *WWW* interface.

4.1.3. *ING*

Around 430 000 observations obtained with the William Herschel Telescope, the Isaac Newton Telescope and Jacobus Kapteyn Telescope is archived by the Isaac Newton Group (http://ing.iac.es). The proprietary time for these data is one year.

4.1.4. *NOAO*

The Save the Bits program started at *NOAO* in 1995. The images are stored in ExaByte tapes in *FITS* format (http://www.noao.edu). The proprietary period for *KPNO* and *CTIO* data is 18 months.

4.2. *Other Ultraviolet Satellites*

4.2.1. *International Ultraviolet Explorer (IUE)*

The International Ultraviolet Explorer (*IUE*) launched on January 26, 1978 had an extremely successful 18+ years of operations. Its detectors could obtain spectra from 1150Å to 3200Å. The telescope consisted of an f/15 45-cm beryllium mirror. The spectrographs obtained data in a low- or high-resolution mode (6 and 0.1Å respectively). At high resolution it was capable of observing sources up to 12mag, while at low resolution the limiting magnitude was 17. Some characteristics of the spectrographs are shown in **Table 4**.

IUE data has been widely distributed, in particular the Uniform Low Dispersion Archive (*ULDA*).

4.2.2. *IUE data*

Spectra obtained with *IUE* consists of a 1-D file with 768 pixels. All reduced images contain a photometrically corrected image file and an extracted spectrum file. All short

	short	medium	long
Bandpass (Å)	70-190	140-380	280-760
Resolution element (Å)	.367-.636	.731-1.27	1.46-2.54
Filter material	Lexan/boron	Al/carbon	aluminium
Transmission (Å)	1588/1172	1566/431 1553	

TABLE 5. *EUVE* spectrometer characteristics

(*SWP*) and long (*LWP*) wavelength data has been reprocessed. The Space Telescope Science Institute is the repository of *IUE*'s final archive. Retrievals can be made at `http://archive.stsci.edu/iue`. European users might prefer to use the Spanish *IUE* node `http://www.vilspa.esa.es/iue`. The Space Telescope Science Institute will be the repository of *IUE*'s final archive.

4.2.3. *Extreme Ultraviolet Explorer*

The Extreme Ultraviolet Explorer (*EUVE*) was launched on June 7, 1992. The satellite has 5 instruments: four photometric imaging systems and a three-channel *EUV* spectrometer capable of obtaining astronomical observations between 70 and 760Å.

During the first phase of the mission a four-band (at 100Å, 200Å, 400Å and 600Å) sky survey in the extreme ultraviolet and a long exposure "deep survey" were conducted. The catalogues and lists resulting from this survey are published in Malina *et al.* (1994) and Bowyer *et al.* (1996), and include 400 *EUVE* sources.

During the on-going second part of the mission a Guest Observer (*GO*) program consisting of pointed spectroscopic observations is carried out.

Each detector has an area of 72.44-cm^2; other characteristics of the *EUVE* spectrometers used in the *GO* phase are shown in **Table 5**.

4.2.4. *EUVE Right Angle Program*

This is a public program that makes data immediately available to the community. The four imaging telescopes obtain these images while the spectrometers are integrating for a Guest Observer program. *RAP* data has been available since the summer of 1996. The exposure time for these observations is such that allows the detection of sources ten times fainter than in the original survey.

4.2.5. *EUVE Data*

After the proprietary period expired (usually 1 year after the observations were made), the data is available through *ADS* and the *EUVE* public data archive. The data is distributed as a set of *FITS* files and tables. Its content is described in the *EUVE GO* Handbook available on-line at *CEA* (`http://www/cea.berkeley.edu/`)

The *CEA* now maintains a dedicated *ADS* node to provide sky survey data, *EUVE* calibration data, and pointed observations.

The on-line data available consists of:
- Surveys:
  - (*a*) sky maps from the survey observations made wit the scanners
  - (*b*) deep survey sky maps
  - (*c*) catalogues of *EUV* sources
- *GO* data:

(*a*) raw data in 3-D *FITS* tables

(*b*) binned *FITS* images

(*c*) 1-d extracted spectra

(*d*) 1-d extracted spectra for calibration sources

A list of the pointed observations can be found at:

`http://www/cea/.berkeley.edu:80/~science/data/dynamic/exposures.html`

A very handy Spectral Data Browser:

(`http://www.cea.berkeley.edu:80/cgi-bin/archive-select`)

allows the user to select and retrieve public data directly from the *WWW* The selection can be done by spectral type or by position.

Data is also periodically released on CD-ROM. The issued volumes include:

• Volume 1.1: all sky survey images and pointed spectroscopic calibration observation data for AT Mic.

• Volume 2.1A: pointed spectroscopic observations of 4 white Dwarfs (WD1845+019, Feige 24, G191−B2B and WD0549+158).

• Volume 2.1B pointed spectroscopic observations of late stars (Capella and Procyon).

• Volume 2.1C pointed spectroscopic observations of the white dwarf WD1620−391, the cataclysmic variable RE1938−461, the BL-Lac object PKS2155−304, the late type star HR1099 and the Moon.

• Volume 2.2A pointed spectroscopic observations of three white dwarfs (WD2309+105, WD1254+223 and HZ43).

• Volume 2.2B pointed spectroscopic observations of three late-type stars (σ Gem, AD Leo and α Cen A).

• Volume 2.2C pointed spectroscopic observations of the B star ϵ CMa, the cataclysmic variable AN UMa and the planet Jupiter.

4.2.6. *Astro Missions*

The two Astro missions flew on board of the space shuttle and consisted of spectrographs, photometers and ultraviolet cameras. Astro-1 lasted 9 days in December 1990 and Astro-2 16 days in March 1995.

The instruments on board the Astro missions were:

BBXRT (Broad Band X-Ray Telescope) produced spectra between 0.3 and 10keV with a resolution of 90-200eV FWHM of around 80 sources during the Astro-1 mission

HUT (Hopkins Ultraviolet Telescope) produced spectra in the far UV (912Å-1850Å) with a 3Å resolution and in the extreme UV (415Å-912Å) with a resolution of 1.5Å for more than 300 sources

UIT (Ultraviolet Imaging Telescope) was a 38cm f/9 Ritchie-Chrétien telescope with two selectable cameras with a wavelength range between 1150Å and 3500Å. *UIT* included a set of Near and Far UV filters. The images were obtained on 70-mm film that was scanned with different pixel sizes.

WUPPE (Wisconsin Ultraviolet Photopolarimeter Experiment) obtained simultaneous spectra and polarisation measurement of 329 sources between 1400Å and 3200Å. The limiting magnitudes for these sources were between 0 and 8 mag. *WUPPE* produced images in three sizes:

• field : 320×256 pix (3.3′ ×4.4′), 1-bit for identification and acquisition.

• zoom : 30×40 pix (24″ ×32″) for fine pointing and focus.

• downfield : 320×256 pix (3.3′ ×4.4′), 6-bit for detailed images.

4.2.7. *Astro Data*

The data produced during the two missions is accessible through the Multimission Archive at *STScI* at `http://archive.stsci.edu/mast.html`

The Web Interface for Searching Archival Research Data (*WISARD*) provides a way to retrieve archival data located in different archival sites. *WISARD* can be found at `http://hypatia.gsfc.nasa.gov/wisard/wisard.html`.

4.2.8. *Far Ultraviolet Spectroscopic Explorer (FUSE)*

FUSE (`http://fuse.pha.jhu.edu`) to be launched in 1999 will obtain high resolution between 905Å and 1195Å. It contains four slits:

- high resolution aperture (1.25″ ×20″).
- high throughput aperture (4″ ×20″).
- large square aperture (30″ ×30″).
- pinhole aperture (∼0.5″).

The *FUSE* spectra will complement those obtained by *HST* as they are further in the UV.

5. The Hubble Space Telescope

The shuttle Discovery launched the Hubble Space Telescope (*HST*) on 24 April 1990. It is an f/24 telescope with a 2.4m primary mirror. Its five science instruments can obtain astronomical observations from the far UV to the IR. Each three years, it is visited by a team of astronauts for a Servicing Mission in which some necessary tune-ups, change of instruments and orbital boosters are performed.

At present, the on-board instruments are three cameras (*FOC, NICMOS & WFPC2*), one spectrograph (*STIS*) and three *FGS*s. These instruments are located in radial and axial bays on the satellite. During the first servicing mission in 1993, the Wide Field and Planetary Camera (*WF/PC*) and the *HSP* were replaced by the *WFPC2* and the pseudo-instrument *COSTAR* (Corrective Optics Space Telescope Axial Replacement), respectively. *COSTAR* actually is an optical bench with a series of arms to which mirrors are attached that intercept the beam and correct the spherical aberration of the primary mirror (only *FOC* still uses *COSTAR*, all the new instruments correct internally for the aberration). During the second servicing mission in February of 1997 the astronauts replaced the *FOS* and *GHRS* with *NICMOS* and *STIS*. The *FOC* will be replaced during the next servicing mission with the *ACS*. Finally, during the 2002 servicing mission, *COS* will be installed in the *COSTAR* bay.

To point the telescope, a pair of Guide Stars is observed with two *FGS*s. In the fine lock mode, the *FGS*s keep the telescope fix (with an RMS of 0.007″) during the observation. The *FGS* that is not used for pointing can be used for astrometric observations.

WFPC2 has four CCDs, shaped as an L with a 2.5′ length. With this camera, it is possible to take images between 1200 and 11000Å with a limiting magnitude of V∼28 (during "normal" operations, campaigns like the Hubble Deep Field have pushed the limit to I∼30). The *FOC* can take high-resolution images in the UV with a field of view of 14″ ×14″ and a limiting magnitude V∼26.5. With a larger field and sensitivity *STIS* can also take UV images and spectra. *STIS* can also take images and two-dimensional spectra from 1150Å to 1.1μm. *NICMOS* has three cameras that can obtain images and spectra in the near IR from 0.8μm to 2.5μm and a limiting magnitude of K∼21 in a typical one orbit integration.

The *FOS* obtained low-resolution spectra with its two detectors, one of which had a range further into the "blue" than the other. The *GHRS* was able to obtain high-resolution spectra with R between 2000 and 80 000, the last one with an echelle.

The following sections briefly describe the most used instruments. In the case of the *WFPC2* also the calibration pipeline while in the *STIS* section a description of the data path is included. More information and detailed descriptions can be found in the *HST* Data Handbook and in the *WWW* instrument pages. **These** extremely brief **notes constitute a very succinct description of each instrument and its calibration (correct at the time they were written, but whose exactness will change when more information is known about the instruments and their data). Readers should consult the instrument *WWW* pages** (http://www.stsci.edu/instruments) **and the *HST* data Handbook before any serious work with *HST* data.**

HST data is more complex than ground-based, in particular the data from *NICMOS* and *STIS*, as explained below, is more complicated than for the other instruments. It was decided, then, to spend more time explaining the "simple" *HST* data to let the readers become familiar with it and summarise the characteristics of these more complex data sets.

5.1. *Fine Guidance Sensors (FGS)*

The three *FGS*s† on board the *HST* are primarily used to point the telescope accurately and to control its stability. The third *FGS* can be used for astrometry while the other two are locked on target. These instruments are ideal to study close binary stars, to determine proper motions and parallaxes, as well as to detect companions.

Relative positions between multiple components of a stellar system can be measured to around 3mas up to a limiting magnitude V=17, using one of the operational modes (*POS*). For the *TRANS* mode, the values are 10 mas and V=16, while the difference in component magnitudes is 4.

The *FGS*s are essentially interferometers, with a total *FOV* of 69 square arcminutes and an instantaneous *FOV* of 5″×5″. Each *FGS* sensor includes a filter wheel with 5 filters. In their optical path there is a pick-off mirror, an aspheric collimator, two star selectors, a polarising beam splitter, corrective optics and a Koester's prism. Light is collected in four *PMT* included in each *FGS*.

Three observing modes are available:

• *POS*, used to measure accurately relative position differences (with an error of 0″.003)

• *TRANS* to measure angular separation of multiple systems, or the sizes of extended objects.

• *SCAN* mode is a *TRANS* observation in which the pattern is specified by the user.

To acquire a target, the instrument first searches the region in a spiral pattern. If the search was successful, the *coarse track* phase begins and, finally, the *fine lock*.

FGS data are the so-called *S-curves*, the graphical representation of the count difference of the two photomultipliers (normalised by the sun of these counts) versus the tilt angle of the wavefront at the entrance of the Koester prism. An example of this transfer function is shown in **Figure 1**.

---

† A detailed description can be found in the latest edition of the instrument handbook available on-line at http://www.stsci.edu/ftp/instrument_news/fgs/html/TOPfgs.html

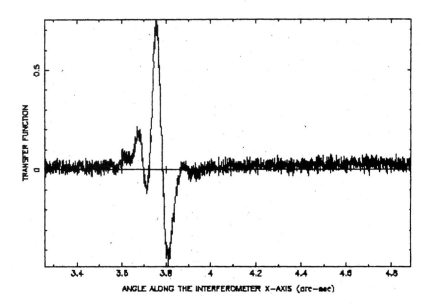

FIGURE 1. The transfer function for a typical *FGS* observation.

5.2. *The Faint Object Camera (FOC)*

The *FOC*‡ is one of the axial instruments on board *HST*. It is a photon counter that obtains images and spectra between 1100Å up to 6500Å. The *FOC* is diffraction limited.

The camera has two relays each with an independent optical path, filter wheels and detectors. They are named from their original (i.e., pre-*COSTAR*) focal length: f/96, f/48 and f/128. When *COSTAR* was installed the focal length changed to f/75.5 and f/151 for the first two. The f/128 relay due to the extremely small size is not used.

A video processor in the camera counts and determines the position of each incident photon that generated a sparkle in the photocathode. This position and number are accumulated in the memory of the instrument's computer. This activity is very sensitive to the speed with which the video-processing unit reacts to the arrival of the photons. The memory unit can store 16 bits per pixel in a 512×512 image and 8 bits in the zoomed format of 512×1024 pixels.

Some characteristics of the *FOC*, for a 512×512 image, are shown in **Table 6**:

The *FOC* has a set of 58 narrow- and broadband filters. In particular the f/96 camera has 3 long-pass filters, 9 broad, 20 medium, 2 narrow-band in addition to the 5 neutral density filters that can be used in different combinations to perform the observations. The camera also has two objective prisms and 3 polarisers. More details about the filters can be obtained in the camera handbook.

The f/48 mode was only used for spectroscopy between shortly after launch until February 1997. It is not now used.

---

‡ A detailed description can be found in the latest edition of the instrument handbook available on-line at http://www.stsci.edu/ftp/instrument_news/foc/html/topfoc.html

	F/48	F/96
Noise	0.002 cts/sec/pix	0.0007 cts/sec/pix
Saturation limit	0.50 cts/sec/pix	0.65 cts/sec/pix
Typical *FOV*	14″ × 14″	7.3″ × 7.3″
Largest *FOV*	28″ × 28″	14.6″ × 14.6″
Scale (post-*COSTAR*)	0.01435 ″/pixel	0.028 ″/pixel
Scale (pre-*COSTAR*)	0.0225 ″/pixel	0.0451 ″/pixel

TABLE 6. *FOC* characteristics.

5.2.1. *Distortion*

As it is essentially a cathode ray tube, the images obtained with the *FOC* detectors suffer from a geometric distortion. This distortion has another component due to the axial position of the camera.

To correct this distortion, *reseaux marks* were engraved in the photocathode. These marks, arranged with a 60-pixel separation, make a 17×17 matrix. These "theoretical" positions are then compared to the observed ones and the image is corrected in such a way as to re-obtain a square grid. This is one of the calibration steps.

5.2.2. *What constitutes an FOC observation?*

An *FOC* "observation" is composed of several files whose name starts with an X and followed by 8 alphanumeric characters. The header is a text file whose extension starts with h. The extension of the binary data starts with d.

The different files (in *GEIS* format) that constitute an "observation" are briefly described in the following table. More details can be found in the *HST* **Data Handbook**.

FOC file extensions

Extension	Contents
Uncalibrated data	
d0h/d0d	uncalibrated scientific data
q0h/q0d	file with data quality information
shh/shd	data packet describing the observation
ulh/uld	history of the observation
Calibrated data	
c0h/c0d	intermediate calibration product
c1h/c1d	calibrated data
trl	history of the pipeline calibration

Among the *FOC* header keywords of interest, we can find:

FOC keywords

Parameter	Description
CRVAL1, CRVAL2	RA and dec. of the (CRPIX1, CRPIX2) pixel
EXPTIME	exposure time
ORIENTAT	position angle
PXFORMT	image format (512x512; 512zoomed x1024; etc.)
PHOTMODE	observing mode (FOC F/96 F486N, for example)
PHOTFLAM	inverse of the sensitivity
	flux = counts x PHOTFLAM/EXPTIME)
PHOTZPT	magnitude zero point in the HST scale

5.2.3. *What can be found in an image?*

An image of the M87 jet obtained in the f/96 mode with the F430W filter and the POL0 polariser is shown in **Figure 2**.

FIGURE 2. A typical *FOC* observation. This is an image of the jet of M87 at f/96, with the *POL0* polariser.

In this image, several common details of *FOC* images can be found:

- *reseaux* marks as mentioned above
- borders somewhat distorted due to the geometric correction
- a vertical band to the right of the image, probably due to a variation in the camera reading speed
- some scratches in the photocathode
- parallel lines close to the brighter parts of the image or details of the electron diffraction in the photocathode
- a variation in the noise (not in the signal) that creates a visual distortion ("finger prints")

5.2.4. *Continuing the reduction*

The *reseaux* can be eliminated from the calibrated image (with the `c1h`/`c1d` extensions) using, for example, `images.imedit` or `rremovex`. The North direction can be determined with the `compass` task or with `foc.focutility.north`.

The approximate positions of the sources in the image can be determined using the `rimcursor` task.

```
   image = ''x0i20a02t.c1h"  image to which coordinates should refer
     wcs = ''world")         output world coordinate system name
wxformat = ''%H")            x coordinate output format
wyformat = ''%h")            y coordinate output format
  cursor = ''")              image cursor
    mode = ''h")
```

To determine better positions, this initial measurement can be complimented with `images.imcntr` and `stsdas.toolbox.imgtools.xy2rd`.

5.3. *Near Infrared Camera and Multi-Object Spectrometer (NICMOS)*

NICMOS† can obtain images and spectra between 0.8 to 2.5 μm. It has three independent cameras (*NIC1*, *NIC2* and *NIC3*) with different magnifications imaging on 256×256 pixel HgCdTe arrays. Each camera has its own collection of broad, medium and narrow-band filters. *NIC1* and *NIC2* have a set of polarisers while *NIC3* has grisms, in addition *NIC2* has a coronographic hole. Some characteristics of the cameras are detailed below :

Most observations are performed in the *MULTIACCUM* mode. In this mode, the images are read at intermediate stages of the integration and at the end. All these intermediate images are then stored and transmitted to the ground. There are some predefined sequences for this sampling, each with 25 readout times as detailed in the Instrument Handbook. The selection of a particular sequence depends on the type of observation and the target.

To eliminate detector non-uniformities, removing the thermal background and to map areas larger than the field of view of *NICMOS* 3 techniques can be used:

- **dithering** small movements of the telescope between the target and an adjacent sky region ($<40''$)
- **chopping** large movements (up to $1800''$) that allow for a proper background subtraction
- **mosaicing** to image an extended source

The Instrument handbook describes several chopping and dithering sequences (linear (along the x- or y-directions of the detector), spiral and square, for example). Each can have between 2 and 40 positions.

† Please refer to the *NICMOS* Instrument Handbook for complete details of the instrument. It can be found at `http://www.stsci.edu/ftp/instrument_news/NICMOS/topnicmos.html`

	NIC1	NIC2	NIC3
Field of View	$11'' \times 11''$	$19.2'' \times 19.2''$	$51.2'' \times 51.2''$
Pixel Size	$0.043''$	$0.075''$	$0.2''$
Spectral Range	0.8-1.8μm	0.8-2.45μm	0.8-2.3μm
Special Filters	[S II], HeI, [FeII], Paα, H$_2$O	Paα, Brγ, H$_2$, HCO$_2$, C$_2$, CO	Paα, [Si VI], H$_2$ CO, HeI, [FeII]
Diffraction limited wavelength	1.0μm	1.75μm	—

TABLE 7. *NICMOS* characteristics.

Suffix	File Contents
_raw	raw science data
_cal	calibrated science data
_ima	intermediate *MULTIACCUM* calibrated data
_mos	Mosaiced data

TABLE 8. Suffixes for *NICMOS* data files.

5.3.1. *What constitutes a NICMOS observation - Associations*

Due to the nature of the data acquisition, it is necessary to combine several exposures to create a useful science observation. These associations are then a collection of exposures that are related and processed (read, calibrated, stored) as a single unit. This new format also implies that *NICMOS* files need to remain in *FITS* format through their reduction and analysis.

These *FITS* files have a three-letter suffix that identifies its contents. Some of the most important are shown in **Table 8**.

Each *NICMOS* science data file is in fact composed of a set of five image extensions:
- the image
- image variance
- data quality map
- number of valid samples at each pixel
- integration time at each pixel

These arrays are stored in five individual *FITS* extensions within the file (see the "File Organisation" chapter in the *NICMOS* section of the *HST* **Data Handbook** for more details).

5.3.2. *NICMOS calibration*

Due to the nature of its data the calibration of *NICMOS* images is done in two parts. The `calnica` program performs the steps that can be done to a single exposure using the header information in a way similar to `calwp2`. In *MULTIACCUM* data, each readout is partially calibrated. The `calnicb` program calibrates and merges the *associations* previously calibrated with `calnica` generating a unique final image from this combination. A third program `calnicc` is used for grism data. Manuals for this program are available from *ST-ECF*.

	CCD	NUV-*MAMA*	FUV-*MAMA*
pixel size (″)	0.05	0.024	0.024
field of view (sq. arcsec)	51×51	25×25	25×25
wavelength range (Å)	2000-11,000	1650-3100	1150-1700

TABLE 9. *STIS* detector characteristics.

5.4. *Space Telescope Imaging Spectrograph (STIS)*

STIS† was installed on board *HST* during the second servicing mission in February 1997. *STIS* can obtain high-resolution imaging and spectroscopy from the UV to the IR. Its echelle gratings are capable of obtaining medium to high spatial resolution spectra in the UV. Long slit or slitless spectroscopy is available for the whole wavelength range at low to medium resolution. As it name implies, *STIS* can also be used to obtain images both in the optical and in the UV.

5.4.1. *Detectors*

STIS uses three large detectors, a "classic" CCD, used for imaging and target acquisition and two Multi-Anode Microchannel Array (*MAMA*) detectors for UV observations. **Tables 9 & 10** summarise some of their characteristics.

STIS has a slit wheel that contains the apertures and slits. It also includes the grating wheel (called the Mode Selector Mechanism - *MSM*) that contains the gratings, the echelles and prisms for spectroscopy and mirrors for imaging. To achieve the full wavelength coverage, there are specific *MSM* positions (grating tilts) at specific central wavelengths that can be specified.

5.4.2. *MAMA detectors*

The *MAMA* detectors are photon-counting detectors sensible to UV radiation. They can be used to take data in an accumulate mode, or as a time series. They offer high spatial resolution and two-dimensional imaging over a large field of view.

Incident photons strike the photocathode creating photoelectrons that pass into the curved Micro-Channel Plates (*MCP*s). They are multiplied to a pulse that is recorded by an anode array behind the photocathode and detected by the detector electronics. In front of the photocathode two different materials have been deposited that make each detector more sensible to near and far-UV radiation.

The peak photocathode response of the FUV-*MAMA* occurs at Lyα while for the NUV is almost flat between 1800Å and 2600Å. Other properties of the detectors are listed in **Table 10**.

It is very important to remember the observational limit brightness when designing *STIS* observations with the *MAMA*s as brighter objects might damage the detectors. If the source exceeds these limits, it is necessary to specify an alternative configuration of the instrument.

MAMA detectors are capable of recording data in *highres* mode producing 2048×2048 images with a 10-30% increase in resolution (but smaller S/N per pixel). This mode also suffers from a lack of flat field reproducibility making it difficult to be of use for high S/N observations.

† Please refer to the *STIS* Instrument Handbook for complete details of the instrument. It can be found at http://www.stsci.edu/instruments

Near-UV *MAMA*	Far-UV *MAMA*	
Photocathode	Cs2Te	CsI
Quantum Efficiency	10% at 2537Å	25% at 1216Å
Dark Count	1.24×10^{-4} cts/pix/sec	6.25×10^{-5} cts/pix/sec
Absolute peak count rate	50 cts/pix/sec	25 cts/pix/sec
Absolute global count rate	300 000 cts/sec	3 000 000 cts/sec

TABLE 10. *STIS MAMA* properties

5.4.3. *Spectroscopic modes*

A detailed description of the 15 spectroscopic modes can be found in Chapter 4 of the Instrument Handbook (Version 2.0 1998, in particular the summary in Table 4.1. The expected limiting magnitude, after integrating an orbit, per spectral resolution element and a S/N=10 for an A-type star is V~21 using the G750L grating, V~14.9 for the G230M and V~11.4 for the E140H echelle (the complete list is in table 4.2).

5.4.4. *Imaging modes*

Selecting *STIS* instead of *FOC* or *WFPC2* to obtain images has several advantages: the detector has a higher throughput over a wider spectral range, a lower read noise and dark current (resulting in higher sensitivity images), true solar blind UV observations with the MAMAs are possible, as well as high time-resolution UV images.

The imaging capabilities with the CCD were designed for target acquisitions. Images can be obtained without a filter, with a longpass, an [O III] λ5007 and an [O II] λ3740 filter. With the *MAMAs*, the filters available include two longpass, two continuum (λ2700, λ1820), a MgII λ2800, a C III] λ1909 and a Lyα filter. More details can be found in the instrument handbook.

5.4.5. *What constitutes a STIS observation?*

All *STIS* data products are *FITS* files and should remain this way. They do not need to be read into *GEIS* or other formats. As any *FITS* file, the header stores the global properties of all the exposures in the files. A series of image extensions contain the data and a specific header to describe it.

Each *STIS ACCUM* data file will contain the following extensions:
- SCI: stores the science values
- ERR: contains statistical errors
- DQ: contains data quality flags

As explained in the section above, *STIS* (and *NICMOS*) *FITS* files will contain multiple science exposures when an associated set is taken. Associations are recognised because they have a zero as the last character in the rootname. The following table summarises the different naming suffixes. Note that the final calibrated spectra are the _x1d, _x2d, _sx1 and _sx2 ones.

Selected STIS File Names

suffix	type	contents
uncalibrated data		
_raw	image	uncalibrated data
_tag	table	timetag event list
_spt	image	planning and telemetry support information
_wav	image	associate wavecal exposure
_asn	table	association file
calibrated data		
_flt	image	flat fielded data
_crj	image	flat fielded and cosmic ray rejected data
_sfl	image	summed flat-fielded data
_x1d	table	aperture extracted, background subtracted, flux and wavelength
calibrated 1-d spectra		
_x2d	image	rectified, wavelength and flux calibrated spectra, or geometrically corrected image
_sx1	table	summed 1-d extracted spectra
_sx2	image	summed 2-d extracted spectra
_trl	table	trailer file

The following table (which as the previous one was extracted from the latest edition of the *HST* **Data Handbook**) details the different extensions that can be found for different observation types. Examples of *STIS* products for different observation types:

Observation type	uncalibrated files	calibrated files
ACQ, ACQ/PEAK	_raw	—
Image *ACCUM* (crsplit or repeatobs)	_raw, _spt, _asn	_flt, _crj, _sx2, _trl
Image *ACCUM* single exposure	_raw, _spt, _asn	_flt, _x2d, _trl
First order spectroscopy, *ACCUM* mode (crsplit or repeatobs)	_raw, _wav, _asn, _spt, _wsp	_flt, crj, _sx2, _trl
First order spectroscopy, *ACCUM* single exposure	_raw, _wav, _asn, _spt, _wsp	_flt, _x2d, _trl
Echelle spectroscopy, *ACCUM* mode	_raw, _wav, _asn, _spt, _wsp	_flt, x1d, _trl
TIMETAG mode (image or spectra)	_tag + *ACCUM* extensions	ACCUM extensions

To determine the different files in an association table the *STSDAS* task `tread` can be used. It will not "read" the *FITS* file, just display the information. The task `infostis` produces an output similar to iminfo for *STIS* data.

Information on the header keywords can be obtained from the *HST* Data Handbook or by querying the on-line Keyword Dictionary (`http://stadtu.stsci.edu/keyword`).

5.4.6. *STIS pipeline*

This section will briefly described how an idea gets transmogrified into a *STIS* spectrum printed in a scholarly paper.

Astronomers from all over the world submit proposals at each *HST*'s cycle Call for Proposals. They are peer-reviewed and the Time Allocation Committee submits the suggested list of scientifically meritorious proposals to the *STScI* Director who makes the final selection.

Successful proposers submit then a Phase II proposal. This file describes in detail the observations that will be performed. They use a graphical editor to create and check this file and *RPS2* to test the feasibility and efficiency of the orbits. Through *RPS2* his Phase II is submitted.

At *STScI* the proposal is ran through an operational version of *RPS2*. It translates the Phase II syntax into other files that will be used to create the commands that the telescope will execute. The structure of the orbits and other information can be analysed in detail at this point. All visits from the submitted proposals are entered into a proposal library upon submission and a program is ran that selects the most appropriate Plan Windows when the observations should be made. After this process (which also entailed the selection of the Guide Stars to be used for the observation) the Project Master Database is updated with the visits to execute for this proposal. These "flight ready" visits are then incorporated in the week-long Science Mission Specification and the commands are loaded to the spacecraft.

Once the observations are executed, they are transmitted to the ground where they are run through Generic Conversion and *FITS* files created. These are the raw data referred to before. During this stage the headers are populated with the calibration information. The program `calstis` is run on the raw data and the calibrated data is generated. After the pipeline the data is compressed and ingested into the optical disks that constitute the *HST* Data Archive. The header information is used to populate the archive tables. After archiving the data is then distributed to the astronomer.

A parallel pipeline exists for the engineering files.

After the astronomer receives the data, it is analysed it is analysed and the final product, a scientific paper, is submitted for publication.

5.5. *Wide Field Planetary Camera 2 (WFPC2)*

WFPC2† was installed on board *HST* during the first servicing mission in December 1993. Since replacing *WF/PC* [1] it became the workhorse of the observatory.

WFPC2 is composed of 4 frontally illuminated CCD cameras. These cameras include optics to correct for the primary's spherical aberration.

A pyramidal mirror divides the incident beam and redistributes it among the four fixed cameras. The F/28.3 Planetary Camera (*PC*) has the highest resolution. The other three, with a F/12.9 and a wider field correspond to the *WFC* (Wide Field Camera). The *WF/PC* had eight cameras and the pyramid was able to rotate, selecting the Planetary or the Wide Field cameras.

† Please refer to the *WFPC2* Instrument Handbook for complete details of the instrument. It can be found at `http://www.stsci.edu/ftp/instrument_news/WFPC2/wfpc2_top.html`

Read Noise	$5e^-$ RMS
Gain	$7e^-$ /ADU or $14e^-$ /ADU
QE at 6000Å, (2500Å)	35% (15%)
WFC2, WFC3, WFC4 Resolution	0.09958, 0.09956, 0.09962 ″/pixel
PC Resolution	0.04553 ″/pixel
WFC Field of view	$2.5' \times 2.5'$ in an L-shape
PC Field of View	$35'' \times 35''$

TABLE 11. *WFPC2* characteristics.

5.5.1. *WFPC2's CCDs*

The camera's CCDs have an array of 800×800, 15μm pixels. Among its characteristics are the ones shown in **Table 11**.

WFPC2 is used generally with the $7e^-$ /ADU gain, the most convenient to observe dim objects. If the target is too bright, it is then possible to switch to the higher gain value and then saturating the well with 53000 e^- instead of 27000 e^-. If the image is saturated, the effects of bleeding or diffraction spikes are observed.

Once a month the camera is heated to eliminate the contaminants that attach themselves to the cool CCDs. In this way the UV sensitivity is regained. This procedure also restores those pixels whose dark counts were abnormally high.

As cosmic rays might be a problem, it is always recommended to split the observations and combine them afterwards, most of them are eliminated in this way. The split images can either be at the same position or one slightly displaced from the other. This second method (dithering) has also the additional advantage of restoring the resolution (severely affected by *WFPC2*'s oversampling).

The pipeline corrects most other anomalies seen in the calibration of the raw data.

5.5.2. *Filters*

The *WFPC2* has 48 filters between the 1200Å and 1μm. The list can be found in the camera Handbook. Please see the Instrument Handbook for a complete detail and spectral coverage of the filters.

The camera does not have a continuum of narrow-band filters. Between 3700Å and 9800Å the ramp filters whose wavelength varies with the position are used. Their width is around 1% or 2% of the central wavelength, being useful then to observe objects no larger than 10″.

The Quad Filters allow observations in [O II] lines with redshifts between 3763Å and 3986Å. The Methane Quads and polarisers with polarisation angles at 0°, 45°, 90° and 135°) are harder to use as some positions are lost due to the L-shape of the camera. The camera has also two solar-blind Wood's filters for far UV imaging.

The broadband filters are similar, but not identical to the Johnson or Cousins ones. Some caution is needed when calculating magnitudes to be sure on what system they are, magnitudes calculated using the header keywords are in *STMAG*.

5.5.3. *What constitutes a WFPC2 observation?*

A *WFPC2* observation is composed of pairs of files whose name starts with U (*WF/PC* files start with W) followed by eight alphanumeric characters. The header is a text file whose extension ends with h, and the binary data's extension with d.

These files are extracted from the Hubble Data Archive in *FITS* format and can be

extension	class	format
d0h/d0d	raw data	integer
q0h/q0d	d0h quality files	integer
x0h/x0d	engineering data	real
q1h/q1d	engineering quality data	integer
shh	standard header packet	ASCII

TABLE 12. *WFPC2* Uncalibrated data files.

extension	class	format
c0h/c0d	calibrated data	realmight be integer)
c1h/c1d c0h/c1h	data quality	integer
c2h/c2d c0h/c0d	histogram	3 lines (original, A-to-D, calibrated)
c3t	photometry table	binary table

TABLE 13. *WFPC2* Calibrated data.

RA_TARG, DEC_TARG	Aperture's RA and Dec
EXPTIME	exposure time
MODE	observing mode e.g., WFALL)
ORIENTAT position angle	
PHOTLAM	inverse of the sensitivity
PHOTZPT	magnitude zero point

TABLE 14. *WF/PC - WFPC2* keywords.

read into *GEIS* for further reduction and analysis. *WFPC2* data sets generally have 4 groups, although sometimes they can have only one if just the image corresponding to the *PC* was stored.

Tables 12 & 13 summarise the different files corresponding to *WFPC2* images. All the files in this table have the same name, only the extension changes.

The only file for which the binary data is not important is the so-called standard header packet. Its extension is shh and contains information about the orbit and the different times associated with an observation.

If the data was extracted from the *HDA* (and not read from a tape) there are two additional files in the data set with extensions dgr and cgr. These are used to populate *StarView* tables. They are not used for any data analysis and can deleted immediately.

The calibration pipeline (or the calwfp task) produces the files shown in **Table 13**:

Some header keywords of interest are listed in **Table 14**:

5.5.4. *Calibrated data*

As mentioned above, the data stored in the Hubble Data Archive has been processed by the calibration pipeline. If needed, they can be calibrated again using the calwp2 task in *STSDAS*. Header keywords determine how the calibration will be done.

When the image was constructed in *GEIS* format (from the original stream received from the spacecraft), these keywords were populated with the names of the best calibration files available at that time. Other keywords were populated with the instructions on how the calibration should be made.

An example of a section of the header of a calibrated file with this information is included below. Only the header of the files is listed, but both the header and the binary data are used for the calibration.

```
                        / RSDP CONTROL KEYWORDS
MASKCORR= 'COMPLETE'        / Do mask correction: PERFORM, OMIT, COMPLETE
ATODCORR= 'COMPLETE'        / Do A-to-D correction: PERFORM, OMIT, COMPLETE
BLEVCORR= 'COMPLETE'        / Do bias level correction: PERFORM, OMIT, COMPLETE
BIASCORR= 'COMPLETE'        / Do bias correction: PERFORM, OMIT, COMPLETE
DARKCORR= 'COMPLETE'        / Do dark correction: PERFORM, OMIT, COMPLETE
FLATCORR= 'COMPLETE'        / Do flat field correction: PERFORM, OMIT, COMPLETE
SHADCORR= 'OMIT       '     / Do shaded shutter correction: PERFORM, OMIT, COMPLE
DOSATMAP= 'OMIT       '     / Output saturated pixel map: PERFORM, OMIT, COMPLETE
DOPHOTOM= 'COMPLETE'        / Fill photometry keywords: PERFORM, OMIT, COMPLETE
DOHISTOS= 'OMIT       '     / Make histograms: PERFORM, OMIT, COMPLETE
OUTDTYPE= 'REAL       '     / Output image datatype: REAL, LONG, SHORT

                        / CALIBRATION REFERENCE FILES
MASKFILE= 'uref\$e2112084u.r0h' / name of the input DQF of known bad pixels
ATODFILE= 'uref\$dbu1405iu.r1h' / name of the A-to-D conversion file
BLEVFILE= 'ucal\$u2jc0105t.x0h' / Engineering file with extended register data
BLEVDFIL= 'ucal\$u2jc0105t.q1h' / Engineering file DQF
BIASFILE= 'uref\$e6110347u.r2h' / name of the bias frame reference file
BIASDFIL= 'uref\$e6110347u.b2h' / name of the bias frame reference DQF
DARKFILE= 'uref\$ea71115au.r3h' / name of the dark reference file
DARKDFIL= 'uref\$ea71115au.b3h' / name of the dark reference DQF
FLATFILE= 'uref\$e3914344u.r4h' / name of the flat field reference file
FLATDFIL= 'uref\$e3914344u.b4h' / name of the flat field reference DQF
SHADFILE= 'uref\$e371355iu.r5h' / name of the reference file for shutter shad
PHOTTAB = 'ucal:u2jc0105t.c3t' / name of the photometry calibration table
GRAPHTAB= 'mtab\$e8210190m.tmg' / the HST graph table
COMPTAB = 'mtab\$eai1341pm.tmc' / the HST components table
```

The file with the x0h extension contains, what in ground-based CCD images is the overscan, is known as the extracted engineering data file. For those observations made in *FULL* mode, it contains 12 columns. The information in this file is used to correct the *BIAS* level if the parameter $BIASCORR = PERFORM$. There is a companion file, with extension .q1h, whose values indicate the quality of the pixels, as in the case of the science data.

Several files used in the calibration process are the ones that determine the quality of each image pixel. Their value come both from the individual observation as well as from the global knowledge of the camera. The file with the q0h extension indicate each pixels' quality. It is important to recognise these potentially "bad" pixels in order to analyse the images correctly. **Table 15** summarises the values that these pixels can have in this file, which will help in their identification.

The calibration process generates several files. The most important file is the one with the c0h/c0d extensions, as it is the final result. This file is also paired with quality data files that have c1h/c1d extensions. This last one is the result of the combination (using the logical operator *AND* of the similar files in the raw and the reference data.

If $DOHISTOS = PERFORM$ a file with extension c2h is created. It contains a histogram of all the good values in the calibrated data. This file contains 3 rows and the same number of groups as the raw file. The first row includes the histogram of the raw data, the second row of the data after the A-to-D conversion and the final one, the histogram of the calibrated data.

value	name	comments
0	goodpixel	"good" pixel will be included in the bias determination and the statistics of the calibrated image
1.	softerror	error generated during the transmission of the data
2	calibdefect	defect that emerged during the calibration process
4	staticefect	defect that is common to all *WFPC2* images (e.g., dead pixels)
8	atodsat	problem that emerged during the A-to-D conversion
16	datalost	the value of the pixel was lost during the data transmission
32	badpixel	bad pixel that didn't fall in the previous categories
64	overscan	pixels outside of the region corrected for spherical aberration

TABLE 15. *WFPC2* Data quality pixel values

The last file created by the calibration process is a table whose extension is c3t. It contains the information needed to populate the photometry parameters in the header (*PHOTLAM, PHOTZPT, PHOTPLAM, PHOTBW*). By multiplying the image by the *PHOTLAM* value, its units are changed from *DN* (counts) to flux units. *PHOTZPT* is the zero point of the system used for the photometry. The other parameters are the wavelength at which the flux was calculated (*PHOTPLAM*) and the width of the filter (*PHOTBW*).

5.5.5. *The calibration process*

The† parameters that control the calibration process are part of the image header, as explained above. They can be modified using the *STSDAS* task stsdas.ctools.chcalpar. If the keyword is set to *PERFORM* that step of the calibration will be performed. Upon successful completion of this step, the keyword is set to *COMPLETE*. If the calibration step is skipped, the keyword value is *OMIT*.

The calibration files which are appropriate for a given observation can be obtained from *StarView*'s calibration screen (see below).

If *MASKCORR = PERFORM* the mask flagging defective pixels is included in the c1h file. These bad pixels, bad rows or bad columns, can be studied by displaying this file.

The next calibration step is the analogue to digital (A-to-D) conversion, done if *ATOD-CORR = PERFORM*. In this case each pixel is replaced by the appropriate value as determined by the corresponding table listed in the header. This table has 3 columns and 8192 rows. One column corresponds to the temperatures in *HST*'s bay 3, the others to the corresponding pixels. For example, the second pixel in the first row is the temperature associated with the second row of the A-to-D table.

† Please refer to the HST Data Handbook for a complete description of the calibration process for *HST* data.

If $BLVCORR = PERFORM$ the level of bias is calculated in the x0h file and subtracted from each pixel in the science image. The value for the odd columns is calculated and stored in the $BIASODD$ keyword, while the value for the even ones is stored in $BIASEVEN$. Each pixel identified as bad is incorporated to the quality file c1h.

The bias image is then subtracted from the science image. This is the calibration file with the r2h extension (and the associated quality file b2h).

PI (last name): SPARKS   Proposal ID: 6967

Target Name: 3C152.0   Release Date: 09/26/98 03:

Dataset Name: U4496302R   Filter1: F555W   Serials: OFF   Mode: FULL

A-D Gain: 7.000   Filter2:   Shutter: A

Orient. 1: -44.561   Orient. 2: 44.922   Orient. 3: 135.237   Orient. 4: -134.20

| | | | LEVEL OF | |
	USED	RECOMMENDED	CHANGE	PERFORMED
A-to-D Correction:	DBU1405IU.R1H	DBU1405IU.R1H	NO CHANGE	COMPLETE
Bias Correction:	H161240AU.R2H	H161240AU.R2H	NO CHANGE	COMPLETE
Dark Current Correction:	H9I1825IU.R3H	H9P14402U.R3H	SEVERE	COMPLETE
Flat Field Correction:	G640925NU.R4H	G640925NU.R4H	NO CHANGE	COMPLETE
Static Pixel Mask:	F8213081U.R0H	F8213081U.R0H	NO CHANGE	COMPLETE
Shutter Shading File:	E371355EU.R5H	E371355EU.R5H	NO CHANGE	OMIT
Engineering File:	U4496302R.X0H			COMPLETE
Photometry Cal. Table:	U4496302R.C3T			COMPLETE
Graph Table:	G8T0918NM.TMG	G8T0918NM.TMG	NO CHANGE	
Components Table:	H7I13018M.TMC	H7I13018M.TMC	NO CHANGE	

In this example, the suggested flat-field files is different. Please note the flag about the actual difference between these files that indicates the need to recalibrate this image.

FIGURE 3. The calibration screen for StarView in the HST data reduction process.

The dark is made as a file with an exposure time of 1 second. This file is multiplied by the header keyword $DARKTIME$, the value of the accumulation time. This resulting file is subtracted from the science image. The reference file, indicated in the header by the $DARKFILE$ keyword has the r3h extension (the file with b3h extension is the quality file; keyword $DARKFIL$). If the observation was longer than 30 minutes or more than 4 images were combined so as to obtain a large SNR, it is advisable to use a "superdark". These special files are the combination of 100 individual darks. If a superdark was used, the hot pixels need to be removed using a dark file obtained close to the observation date or creating a mask from the bad pixel lists.

The resulting image from these last operations is multiplied by the flat-field (identified in the header by the *FLATFILE* keyword and with a `r4h` extension). The flat has already been normalised and its inverse is use to avoid a division by zero.

Finally if *SHADCORR = PERFORM* the errors caused by the different illumination due to the shutter blades at the entrance pupil. This step is only needed if the observations are shorter than 10 seconds.

As mentioned above if *DOPHOTOM = PERFORM* the photometry parameters are populated. This is done using the throughput values of the different optical components. These values are calculated using different synphot subroutines.

5.5.6. *How to select the "best" reference files*

The images obtained with *WFPC2* are distributed and archived already calibrated. This calibration was performed with the more adequate reference files available at the time. It is frequent to find out that these are not the "optimal" ones. An example was mentioned above: the "superdarks". The most common one, though, is the existence of new (and probably better) bias and flat field files. Also, as more details are learned about the instrument, the calibration algorithms are modified. Changes to `calwf2` are announced in the Space Telescope Analysis Newsletters and in the *WFPC2 WWW* pages.

It is always convenient then to be sure that the images are adequately calibrated. The easiest way to check that the appropriate files were used, is to use *StarView*'s calibration screen. As shown in **Figure 3**, this screen lists the used and suggested files.

5.5.7. *Changing the calibration parameters*

As described above, the *STSDAS* task `chcalpar` can be used to change the calibration parameters. This task allows to change the name of the files as well as the calibration controls.

The `chcalpar` parameters are, for example:

```
    images = ''u27L0402T.d0h" List of images to modify
    template = ''")              Image to read header from
    keywords = ''")              Pset to use if not reading from an image
        add = yes)              Add keywords if not present in header?
    verbose = yes)              Print out files as they are modified?
    Version = ''25Mar94")        Date of Installation
        mode = ''al")
```

While the `pset` keywords are:

```
maskcorr = ''complete")    >do mask correction?
atodcorr = ''complete")    >Do A-to-D correction?
blevcorr = ''complete")    >do bias level correction?
biascorr = ''complete")    >do bias correction?
darkcorr = ''complete")    >do dark correction?
flatcorr = ''complete")    >do flat field correction?
shadcorr = ''omit")        >Do shaded shutter correction?
dophotom = ''complete")    >fill photometry keywords?
dohistos = ''omit")        >Make histograms?
outdtype = ''real")        >output image data type?
maskfile = ''uref\$e2112084u.r0h") >name of the input DQF of known bad pixel
atodfile = ''uref\$dbu1405iu.r1h") >name of the A-to-D conversion
blevfile = ''ucal\$u2jc0101t.x0h") >engineering file with extended register
blevdfil = ''ucal\$u2jc0101t.q1h") >engineering file DQF
biasfile = ''uref\$e6110347u.r2h") >name of the bias frame reference file
biasdfil = ''uref\$e6110347u.b2h") >name of the bias frame reference DQF
darkfile = ''uref\$ea71115au.r3h") >name of the dark reference file
```

```
   darkdfil = ''uref\$ea71115au.b3h") >name of the dark reference DQF
   flatfile = ''uref\$e3914344u.r4h") >name of the flat field reference file
   flatdfil = ''uref\$e3914344u.b4h") >name of the flat field reference DQF
   shadfile = ''uref\$e371355iu.r5h") >name of the reference file for shaded sh
    phottab = ''ucal:u2jc0101t.c3t") >name of the photometry calibration table
   graphtab = ''mtab\$e8210190m.tmg") >the HST graph table
    comptab = ''mtab\$eai1341pm.tmc") >the HST components table
 instrument = ''wfpc2")          Instrument represented by this pset
    Version = ''14Feb94")        Date of Installation
       mode = ''al")
```

The control parameters can be changed to *PERFORM* to execute, or *OMIT* if that part of the calibration will not be run. The recommended flat-field needs to be extracted from the *HDA* and its name entered in the pset before proceeding with the calibrations.

5.5.8. *Continuing the reduction*

The reduction process can continue after the calibration. Several steps can be taken depending on the desired final results.

5.5.9. *Cosmic rays*

Probably the next step is the elimination of the cosmic rays. This is not an easy task as, due to the undersampling, point sources can be confused with a cosmic ray.

To eliminate the cosmic rays in an image, the stsdas.hst_calib.wfpc2.crrej task can be used to combine a pair of *CR-SPLIT* images. An example of the parameters of this task is:

```
  infile = ''u*.c0h"       >input images
    outfile = ''3c353"      >output image name
    outfile2 = '' "         >output (number of good points) image name (opt
     sigmas = ''5,5,4,3"  >rejections levels in each iteration
     radius = 0.            >CR expansion radius in pixels
    pfactor = 0.            >CR expansion discriminator reduction factor
  hotthresh = 4096.         >hot pixel threshold in DN
     minval = -99.          >minimum allowed DN value
    initial = ''min"        >initial value estimate scheme (min, med, or fi
   noisepar = '' ")         >Noise model parameters (pset)
    fillval = INDEF)        >fill value for pixels having CR in all input i
    verbose = yes)          >print out verbose messages?
       mode = ''al")
```

The pset in this case is:

```
(readnoise = 0.85)      Read noise (in DN) from noise model
      gain = 7.)        Detector gain in electrons/DN
 scalenoise = 5.)       Multiplicative term (in percent) from noise mod
      mode = ''al")
```

5.5.10. *Orientation*

The position angle is stored in the *ORIENTAT* header keyword. Although the image can be easily rotated so as the North direction is up (and East is left), it is easier to start the analysis using the stsdas.graphics.sdisplay.compass task. It draws an arrow pointing North.

To combine the four images (each in one group) and obtain an image with the full *WFPC2* field of view, it is possible to use the stsdas.hst_calib.wfpc2.wmosaic task. The resulting image has not been rotated to have North at the top.

5.5.11. *Positions*

Images obtained with the *WFPC2* suffer from a geometric distortion due to the optics and the physical characteristics of the camera. This distortion varies with the position of the pixel, being smaller close to the centre of the chip and around 3-4% at the edges. To measure absolute positions in a *WFPC2* image, it is necessary to account for this effect. The task `stsdas.hst_calib.wfpc2.metric` is the recommended one to perform position measurements. The task `rimcursor` can be used to measure positions in images resulting from `wmosaic`. The precision of the relative measurements is between 4 and 10 milliarcseconds. The absolute measurements are as "good" as the ones from the guide stars used to point the telescope while obtaining the image. The nominal error of the Guide Star Catalogue (*GSC*) is $0.7''$.

In the case of the so-called early acquisition images it is necessary first to locate a common star in the *WFPC2* and *GSC* plate images. The offset, if existent between these two positions can be determined and then applied to the absolute measurement of the target. This position is then the one used to point the spectrograph.

5.5.12. *Photometry*

To convert the *DN*s into magnitudes, in principle, only the following formula can be used:

$$m = -2.5 \times \log(\text{counts}/\text{EXPTIME}) + \text{ZEROPT} \qquad (5.1)$$

where

$$\text{ZEROPT} = -2.5 \times \log(\text{PHOTFLAM}) + \text{PHOTOZPT} \qquad (5.2)$$

PHOTOZPT is the magnitude zero point. Its value (and the corresponding *PHOTLAM*) are included in the header. This system is based in a spectrum with a constant flux per wavelength. In this case the flux density is simply:

$$\text{fluxdensity} = \text{counts} \times \text{PHOTFLAM}/\text{EXPTIME} \qquad (5.3)$$

The values obtained in this way, need to be corrected still for temporal and spatial defects, as for example those due to:
- contamination of the entrance pupil (especially in the UV)
- change in focus due to distortion or changes in temperature during the different parts of the orbit
- geometric distortion and border effects
- *CTE*
- aperture corrections
- differences between CCDs
- colour terms

A detailed description can be found in Holtzman *et al.* (1995). For more details and examples please see the *WFPC2* photometry page at:

`http://www.stsci.edu/ftp/instrument_news/WFPC2/WFPC2_doc.html#Phot`

With the task `synphot` the different magnitude values can be estimated.

Without much additional work, it is possible to obtain magnitudes with a 5% error. By using more precise models and performing all the necessary corrections the errors are reduced to 2% or 3%.

	U-F336W	B-F439W	V-F555W	R-F675W	I-F814W	R-F675W	I-F814W
O5 V	0.53	0.67	0.05	0.67	1.11	0.71	1.22
B0 V	0.46	0.66	0.05	0.67	1.13	0.70	1.22
A0 V	0.08	0.67	0.02	0.68	1.22	0.67	1.21
F2 V	0.03	0.62	0.00	0.69	1.28	0.63	1.22
G0 V	0.02	0.58	0.01	0.70	1.31	0.60	1.23
K0 V	0.10	0.53	0.01	0.69	1.32	0.58	1.23
M0 V	0.04	0.43	0.00	0.78	1.48	0.54	1.22
M6 V	0.05	0.29	0.03	1.05	1.67	0.56	1.21

TABLE 16. Conversion values between *STMAG* and UBVRI or RI for *WFPC2* from the WFPC2 **Photometry Cookbook**.

5.5.13. *Conversion from STMAG to Johnson's UBVRI and Cousins' RI*

The following table can be used to convert the magnitudes obtained with the formula above to the Johnson or Cousins' systems. The values listed in the table were obtained using synphot and have errors of ∼5% (probably larger in the U band). Different Bruzual models were used (see the synphot manual) to create this table as the correction depends on the class of the object. For more details see the Photometry documents in the *WFPC2* WWW pages

The following tasks can be used to plot the different filter throughputs and compare them:

```
sy> plband ''band(wfpc2,1,a2d7,f814w) left=6000 right=12000 norm=yes
ltype=solid dev=stdgraph
```

```
sy> plband ''band(johnson,i)" normali+ ltype=dashed append+
device=stdgraph
```

```
sy> plband ''band(cousins,i)" normali+ ltype=dashed append+
device=stdgraph
```

To generate one of the entries of the table, the command is

```
sy> calcphot ''band(johnson,u) crgriddbz77\$bz\_1 vegamag
sy> calcphot ''band(wfpc2,1,a2d7, f336w)" crgriddbz77\$bz\_1 stmag
```

Where "band(wfpc2,1,a2d7,f336w)" defines the F336W band; "band(johnson,u)" is Johnson's U band and "crgriddbz77$bz_1" is an O5 V spectrum from Bruzual's model library. The entries in the table are the differences between these two values. The errors are ±0.05mag (probably larger in the UV).

Example

The magnitude of an M6 V star observed with the *WF4* needs to be calculated in Cousins I band. From the header parameters

$$PHOTZPT = -21.1 \text{ and } PHOTLAM = 2.6044 \times 10^{-18}$$

Then,

$$ZEROPT(STMAG) = -2.5 \times \log10(2.6044E - 18) - 21.1 = 22.861 \qquad (5.4)$$

From the previous table, the correction value for this case is −1.21, then

$$\text{ZEROPT}(I_{\text{Cousins}}) = 22.861 - 1.21 = 22.861 \qquad (5.5)$$

The measured counts can then be converted to magnitudes in the Cousins band using

$$\text{mag}(I_{\text{Cousins}}) = -2.5 \times \log10(\text{counts}) + 21.651 \qquad (5.6)$$

If the star had 137 counts/sec, then the magnitude is $I = 16.31$mag.

Note that all these calculations were performed with the counts measured as count rates. Some photometry programs expect counts in units of exposure time. In this case, then the magnitude zero point is

$$21.651 + 2.5\log10(600) = 28.596 \qquad (5.7)$$

5.6. *Faint Object Spectrograph (FOS)*

The *FOS*†‡ was one of the original spectrographs on board *HST*. It obtained spectra between 1150 and 8500Å. It had two detectors with independent optical paths. One detector was more sensitive towards the blue (*FOS*/BL, 1150-5400Å) and the other one towards the red (*FOS*/RD, 1620-8500Å).

The incident light goes through one of the apertures, a polariser (in the case of spectropolarimetric observations) and a filter (if necessary), it is then dispersed (by a dispersion grating or a prism) or reflected by a mirror (to create an image). An array of 512 diodes is at the end of the optical path. These diodes have a width of 0.31″ in the dispersion direction (X) and 1.21″ in the cross-dispersion direction (Y). The last one is measured in units called Ybase. The diodes have approximately 256 y-base units in height.

The different gratings and prisms produce spectra in different regions of the photocathode and on different groups of diodes. **Figure 4** (from the instrument manual) illustrates the different apertures. In this example, the diode array is centred in the circular apertures.

FOS spectra were typically obtained using two techniques:

• substepping: the spectrum is deviated a fraction of a diode in the dispersion direction a value given by $1/NXSTEPS$ (*NXSTEPS* is a header parameter). This process is done *NXSTEPS* times. In this way it is assured that all wavelengths are well represented. Depending on the value of *NXSTEPS*, the spectrum might be multidimensional

• overscanning : in this case, the spectrum is deviated an integer fraction of a diode so that each pixel in the final spectrum originated from several diodes. In this way diode to diode variations are eliminated. This method does not change the size of the spectrum but the contribution of the diodes to it.

5.6.1. *Apertures, gratings, prisms and polarisers*

Tables 17 and 18 list the *FOS* gratings. Note that the name in the header is different from the one in the Handbook.

The values of the width for the red gratings are negative because the wavelength direction runs in the opposite direction. It is the same case for the spectra obtained with the prism.

---

† archival instrument

‡ A detailed description can be found in the *FOS* **Instrument Handbook, available at** http://www.stsci.edu/ftp/instrument_news/FOS/topfos.html, version 6, May 1995

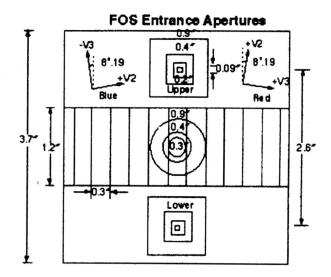

FIGURE 4. The *FOS* apertures in the *HST*.

Grating	Header Name	Wavelength (Å)	Width (Å/diode)
G130H	H13	1140 - 1606	1.00
G190H	H19	1573 - 2330	1.47
G270H	H27	2221 - 3301	2.09
G400H	H40	3240 - 4822	3.07
G570H	H57	4574 - 6872	4.45
G160L	L15	1140 - 2508	6.87
G650L	L65	3540 - 9022	25.11
PRISM	PRI	1500 - 6000	-

TABLE 17. *FOS* Blue Side (*FOS*/BL) gratings.

5.6.2. *What constitutes a "spectrum" obtained with the* FOS

An *FOS* "observation" is composed of several pairs of files whose name start with Y followed by eight alphanumeric characters. The header is a text archive whose extension finishes with the letter h. The binary data extension ends with d.

The following table briefly describes the different files (in *GEIS* format) that constitute an *FOS* "observation". More details about these files can be found in the *HST* Data Handbook.

Note that the calibrated "spectrum" is composed of two pairs of files that need to be combined:

- c0h/c0d: wavelength calibrated
- c1h/c1d: flux calibrated

Grating	Header Name	Wavelength (Å)	Width (Å/diode)
G190H	H19	1590 − 2312	1.45
G270H	H27	2222 − 3277	2.05
G400H	H40	3235 − 4781	3.00
G570H	H57	4569 − 6818	4.37
G780H	H78	6270 − 8500	5.72
G160L	L15	1571 − 2424	6.64
G650L	L65	3540 − 7075	25.44
PRISM	PRI	1850 − 8950	-

TABLE 18. *FOS* Red Side (*FOS*/RD) gratings.

FOS extensions

Extension	Contents
Uncalibrated Data	
d0h/d0d	uncalibrated scientific data
q0h/q0d	file describing the data quality
shh/shd	data packet describing the observation
ulh/uld	observation history
Calibrated Data	
c0h/c0d	wavelength-calibrated data
c1h/c1d	flux-calibrated data
c2h/c2d	statistical errors
c4h/c4d	number of counts
c5h/c5d	"object" spectrum after flat-fielding
c6h/c6d	"sky" spectrum after flat-fielding
c7h/c7d	background spectrum
c8h/c8d	spectrum of the "object" after flat-fielding and sky subtraction
cqh/cqd	quality of the calibrated data

The calibration process produces several intermediate spectra (the output of the flat-fielding, sky subtracting, background spectra, for example) that can be used for further analysis.

Table 19 describes some parameters of interest included in an *FOS* header:

5.6.3. *Number of pixels in the spectrum*

The *FOS* observations obtained in *ACCUM* mode are made deviating the spectrum magnetically in the dispersion direction. The parameters *SUBSTEP* and *OVERSCAN* control this procedure. The most common case is the one with *SUBSTEP=4* and *OVER-SCAN=5*, each pixel in the spectrum (except the edges) is populated by contributions

RA_APER1, DECAPER1	RA and dec. of the aperture
PA_APER	position angle
OBTIME	exposure time
APER_ID	aperture used (e.g., *SAA* - small science aperture)
FGWA_ID	grating used (e.g., G570H)
GRNDMODE	observing mode

TABLE 19. Some parameters of interest included in a *FOS* header.

time	parameter	unit
observation start	EXPTIME	MJD
observation end	FPKTIME (last group)	MJD
exposure time per pixel	EXPOSURE	sec

TABLE 20. *FOS* times.

from 5 diodes. Despite the fact that there are 512 diodes, the number of pixels in a spectrum is:

$$\text{pixels} = (\text{number of diodes} + (\text{OVERSCAN} - 1)) \times \text{SUBSTEP} \tag{5.8}$$

For the values of the example above, the number of pixels is 2064.

5.6.4. *Exposure time*

Spectra obtained in the *ACCUM* mode were read at regular intervals. Each group is the accumulation of the counts of all the previous groups. The last group is the one that contains all the counts accumulated during all the exposure time. It is then of interest to know the different times involved in an *ACCUM* observation. These times are listed in **Table 20**.

5.6.5. *Analysis of FOS spectra*

As detailed above, *FOS* calibrated spectra are divided in two files: one wavelength- and the other flux-calibrated. To combine them and perform the analysis with the tasks in the `onedspec` package, the *STSDAS* task `mkmultispec` is used. This task modifies the header of the wavelength-calibrated data by including information about the flux. The data itself are not changed.

If the analysis requires, for example, the combination of spectra obtained at several different wavelengths, it is then necessary to use the `resample` task. It is recommended, though, to avoid as much as possible to use tasks that modify the data.

5.6.6. *Orientation of the aperture*

The acquisition mode of the spectrum (*ACQUISITION* or *BINARY*) is indicated by the *OPMODE* keyword in the standard header packet (`shh`).

For *OPMODE = ACQ* the Right Ascension of the aperture (in degrees) is given by the *RA_APER1* and the declination (in degrees) by *DEC_APER1*.

To obtain the position of the aperture as it was in the sky, i.e., with North "up" and East to the left, the data can be rotated by the *PA_APER* value. In the *FOS* co-ordinate system, the aperture is aligned with the X axis parallel to the diodes (this is the side that measures 4.3″) and the Y axis is perpendicular to them (this is the side that measures 1.4″) as illustrated in **Figure 5**:

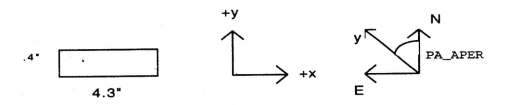

FIGURE 5. The size and orientation of the *FOS* aperture in *HST* observations.

5.6.7. *Flux determination for extended sources*

All the calculations of the flux described above were done supposing that source was a point. It is then necessary to correct these values for extended sources.

The steps to follow are:

(*a*) select the IVS file corresponding to the 4.3″ aperture and closest to the observation date (use the *DATE-OBS* keyword to determine it).

(*b*) For the cases in which the sources completely fills the aperture.

$$\text{flux} = \texttt{c5h} \text{ file} \times \texttt{IVS} \text{ file} \times 0.73$$

the first factor is the result of the division by the flat-field, the third one is the estimate of the flux loss.

For those targets that do not completely fill the aperture it is necessary to include one more factor that can be calculated from the convolution of a *PSF*, for example) and an appropriate model of the source.

This procedure will estimate the flux with a 5 to 10% error.

5.6.8. *Special wavelength calibrations*

To calculate the wavelengths of a comparison lamp or to determine the difference between the positions between the lines in the calibrated spectrum and those of the lamp the linefind and dispfity tasks can be used.

Note that as these calculations are made from a spectrum obtained with an internal calibration lamp, it is necessary to account for the differences between the diode positions between these "internal" and those "external" sources. These differences are +0.176 for the red side and 0.102 for the blue. In this way, for example, a line (obtained with the *FOS/BL*) will appear in an higher index pixel.

5.7. *Goddard High Resolution Spectrograph (GHRS)*

The *GHRS*†‡ was one of the two original spectrographs on board *HST*. It was replaced during the second servicing mission. The *GHRS* obtained high resolution spectra between 1150Å and 1800Å with resolving power between R = 2000 (for the "L") gratings and 25 000 (for the "M") on one of its sides; while for the other the resolving powers ranged between 25 000 and 80 000 and the wavelength range between 1680Å and 3200Å. Each

† Archival instrument

‡ A detail description can be found in the *GHRS* **Instrument Handbook**, version 6, May 1995. Available at http://www.stsci.edu/ftp/instrument_news/GHRS/topghrs.html

Name	Range (Å)	Å/diode	Width (Å)	Side
G140L	1100-1900	0.572-0.573	286-287	1
G140M	1100-1900	0.056-0.052	28-26	1
G160M	1150-2300	0.072-0.066	36-33	2
G200M	1600-2300	0.081-0.075	41-38	2
G270M	2000-3300	0.096-0.087	48-44	2

TABLE 21. *GHRS* gratings.

OBSMODE	exposure type	TARGNAME
DEFCAL	centring of the source, no spectrum	source name
SPYBAL	calibration lamp spectrum	WAVE
ACQ/TARG	acquisition	source name
IMAGE	image of the photocathode	source name
ACCUM	spectrum count integration	source name
RAPID	rapid acquisition without accumulation	source name

TABLE 22. *GHRS OBSMODE* types.

of these sides had its own detector (a Digicon) with 512 diodes each. The central 500 diodes were the ones used to obtain the spectra, while the first and last six (the so-called "special diodes") collect information about the observation itself.

The *GHRS* had two apertures: the Large Science Aperture (*LSA*) with a pre-*COSTAR* size of 2.0″ (1.74″ post-*COSTAR*) and the Small Science Aperture (*SSA*) with a pre-*COSTAR* size of 0.25″ (0.22″ post-*COSTAR*). Almost all the light from a point source was captured by the *LSA*. The *LSA* was used to obtain precise fluxes, while for the determination of radial velocities or line profiles it was the *SSA*.

The seven gratings, echelles and mirrors (used for the acquisitions) are mounted on a movable carrousel. Depending on the configuration, either an image of the aperture, a one or two-dimensional spectrum was obtained.

Due to the granularities of the photocathode (that limit the signal to noise) spectra were obtained divided in two or four parts, moving slightly the carrousel each time. This procedure is called FP-SPLIT. The final spectrum was obtained by combining them.

The size of the *GHRS* diodes is larger than the *PSF*. To re-gain the actual resolution it was necessary then to oversample the spectrum, i.e. to allow the light to lay over several diodes. To do this, the beam was diverted a quarter or half a diode and reading the spectrum after each movement. The number of pixels in the calibrated spectrum depended how these deviations were made. The raw spectrum always has 500 (no deviations made), 1000 (0.5 diode shift) or 2000 (0.25 diode shift) pixels.

5.7.1. *Gratings and echelles*

The first order *GHRS* gratings are listed in **Table 21**. The width listed corresponds to the number of Å that were obtained per observation.

The width listed corresponds to the number of Å that were obtained per observation.

5.7.2. *What constitutes a GHRS observation*

An "observation" with the *GHRS* is composed of several pairs of archives, whose name always start with z. The header, as usual, is a text archive whose extension ends with h and the binary data's extension with d.

The different files, in *GEIS* format, that constitute an "observation" are briefly described in the following table. Please refer to the *HST* Data Handbook for more details.

Note that a calibrated "spectrum" is composed of two pairs of files:

- c0h/c0d wavelength calibrated data
- c1h/c1d flux calibrated data

A list of the different *GHRS* filename extensions is given below:

GHRS extensions

Extension	Contents
Uncalibrated Data	
d0h/d0d	uncalibrated science data
q0h/q0d	data quality file
x0h/x0d	data from the special diodes
d1h/d1d	acquisition data (return to brightness mode)
Calibrated Data	
c0h/c0d	wavelength-calibrated data
c1h/c1d	flux-calibrated data
c2h/c2d	statistical errors
c3h/c3d	special diodes calibrated data

The different exposure types are described by the *OBSMODE* header parameter, which are shown in **Table 22**:

TARGNAME is the name with which the source is referred to in the header. If a special calibration was requested, the keywords are *DATE-OBS = ACCUM* and *TARGNAME = WAVE*.

Table 23 describes some of the header parameters of interest from a *GHRS* observation:

5.7.3. *Working with GHRS spectra*

If the spectra were obtain with an *FP-SPLIT* before starting the analysis, it is necessary to combine them using the *STSDAS* tasks poffsets and specalign. The spectrum is still divided in two: the wavelength and flux calibrated parts. To combine them and proceed with the study, the mkmultispec task is used.

The wavelength calibration has an RMS error of 1.25 diodes. If additional data was obtained (like *SPYBALS*, for example) it is possible to reduce this number. Additionally the task wavecal can be used to obtain new dispersion coefficients and zero-point differences.

APERTURE	aperture used (*LSA*, *SSA*)
RA_APER1, DECAPER1	aperture position
EXPOSURE	exposure time per group
AEXPSTART	exposure time in MJD
EXPTIME	total exposure time
FP_SPLIT	FPSPLIT mode (2 or 4)
GCOUNT	number of groups
NAXIS1	number of pixels in the spectrum
RPTOBS	number of repeated observations
STEPPAT	substepping mode
OBSMODE	observing mode

TABLE 23. *GHRS* keywords.

	Wide Field Camera	High Resolution Camera	Solar Blind Camera
Field of View	$200'' \times 200''$	$26'' \times 26''$	$26'' \times 26''$
Number of Pixels	4000×4000	1000×1000	1000×1000
Pixel Size	$0.05''$	0.025 0.025	
Bandwidth	$4000\text{Å} - 1\mu m$	$2000\text{Å} - 1\mu m$	$1150\text{Å} - 1700\text{Å}$

TABLE 24. *ACS* characteristics.

5.8. *High Speed Photometer (HSP)*

The High Speed Photometer† was able to take measurements in the optical and UV at rates up to 10^5 Hz. It could also measure very low amplitude variability and linear polarisation. It was the only instrument on board *HST* without moving parts. *HSP* had four Image Dissector Tubes (*IDT*), three were used for photometric measurements and one for polarimetry. The photomultiplier tube was used simultaneously with one of the *IDT*s. The time resolution was around 1ms. *HSP* had 23 filters covering from 1200Å to 7500Å.

Three science collecting modes were available: single colour photometry, star-sky photometry and area scans.

During the first servicing mission the *HSP* was replaced by *COSTAR*.

5.9. *The Advanced Camera for Surveys (ACS)*

The *ACS* will be installed during the third servicing mission in the bay currently occupied by the *FOC*. It has been designed as an optimal camera for all faint ultraviolet targets. The *ACS* has three cameras: the Wide Field Camera (*WFC*) has high throughput, a large field and will mostly be obtaining images in the I band; the High Resolution Camera (*HRC*) is critically sampled at 5000Å and includes a coronograph; the Solar Blind Camera is optimised for far UV observations. Some characteristics of these instruments are included in **Table 24**.

The set of filters have been chosen as to complement those of the *Sloan Digital Sky Survey*. With the *ACS* it will be possible to observe multiple Lyman α forest systems with sufficient resolution as to separate the D and H contributions. The position of the

† Archival instrument.

	Wavelength Coverage	$\Delta\lambda/\lambda$
High Dispersion Channel 1 (HDC1)	1150Å - 1450Å	20 000 - 24 000
High Dispersion Channel 2 (HDC2)	1405Å - 1775Å	20 000 - 24 000
High Sensitivity Channel (HSC)	1230Å - 2000Å	2500 - 3000

TABLE 25. *COS* characteristics.

WFC and *HRC* in *HST*'s focal plane (parallel to the *V2* and *V3* axis) makes it possible to obtain mosaiced images while the Guide Stars are stepping along the directions of the *FGS*.

5.10. *Cosmic Origins Spectrograph (COS)*

This new spectrograph will be installed in the bay currently occupied by *COSTAR* during the fourth servicing mission in 2002. *COS* is a large-format slitless UV spectrometer with three gratings:

The detector is a windowless microchannel-plate array, with an opaque CsI photocathode, and a double delay-line readout. It is optimised for UV and far UV spectroscopy.

COS most frequent use with be in time-tag photon counting mode. For brighter target an image accumulation mode can be used too.

6. The Hubble Data Archive

Since the beginning of the Hubble Space Telescope project, an archive of its data was envisioned. In this way the astronomical community was able to immediately access these invaluable observations.

The Hubble Data Archive (*HDA*)† consists of:
- a catalogue, containing the information about the data
- a system to write and read the data stored in optical disks
- an interface to query the contents of the catalogue

The *HDA* includes among others, besides the observations obtained with past and present instruments on board *HST*, several other astronomical catalogues (like the *HD*, *SAO*, *PKS*, *IRAS*, *RC3* and *UGC*, for example). It also has the *VLA* radio maps obtained with *FIRST* (Faint Images of the Radio Sky at Twenty-cm). Additionally, there are the Digitised Sky Survey images obtained from the scanning of the Palomar Observatory Sky Survey *XE* plates and the *ESO-SERC* Southern Sky Survey (see the description in these notes), the second epoch plates of *POSS*. Finally, the reduced *FOS* spectra obtained as part of the QSO absorption line *HST* key project are also included. The Multimission Archive at *STScI* is also available through the *HDA* page.

6.1. *The Data Archive and Distribution Service (DADS)*

The engineering data (temperatures and voltages in different places of the spacecraft, orbital data, jitter), ephemerides, observation plans are also written to optical disks. Finally, also included in the Archive are the reference files used to calibrate the science data.

† Based on documentation produced by *STScI*'s data Archive Branch. A more detailed explanation can be found on-line

As of 15 February 1999 there were 6.08 Tbytes stored in the *HDA* corresponding to more than 10000 sources in 230000 data sets. Most of this data is already public. Twenty percent of the data volume correspond to *NICMOS* observations which in 2.5 years of operations produced more than 103 000 observations.

The different tables that are part of *DADS* are populated by the parameters in the headers of the images, the spectra and the tables that are part of the data sets.

6.2. *Browsing the HDA*

StarView a stand-alone X-windows interface or *WWW* pages can be used to browse the Hubble Data Archive. *StarView* can installed locally in the user's machine. More information is available at `http://archive.stsci.edu/hst/starview.html`. The queries via the *WWW* can be done at `http://archive.stsci.edu/cgi-bin/nph-hst`, a page just below the archive home page.†

The interaction with *StarView* is done through several screens. Each one has fields with information about the properties of the object, the observational parameters and others. These fields are filled by the values in the *DADS* tables after the initial query. The results of the *WWW* query are similarly populated.

It is possible to preview the public images and spectra without actually extracting the files. In this way the user can assess the utility of the files for the research program in question. Another useful utility for the preparation of proposals is the duplication checking facility.

A very common issue in trying to find observations of a certain object is doing it by its name. Each user calls it in his or her preferred way (an observer calls a galaxy 3C 84, while another one NGC 1275; the star HD 137603 can also appear as WR 70 or MR 57, for example). It is very easy to solve this problem using a name resolver. By entering a name, the resolver queries either the *NED* or *SIMBAD* databases and retrieves its co-ordinates. The catalogue query is then performed around this position and not using a particular name.

Another common query is the one in which the user tries to find all the observations of a certain type of objects (e.g., supernovae remnants, pulsars, gravitational lenses, etc.). The possible keywords are on-line by pressing *StarView*'s Strategy button.

The easiest way to start a query with *StarView* is to select the Quick Search screen. Most of the basic information about the observations can be found in it. Remember to use the name resolver and the strategy button to find suggestions on how to proceed.

The data of interest are "marked" and extracted from the *HDA*. To extract observations it is necessary to register with the Archive. The information and the form can be found at `http://archive.stsci.edu`.

6.3. *How to use the HDA*

6.3.1. *What to do after querying the database and identifying the images or spectra of interest?*

First, check the instrument's calibration screen to see if it is necessary to recalibrate the data with newer (and possibly better) calibration files. Not only the reference files might have changed, but also the algorithms used to create the calibration code. If the program had indeed changed, it will be necessary to decide the impact of this change based in the use you will give to the observations. This information can be found in the

---

† Users in Europe can browse the catalogue at the *ST-ECF* (`http://archive.eso.org`). Those in Canada at the *CADC* (`http://cadcwww.dao.nrc.ca/`) and those in the United Kingdom at the UK *HST* Support Facility (`http://www.ast.cam.ac.uk/hst/home.html`).

information	can be found at:
WF/PC and *WFPC2*	
do superflats exist?	list in the *WWW* instrument pages
precise exposure time	header, *HST* Data Handbook (*DHB*)

FOS	
light diffusion problem?	ISR *FOS*-127, *HDB*, Instrument Handbook (*IHB*)
background problem?	ISR *FOS*-114, *IHB*, *DHB*
GIMP correction applied?	header (observation date)
precise value of *FPKTTIME*	*IHB*

GHRS	
precise value of *AEXPSTART*	header, *DHB*
wavelength calibration	*SPYBAL* information (header or Phase II listing)

TABLE 26. Where to find information about *HST* exposures (based on a presentation by M. Rosa at the Second *HST* Calibration Workshop).

HST **Data Handbook**, the instrument manual or the *WWW* pages with instrument information (http://www.stsci.edu/instruments.html.)

The absolute calibration of all observations taken with *HST* uses a scale based in the white dwarf scale.

In the case of the very early observations (between 1990-92) the aperture position listed in the header is not accurate. To verify and, if necessary, correct them it is possible to use the jitter files or using an image from the Digitised Sky Survey.

Other information of interest about the data is listed in **Table 25**.

Several tasks need to be completed before analysing the data. It is recommended to extract all files related to an observation, even the engineering ones; i.e., the whole data set. With all this information it is possible to reconstruct what happened during the observation. For example, the number of re-centring events, which guide stars were used to point the telescope, where there parallel observations obtained at the same time and the comments about the calibration and quality of the data. More information about these files can be found in the *HST* Data Handbook.

As explained above, it is very desirable to always check *StarView*'s calibration screen and the instrument pages (http://www.stsci.edu/instruments). The Advisories section in the last ones is a must read! It contains invaluable information about certain characteristics of the instruments and their data. Other items that can be found in the instrument pages are the latest edition of the instrument's handbook, Frequently Asked Questions, Instrument Science Reports and all the Space Telescope Analysis News issues. Other information of interest about the data can be found in the Instrument Science Reports. **Table 26** is an example of this information.

Finally, it is also wise to read the contents and latest news posted in the *STSDAS* page (http://www.stsci.edu/stsdas.html). In these pages, not only the latest version of the software can be found, but also several other documents, handbooks and user's guides. This page has a link to the *IRAF* page, where the latest versions and documents can be found.

7. Surfing for data

There is nothing more thrilling that the sense of discovery. Of being out there in the middle of the night and finding an answer. The path was not easy and what we have talked about and discussed these past two weeks are part of it.

We just gave additional tools to explore and test your ideas, to see if someone thought of it before, to look at their results, to find your way around the maze.

Don't cease to explore and question. Start firing your browsers!

I thank the organisers of this IX Canary Islands Winter School for inviting and giving me the opportunity to share this tingling sensation with a new generation of students, which like me some years ago, attended the School to broaden our horizons.

8. Acronyms

A complete list of the acronyms employed in this review is given in **Table 27**.

ADC:	Astronomical Data Center
ADS:	Astrophysics Data System
AEC:	Archival Exposures Catalogue
AR:	Archival Research
AURA:	Association of Universities for Research in Astronomy, Inc.
CADC:	Canadian Astronomical Data Center
CCD:	Charge Coupled Device
CDBS:	Calibration Data Base System
CDS:	Centre des Données Stellaires
CEA:	Center for Extreme Ultraviolet Astronomy
CFA:	Center for Astrophysics
CFHT:	Canada France Hawaii Telescope
COSTAR:	Corrective Optics Space Telescope Axial Replacement
CP:	Call for Proposals
CTE:	Charge Transfer Efficiency
CVZ:	Continuous Viewing Zone
DADS:	Data Archive and Distribution System
ESA:	European Space Agency
ESO:	European Southern Observatory
EUVE:	Extreme Ultraviolet Explorer
FGS:	Fine Guidance Sensors
FITS:	Flexible Image Transport System
FOC:	Faint Object Camera
FOS:	Faint Object Spectrograph
FOV:	Field of View
GASP:	Guide Star Selection System Astrometric Support Program
GEIS:	Generic Edited Information Set
GHRS:	Goddard High Resolution Spectrograph
GIMP:	Geomagnetically-Induced Image Motion Problem
GO:	General Observers
GSC:	Guide Star Catalogue
GSFC:	Goddard Space Flight Center
GSSS:	Guide Star Selection System
GTO:	Guaranteed Time Observer
HDA:	Hubble Data Archive
HSP:	High Speed Photometer

HST:	Hubble Space Telescope
IDT:	Instrument Definition Team
IPAC:	Infrared Processing and Analysis Center
IRAF:	Image Reduction and Analysis Facility
IRAS:	Infrared Astronomical Satellite
ISO:	Infrared Astronomical Observatory
ISR:	Instrument Science Report
IUE:	International Ultraviolet Explorer
JPL:	Jet Propulsion Laboratory
Leda:	Lyon-Meudon Extragalactic Database
LSA:	Large Science Aperture
LSF:	Line Spread Function
MAMA:	Multi-Anode Microchannel Array
MJD	Modified Julian Date (JD-2400000.5)
NASA:	National Aeronautics and Space Administration
NOAO:	National Optical Astronomical Observatory
NED:	*NASA* Extragalactic Database
NICMOS:	Near Infrared Camera and Multi-Object Spectrograph
NSSDC:	*NASA* Space Science Data Center
OFAD:	Optical Field Angle Distortion
OIF:	Old *IRAF* Format
OMS:	Observatory Monitoring System
OPUS:	*OSS* and *PODPS* Unified System
OSS:	Observatory Support System
OTA:	Optical Telescope Assembly
PC:	Planetary Camera
PODPS:	Post-Observation Data Processing System
PRESTO:	Project to Re-Engineer Space Telescope Observing
PSF:	Point Spread Function
QE:	Quantum Efficiency
ROSAT:	Röentgen Satellit
RPS(2):	Remote Proposal Submission System (2)
RSDP:	Routine Science Data Processing
SAA:	South Atlantic Anomaly
SI:	Scientific Instrument
SIMBAD:	Set of Information, Measurements and Bibliography for Astronomical Data
SMOV:	Servicing Mission Orbital Verification
SMS:	Science Mission Specification
SPYBAL:	SPectrum Y BALance
SSA:	Small Science Aperture
ST-ECF:	Space Telescope - European Co-ordination Facility
STAN:	Space Telescope Analysis News
STEIS:	Space Telescope Electronic Information System
STIS:	Space Telescope Imaging Spectrograph
STOCC:	Space Telescope Operations Control Center
STScI:	Space Telescope Science Institute
STSDAS:	Space Telescope Science Data Analysis Software
TAC:	Time Allocation Committee
TDRSS:	Tracking and Data Relay Satellite System
TIM:	Telescope Image Modelling
URL:	Universal Resource Locator

WFC:	Wide Field Camera
WF/PC:	Wide Field and Planetary Camera [1]
WFPC2:	Wide Field Planetary Camera 2
WWW:	World Wide Web

TABLE 27. A complete list of acronyms which have been used in this text. Many of these are of common-usage in the Internet and in astronomical documentation.

9. Astronomical CDs

A selection of interesting astronomical CDs is shown in **Figures 6-8**.

SELECTED ASTRONOMICAL CD-ROMs		
TITLE	**CONTENTS**	**AVAILABLE FROM**
CATALOGS		
Selected Astronomical Catalogs - Volumes1- 3	several galactic and extragalactic catalogs	request@nssdca.gsfc.nasa.gov
The Guide Star Catalog	Astrometric position for more than 18M objects	STScI sturch@stsci.edu
PGC-ROM 1996	Principal Galaxy Catalog - LEDA	Observatoire de Lyon cdrom@obs.univ-lyon1.fr
Astrophysics on disk	tables from ApJ Supp	ApJ subscription
ASTROMETRY		
Hipparcos Input Catalog	astrometric data	INCA consortium
Hipparcos and Tycho Catalogs	astrometric data	ESA
Astrographic Catalog as reduced to the ACRS	Astrometric data	U.S. Naval Observatory
RADIO		
Images from the Radio Universe	radio data from QSOs galaxies, planets and sky maps	NRAO jcondon@nrao.edu dwells@nrao.edu
The Green Bank sky maps and radio source catalog	Green Bank 1400MHz, 4.85GHz and 87GB maps	NRAO jcondon@nrao.edu
X-RAY		
The Einstein Observatory database of IPC X-ray images in event list format	database of X-ray sources	CfA edpo@cfa.harvard.edu

FIGURE 6. A listing of just a few of the interesting astronomical CDs.

SELECTED ASTRONOMICAL CD-ROMs		
TITLE	**CONTENTS**	**AVAILABLE FROM**
The Einstein Observatory SSS, MPC and FPCS data products	X-ray images and databases	CfA edpo@cfa.harvard.edu
The Einstein Observatory database of HRI X-ray images in envent list format	database of X-ray sources	CfA edpo@cfa.harvard.edu
The Einstein Observatory database of HRI x-ray images	X-ray images and databases	CfA edpo@cfa.harvard.edu
The Einstein Observatory catalog of IPC X-ray sources	database of X-ray sources	CfA edpo@cfa.harvard.edu
The Einstein Observatory SLEW Survey	Catalog of sources and X-Ray data	CfA edpo@cfa.harvard.edu
ROSAT- The Images (5 vol.)	X-ray images from ROSAT	NSDDC
IR		
IRAS Sky Survey Atlas Vol II	coadded IR images and individual HCON for b<50	IPAC
IRAS Sky Survey Atlas	IR images	IPAC
COBE -DMR data	sky maps and other data	NSSDC
COBE- DIRBE data (3 vol.)	DIRBE weekly sky maps	NSSDC
UV		
EUVE Science Archive Vol1.1; 2.1A-C; 2.2A-B	EUV images and spectra	CEA
COMMERCIAL PLANETARIUMS		
Almagest	software for displaying sky maps and images from DSS	PASP (commercial)
TheSky	software for displaying sky maps and images	Software Bisque (commercial)

FIGURE 7. A listing of just a few of the interesting astronomical CDs. (cont.)

REFERENCES

BOLTON, J.G., & SAVAGE, A., 1977 *Aust. J. Phys Astroph.* **44**, 21

BOWYER, S., LAMPTON, M., LEWIS, J., WU, X., JELINSKY, P., & MALINA, R.F., 1996 *ApJS*, **102**, 129

EGRET, D. & ALBRECHT, M.A., 1995 In "Information and On-Line Data in Astronomy", Dordrecht: Kluwer

GRØSBOL, P.J., HARTEN, R.H., GREISEN, E.W., & WELLS, D.C., 1988 *A&AS*, **73**, 359

HARTEN, R.H., GRØSBOL, P.J., GREISEN, E.W., & WELLS, D.C., 1988 *A&AS*, **73**, 365

HELOU, G. *et al.* , 1995 In "Information and On-Line Data in Astronomy" D. Egret & M. Albrecht (editors), Dordrecht: Kluwer

HOLTZMAN, J., BURROWS, C.J., CASERTANO, S., HESTER, J.J., TRAUGER, J.T., WATSON, A.M., & WORTHEY, G., 1995 *PASP*, **107**, 1065

JENKNER, H., LASKER, B.M., STURCH, C.R., MCLEAN, B.J., SHARA, M.M., & RUSSELL,

SELECTED ASTRONOMICAL CD-ROMs		
TITLE	**CONTENTS**	**AVAILABLE FROM**
Starry night deluxe	software for displaying sky maps and images	Sienna Software (commercial)
Voyager II	software for displaying sky maps and images	Carina Software (commercial)
DIGITAL SKY SURVEY IMAGES		
RealSky	POSS plates 100X compressed	PASP
Digitized Sky Survey	POSS and ESO/SERC plates 10X compressed	PASP
SOLAR SYSTEM		
Mission to Mars	Viking images of Mars	NSSDC
Voyager to the outer planets	Voyager images of Jupiter, Saturn and Uranus	NSSDC
Clementine EDR Image Archive (88 vol.)	Images of the Moon obtained by the Clementine satellite	NSSDC
Magellan	Images of Venus	NSSDC
Planetary images on CD-ROM	200 images of Solar System objects	NSSDC
Welcome to the Planets	multimedia introduction to the Solar System	JPL
MISCELANEOUS		
Ground-based CCD images in support of the Astro-1/UIT mission (3 vols.)	UBVR wide field CC photometric images	NSSDC
Gems of Hubble	Images obtained with HST	ASP
Kurucz models	stellar models	CfA

FIGURE 8. A listing of just a few of the interesting astronomical CDs. (cont.)

J.L., 1990 *AJ*, **99**, 2082

LASKER, B.M., STURCH, C.R., MCLEAN, B.J., RUSSELL, J.L., JENKNER, H., & SHARA, M.M., 1990 *AJ*, **99**, 2019

LASKER, B.M., DOGGETT, J., MCLEAN, B., STURCH, C., DJORGOVSKY, S., DE CARVALHO, R.R., & REID, I.N. 1996 *IAU Commission 9, Newsletter* **8**, 20

LORTET, M.-C., BORDE, S. & OCHSENBEIN, F., 1994 *A&AS*, **107**, 193

MACCACARO, T., WOLTER, A., MCCLEAN, B., GIOLA, I.M., STOCKE, J.T., DELLA CECA, R., BURG, R., & FACCINI, R., 1994 *Astrop. Lett. Comm.*, **29**, 267

MALINA R.F., MARSHALL, H.L., ANTIA, B., CHRISTIAN, C.A., DOBSON, C.A., FINLEY, D.S., FRUSCIONE, A., GIROUARD, F.R., HAWKINS, I., JELINSKY, P., LEWIS, J.W., McDON-ALD, J.S., MCDONALD, K., PATTERER, R.J., SABA, V.W., SIRK, M., STROOZAS, B.A., VALLERGA, J.V., VEDDER, P.W., WIERCIGROCH, A., BOWYER, S., 1994 *AJ*, **107**, 751

PONZ, J.D., THOMPSON, R.W., & MUÑOZ, J.R., 1994 *A&AS*, **105**, 53

RUSSELL, J.L., LASKER, B.M., MCLEAN, B.J., STURCH, C.R., & JENKNER, H., 1990 *AJ*, **99**, 2059

SCHNEIDER, D.P., BAHCALL, J.N., SAXE, D.H., BAHCALL, N.A., DOXSEY, R., GOLOMBEK, D., KRIST, J., MCMASTER, M., MEAKES, M., & LAHAV, O., 1992 *PASP*, **104**, 678

WELLS, D.C., GREISEN, E.W., & HARTEN, R.H., 1981 *A&AS*, **44**, 363

ZOMBECK, M.V., 1990 *"Handbook of Space Astronomy and Astrophysics"*, Second Edition, New York: Cambridge University Press

10. General References

Baade, D. & Grøsbol, P.J. (editors), 1990, "Second *ESO/ST-ECF* Data Analysis Workshop", Garching bei München: European Southern Observatory

Benvenuti, P. & Schreier, E. (editors), 1992, "Science with the Hubble Space Telescope", Garching bei München: European Southern Observatory

Blades, J.C. & Osmer, S.J. (editors), 1994, "Calibrating Hubble Space Telescope", Baltimore: Space Telescope Science Institute

Casertano, S. *et al.* (editors), 1997, "The 1997 HST Calibration Workshop - With a New Generation of Instruments", Baltimore: Space Telescope Science Institute

Crabtree, D., Hanisch, R.J. & Barnes, J. (editors), 1994, "Astronomical Data Analysis Software and Systems - III", San Francisco: Astronomical Society of the Pacific

Grøsbol, P.J., & Ruijsscher, R.C.E. (editors), 1993, "Fifth *ESO/ST-ECF* Data Analysis Workshop", Garching bei München: European Southern Observatory

Grøsbol, P.J., & Ruijsscher, R.C.E. (editors), 1992, "Fourth *ESO/ST-ECF* Data Analysis Workshop", Garching bei München: European Southern Observatory

Grøsbol,P.J. & Warmels, R.H. (editors), 1991, "Third *ESO/ST-ECF* Data Analysis Workshop", Garching bei München: European Southern Observatory

Grøsbol, P.J., Murtagh, F. & Warmels, R.H. (editors), 1989, "First *ESO*/ST- ECF Data Analysis Workshop", Garching bei München: European Southern Observatory

Hanisch, R.J. & White, R.L., 1994, "The Restoration of *HST* Images and Spectra - II", Baltimore: Space Telescope Science Institute

Hanisch, R.J., Brissenden, J.V. & Barnes, J. (editors), 1993, "Astronomical Data Analysis Software and Systems - II", San Francisco: Astronomical Society of the Pacific

Howell, S.B. (Editor), 1992, "Astronomical CCD Observing and Reduction Techniques", San Francisco: Astronomical Society of the Pacific

Hunt, G. & Payne, H.E. (editors), 1997 "Astronomical Data Analysis Software and Systems - VI", San Francisco: Astronomical Society of the Pacific

Jacoby, G.H. (editor), 1990, "CCDs in Astronomy", San Francisco: Astronomical Society of the Pacific

Kitchin, C.R. 1991 "Astrophysical Techniques", 2nd. Ed. Bristol: Adam Hill

Kortaker, A, & Leitherer, C. (editors), 1995, "Calibrating Hubble Space Telescope: Post Servicing Mission", Baltimore: Space Telescope Science Institute
Léna, P. 1988, "Observational Astrophysics", Berlin: Springer-Verlag

Payne, H., Hanisch, R.J. & Barnes, J. (editors), 1995, "Astronomical Data Analysis Software and Systems - IV", San Francisco: Astronomical Society of the Pacific

Russ, J.C., 1995, "The Image Processing Handbook", Boca Raton: CRC Press

Shaw, R.A., Payne, H.E. & Hayes, J.J.E. (editors), 1995 "Astronomical Data Analysis Software and Systems IV", San Francisco: Astronomical Society of the Pacific

White, R.L. & Allen, R.J., 1990, "The Restoration of Hubble Space Telescope Images and Spectra" - I, Baltimore: Space Telescope Science Institute

Worral, D.M., Biemesderfer, C. & Barnes, J. (editors), 1992, "Astronomical Data Analysis Software and Systems - I", San Francisco: Astronomical Society of the Pacific

X-ray Astronomy

By ANDREW FABIAN

Institute of Astronomy, Madingley Road, Cambridge CB3 0HA

This article reviews X-ray astrophysics from a theoretical viewpoint. We treat how X-rays are generated, looking at the different astrophysical processes that can lead to X-ray emission. We then examine a range of astronomical objects which can generate X-ray and high energy emission (compact objects, supernova remnants, Active Galactic Nuclei, clusters of galaxies and the X-ray background), examining the characteristics of the emitted radiation and the processes which lead to its generation in each case. Finally, we examine the relevance of X-ray studies to the cosmic baryon fraction and the problem of closure of the Universe.

1. Introduction

X-rays have energies of between about 100 and 100 000 times the energy of an optical photon and interact strongly with matter. This interaction is commonly made with the inner or K-shell electrons of many elements and results in the ejection of an electron; the process is called photoelectric absorption. Typical cosmic X-rays have an energy of about 1 keV and a pathlength of only a few cm in air. Consequently the surface of the Earth is completely dark to an X-ray detector (unless certain radioactive sources or an X-ray machine are very nearby) and to view the Universe in X-rays the detector has to be placed above the atmosphere, preferably orbiting in a satellite.

X-rays were detected from the Sun using rocket-borne detectors in the late 1940s and 1950s. Non-solar X-ray astronomy, which is the subject of this chapter, began in 1962 with a sounding rocket flight led by R. Giacconi in which the bright X-ray source Sco X-1 and the X-ray Background were discovered. Later rocket experiments in the 1960s found the Crab Nebula and pulsar and a few tens of other sources before the first all-sky survey was performed by the Uhuru satellite (also led by Giacconi), launched in 1970. The subject has grown enormously since then, with over 100 000 sources now found from the *ROSAT* All-Sky Survey, which took place in 1990, and detailed maps and spectra available from imaging telescopes and charge coupled device (CCD), and other, detectors.

Gamma-ray astronomy has similarly grown from a handful of sources in the early 1970s from balloon, rocket and satellite instruments to all-sky maps from the Compton Gamma-Ray Observatory, launched in 1990.

The X-ray band is usually subdivided into a soft band below 1 keV, a band between about 1 and 10 keV, and a hard band at higher energies. Low energy gamma-rays occur between 500 keV (the rest mass energy of an electron) and say 10 MeV and high energy gamma-rays above a GeV (the rest mass energy of a proton). Soft X-rays can easily photoionize hydrogen, which has an ionisation potential of 13.6 eV and is the main consituent of our interstellar medium. Consequently much of the extragalactic sky is inaccessible to extreme ultraviolet radiation at energies below about 0.1 keV and there is a natural observational gap between the UV and X-ray bands. Nearby objects can and have been detected in this gap and some extragalactic objects along directions where there is little interstellar gas. The cross section for photoelectric absorption drops off approximately as the cube of the photon energy above threshold, so the interstellar medium rapidly becomes transparent to X-rays as the energy increases above 0.1 keV, until the whole Galactic disk is transparent above about 5 keV.

This overview is meant to give a broad introduction into X-ray astronomy and does not contain detailed references into the research literature. That is best carried out, and will be more up-to-date, using the Astrophysics Data System (http://adsabs.harvard.edu/) with appropriate title words or keywords. Where applicable, references to major review articles and books are given at the end of sections.

1.1. *Where cosmic X-rays occur*

To produce an X-ray of energy ϵ_x, requires an electron with at least that energy to be accelerated (or decelerated). For a thermal gas this means particle energies of

$$\frac{1}{2}m_P v^2 \sim kT > \epsilon_x.$$

Galaxies, groups of galaxies and clusters of galaxies have internal motions of $\sim 250, 500$ and $1000\,\mathrm{km\,s^{-1}}$ respectively so protons in these objects have kinetic energies of $0.4 - 6\,\mathrm{keV}$. If the electrons are in equipartition then X-ray emission can be seen as natural to these objects (the temperatures are then $kT \sim 0.4 - 6\,\mathrm{keV}$).

Particle collisions are responsible for the pressure in a gaseous atmosphere, so if such an atmosphere is extensive with a scale height kT/mg of order the radius, R, (g is the local gravitational acceleration) then

$$kT \sim GMm_p/R$$

(cf the Virial Theorem). For the Sun, a white dwarf, and a neutron star/black hole, this corresponds to ~ 2, $200\,\mathrm{keV}$ and $200\,\mathrm{MeV}$, respectively. (Note that a long lived atmosphere requires that the scale height is about a tenth or less of the radius, so the temperatures are proportionately smaller.)

The net result of this is that X-ray production is common in the extensive atmospheres in clusters and groups, and haloes of galaxies, and on stars and compact objects (*WD*, *NS* and *BH*). In this list, the gas around objects down to normal stars is optically thin to X-rays (photons can freely escape) and they are in collisional temperature and ionisation equilibrium. For the compact objects it is often optically thick (photons interact with the matter by scattering and/or absorption) and the conditions are then determined, or at least influenced, by photon-matter interactions. At the limit when these interactions are maximum (complete equilibrium between photons and matter) then the resultant emission is blackbody radiation;

$$L = 4\pi R^2 \sigma T^4 \approx= 0.6 L_{36}^{1/4} R_1^{1/2}\,\mathrm{keV},$$

where $L = 10^{36}\,\mathrm{erg\,s^{-1}}$ and $R = 10R_1$ km, appropriate for a neutron star. This is the minimum temperature that a neutron star surface of this luminosity can have and it must thus be an X-ray source.

A further way to write the condition for an extensive atmosphere is

$$kT = (2GM/Rc^2)(m_p c^2/2) = (R_s/R)(m_p c^2/2) \sim 10^{12}(R_s/R)\,\mathrm{K},$$

where $R_s = 2GM/c^2$ the Schwarzschild radius of a black hole with a mass equal to that of the object. What matters is the size of the object in terms of R_s or in other words, its size in gravitational units. Surprisingly (see **Figure 1**), galaxies and hydrogen-burning stars have a similar size in these units and clusters of galaxies are smaller, despite being 100 times larger in physical units!

Plotted on Figure 1 are lines of unit optical depth for Thomson scattering ($n\sigma_T R = 1$; solid line) and for a bremsstrahlung cooling time of a Hubble time ($2 \times 10^{11} T^{1/2}/n = 10^{10}$ yr; dashed line). It has been assumed in estimating the density n that all the mass is in baryons. The inclusion of dark matter shifts the lines very slightly to the left

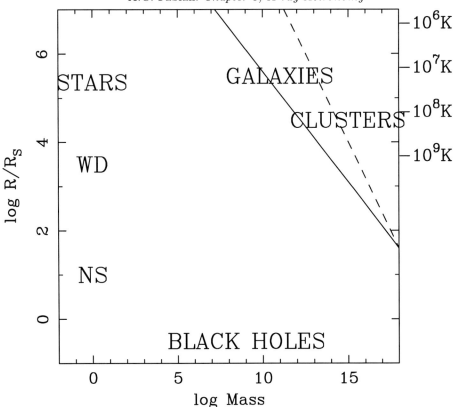

FIGURE 1. Radius-mass relation for objects in the Universe, where radii are measured in gravitational radii and masses in Solar units. If most of the mass is gaseous then objects are optically thin to Thomson scattering above the solid line and cannot cool in a Hubble time above the dashed line. An object above the line can therefore have a substantial hot, optically thin, atmosphere.

in the diagram. The important issues are that objects are optically thick to electron scattering below and to the left of the solid line and optically thin above and to the right of it. Similarly the gas will have cooled without a heat source for objects to the left of the dashed lines and will remain hot to the right. Opacity means that X-ray emission from stars, white dwarfs and neutron stars only emerges from a thin surface atmosphere and particle-photon interactions are commonplace. Clusters of galaxies can retain a significant optically-thin atmosphere which radiates X-rays. Atmospheres in galaxies are probably optically thin and most of the gas has probably cooled (this is the process of galaxy formation!).

1.2. *The astrophysical significance of X-ray observations*

The X-ray sky gives direct insight into accretion onto neutron stars and black holes which is the most efficient, in an $E = mc^2$ sense, process known (ignoring matter-antimatter annihilation which appears to be uncommon in our Universe). We can learn about the physical properties of neutron stars (mass, moment of inertia, radius, magnetic field etc) and the near environment of black holes and strong gravity. Through supernova remnants it shows us the physics of strong shocks and the metal enrichment of the interstellar medium. For elliptical galaxies and clusters, we can determine the profile of the dark matter there, the enrichment history of the abundant intracluster gas, observe cooling flows and use the baryon fraction to estimate the mean density of the Universe.

2. X-ray Photon Detectors and Telescopes

X-rays and gamma-rays are usually detected by the absorption, or interaction, of the photons with matter, giving rise to an electron which is then amplified to yield a current pulse. The absorption/interaction process depends on the photon energy; the photoelectric effect dominates up to several 100 keV, the Compton effect to several MeV and pair production in the presence of a nucleus at the highest energies.

X-ray detectors therfore depend on the photoelectric effect. The matter is either a gas or a solid (liquids would cause problems in space) and the interaction causes a photoelectron to emerge with an energy $\omega = \epsilon - I$ the ionisation potential. The cross-section is comparable to the orbital size of the ejected electron at threshold (ie $\omega = 0$) and drops off as ϵ^{-3}. The atom readjusts by an electron dropping into the vacant space and emitting a photon (fluorescence). If however the atomic number is low, another (Auger) electron is likely to be ejected (the yield of fluorescent photons varies as Z^4.)

The energy resolution of the detector is usually determined by how the energy of the photoelectron is converted into an electrical current. Typically it collides with neighbouring atoms giving rise to secondary electrons each with an energy loss ΔE of about 30 eV in a gas and 1 eV in a solid. So the total number of electrons is roughly $\epsilon/\Delta E$. The energy resolution is then proportional to the uncertainty in this number, which from Poisson statistics is $\Delta\epsilon/\epsilon \sim \sqrt{\Delta E/\epsilon}$. (It is not strictly equal to this but smaller by a the Fano factor, which takes into account that there is a fixed amount of energy to be shared between all the electrons.) This means that a solid detector (such as a CCD) has about 5 times better (smaller) resolution than a gas one (such as a gas-filled proportional counter). The resolution can be further improved in solid detectors by relying on phonons (in the bolometers) and superconducting pairs (in an STJ) which take considerably less energy to produce and so occur in far greater numbers for a given photon.

The bandwidth of the detctor is determined by the thickness of the window (or any protective shield) on the low energy side and by the thickness of the active region of the detector on the high energy side. The quantum efficiency between these bounds is often 90 or more per cent.

The main gas detectors are proportional counters which contain an inert gas (A or Xe) and a central anode for collection of the electrons. The electric field accelerates the electrons near the anode causing multiplication of the electron number. These detectors can be made position sensitive (as in the *ROSAT* Position Sensitive Proportional Counter or *PSPC*). Apart from *ROSAT* they have recently been used in the Ginga Large Area Counter or *LAC*, and the Rossi X-ray Timing Explorer, *RXTE*. Light flashes can be produced from collisionally excited atoms in such devices and can be detected instead of the electrons, yielding a factor of 2 improvement in energy resolution. The *ASCA* Gas Imaging Spectrometer, *GIS* and the *BeppoSAX*, *LECS*, and *MECS* work on this principle.

The main solid state detectors are microchannel plates (the *ROSAT* High Resolution Imager *HRI*, *Chandra (AXAF)* High Resolution Camera *HRC*), which have poor energy resolution, and Charge Coupled Devices or CCDs (*ASCA* Solid state Imaging Spectrometer *SIS*, *Chandra ACIS*, *XMM EPIC*) which are becoming the workhorse for spectral imaging. The latter are an array of metal-oxide silcon capacitors in which the charge created by the X-ray is transferred from pixel to pixel as it is read out. That charge is proportional to the energy of the incoming photon ($\Delta\epsilon \sim 100$ eV) provided that another photon does not arrive before the charge is read out (the limits the brightness of a source which can be observed). The Bolometer on *Astro-E* works at 60 mK and relies on phonon production in a small crystal. This yields $\Delta\epsilon \sim 10\,\text{eV}$.

Telescopes and imaging are required to improve the signal-to-noise ratio for point sources and to discern detail in extended ones. Noise can be unwanted signal from other sources, the diffuse X-ray background or the particle background caused by cosmic rays passing through the detector. This is of course reduced as the detector is made physically smaller which means that some sort of focussing is required if a large collecting area is required.

Bright sources can be studied with simple collimators, shielding out unwanted other sources if the detector can be physically large. This approach has been applied from Uhuru through Ginga to *RXTE*. A more complex system involving coded masks can also be used (particularly at high energies as for *INTEGRAL*).

But for faint objects and to resolve detail, a focusing telescope must be used. Owing to the large interaction with matter, simple normal incidence mirrors cannot be used (normal incidence multilayer mirrors relying on interference effects can be used in the soft X-ray band) and grazing incidence reflection must be used instead. This works because the refractive index of metals is just less than unity. There is thus total internal reflection taking place as an X-ray grazes a metal surface. The angle between photon and surface has to be a few degrees at energies less than 1 keV droppping to only a few arcmin at 10 keV. (The simple analogy is with a stone being skimmed on water.) The mirror surface used is typically the outer parts of a steep paraboloid. Since the reflection takes place well off axis the image would be comatic and a second reflection off a hyperboloidal mirror is employed to correct for this. Such Wolter mirrors have been used on the Einstein Observatory, *ROSAT* and will reach the zenith with *Chandra*, for which 70 per cent of the power will appear within a 0.5 arcsec radius. At grazing incidence, surface irregularities are a problem and the mirrors must be polished to within a few Angstrom.

The surface can be approximated by conical, or at least more simple surfaces and so made out of rolled Al foil, as has been used for the *ASCA* mirrors. Such Serlemitsos mirrors are relatively inexpensive when compared with full mirrors.

The detectors mentioned earlier are non-dispersive in their measurement of ϵ. Dispersive spectrometers do give higher resolution for point sources, although with lower efficiency. Bragg Crystals were used in the *FPCS* of Einstein, transmission gratings will be used in *Chandra* and reflection gratings in *XMM*.

Further details of the processes discussed in this section may be found in Fraser (1989).

3. Radiation Processes

There are three principle interactions which lead to X-ray emission; electron + proton (or nucleus) which gives bremsstrahlung, electron + ion which leads to line and recombination radiation, and electron + magnetic field which gives cyclotron and synchrotron radiation.

The outgoing spectrum can then be modified by interaction with atoms and ions (photoelectric absorption), electrons (Comptonisation) and other photons if they are energetic enough (photon-photon pair production).

It is often of importance to know whether the energy distribution of the electrons is thermal or not. If it is, then particle-particle collisions will be important and the distribution is characterized by one number – the temperature. If not then collisions may not be important and some information on the acceleration process may be preserved. The energy distribution may be a power-law.

3.1. *Hot thermal plasmas*

The continuum from a hot thermal plasma is usually dominated by bremsstrahlung, in which the radiating electron is accelerated by the Coulomb field of an atomic nucleus. The process is intrinsically 2-body so the emissivity (energy emitted per unit volume per unit time) is proportional to the product of the electron and ion densities. In a cosmic plasma dominated by hydrogen, these densities are almost equal and the emission is proportional to the square of the density. The spectrum of bremsstrahlung is flat up to $\epsilon \sim kT$, above which it drops exponentially with frequency.

As the temperature of a cosmic plasma increases so does the energy of the electrons and ions and their ability to be collisionally ionised (i.e. $kT > I$). This means that above temperatures of a few million K most abundant elements apart from iron have lost all but their 2 innermost (K shell) electrons and by 100 million K iron, and all the lighter elements, are mostly fully stripped of all electrons. Collisional excitation of ionic electrons, leading to line radiation, dominates the energy loss of hot plasma below about 10 million K and continuum emission dominates above.

Precise calculations of line emission from a hot plasma require first that the ionisation balance be determined, which may be complicated. For multi-electron ions the excitation or recombination of an electron may cause the other electrons to be ejected or excited. Such processes (autoionisation and dielectronic recombination) are very important for large atomic number such as iron. Most X-ray astronomers rely on the Raymond-Smith and Mewe-Kaastra-Liedahl (*MEKAL*) codes (**Figure 2**), which incorporate many lines and are continuously being updated (the first by Brickhouse). (The *MEKAL* and other codes are incorporated into the spectral-fitting package *XSPEC*; this and much other software can be obtained from `http://legacy.gsfc.nasa.gov/`)

When individual lines complexes can be resolved, line ratios (eg Ly α/Ly β) can yield temperatures and (eg forbidden/intercombination) densities. The ratio of intercombination + forbidden to resonance lines for a given transition can also show whether the medium is under- or over-ionised.

So far I have assumed that the gas is optically thin and collisional ionisation alone is important (this is known as the coronal model). There is a unique ionisation state for the gas at each temperature. When gas is close to a powerful source of ionizing radiation such as an X-ray source, then photoionisation may dominate. The ionisation state, and probably the gas temperature, then depend upon the source flux and must be computed for that state (using say Ferland's *CLOUDY* code or Kallman's *XSTAR*).

3.2. *Comptonisation*

X-rays scatter off electrons in many cosmic sources. The appropriate cross section is the Thomson one, reducing as the more complex Klein-Nishina one as ϵ approaches and exceeds $m_e c^2$. When the energy of the photon exceeds that of the electron, the fractional mean change in energy of the photon from the Compton effect, $\Delta\epsilon/\epsilon \approx -\epsilon/m_e c^2$. This is Compton down-scattering (of the photon) or Compton heating (of the electron). For the other case when the electron energy exceeds that of the photon, $\Delta\epsilon/\epsilon \approx v^2/c^2$, or for a thermal gas $\Delta\epsilon/\epsilon = 4kT/m_e c^2$. This is Compton up-scattering (photon) or Compton cooling (electron). Overall,

$$\Delta\epsilon = \epsilon m_e c^2 (4kT - \epsilon).$$

If the Thomson depth of the medium $\tau = n_e \sigma_T r$ is large and $\epsilon \ll kT \ll /m_e c^2$, then

MEKAL models

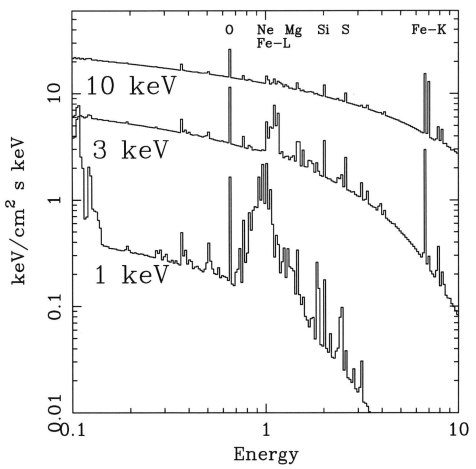

FIGURE 2. Energy spectra of a 1 keV, 3 keV and 10 keV plasma of cosmic abundance, computed using the *MEKAL* code.

there will be $\sim \tau^2$ scatterings before the photon exits the medium at each of which the photon energy is boosted by a factor $1 + \epsilon 4kT/m_ec^2$, so the ratio of final to initial photon energies, $\epsilon_f/\epsilon_0 \sim \exp y$, where the Compton y-parameter is $y = (4kT/m_ec^2)\tau^2$. The emergent spectrum can be calculated from the Kompaneets equation for the diffusion of photons in energy space and is a powerlaw of index $\alpha = \sqrt{(2.25 + 4/y)} - 1.5$.

Often, however $\tau \sim 1$ and $kT/m_ec^2 = \Theta \sim 1$. In this case diffusion is not a good description of what is happening. The mean photon amplification factor per scattering is $A = \ln(1 + 4\Theta + 16\Theta^2)$ (**Figure 3**), so after k scatterings the photon energy $\epsilon_k \sim A^k$. The probability that the photon is still in the medium to be scattered is τ^k and it is this which determines the intensity at ϵ_k. This leads to a powerlaw of index $\alpha = \ln \tau/((1 + 4\Theta + 16\Theta^2)$.

3.3. *Inverse Compton radiation*

Compton upscattering by an energetic electron is called the inverse Compton process, particularly if the electron is relativistic. If the interaction is with an electron of Lorentz

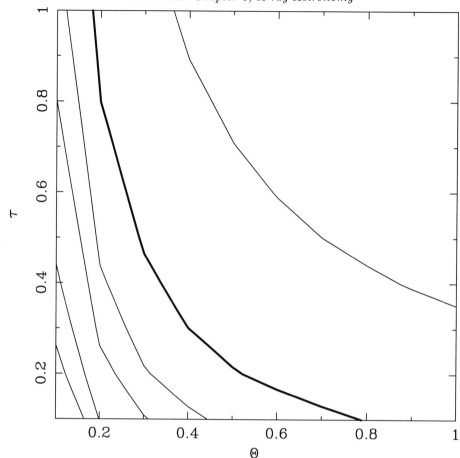

FIGURE 3. Contours of constant energy index α in the $\tau - \Theta$ plane produced by Comptonization in a gas with Thomson depth τ and temperature (in units of $m_e c^2$) of Θ.

factor γ then the incoming photon changes its energy from ϵ_0 to $\epsilon_1 = \frac{4}{3}\gamma^2\epsilon_0$. The factor of γ^2 comes about because of the need to transform into the electron rest frame where the scattering is a Thomson one, then back to our frame, each transformation introducing a factor of γ.

An electron passing through a radiation field of photon density n_γ loses energy in lumps of $\frac{4}{3}\gamma^2\epsilon_0$ at rate $n_\gamma\sigma_T c\frac{4}{3}\gamma^2\epsilon_0$. Now the energy density of the radiation field is $\mathcal{E}_{\text{rad}} = n_\gamma\epsilon_0$ so $\dot{E} = \frac{4}{3\sigma_T c\mathcal{E}_{\text{rad}}\gamma^2}$. The energy loss time of the electron is then $E/\dot{E} \propto \gamma\mathcal{E}_{\text{rad}}^{-1}$.

This process may be important in radio sources where there is a mixture of relativistic electrons and radiation, either from the active nucleus itself or from the microwave background.

3.4. *Cyclotron and Synchrotron radiation*

The acceleration (orbital motion) of an electron in a magnetic field B leads to a cyclotron power $\propto B^2 v^2$ when the electron is non-relativistic and a synchrotron power $\propto \gamma^2 B^2$ when it is relativistic. The formulae are similar to inverse Compton scattering since the process may be seen as Compton scattering of virtual cyclotron photons, replacing $\mathcal{E}_{\text{rad}}$ by $B^2/8\pi$.

Self absorption commonly affects the lowest frequencies of cyclotron or synchroton spectra.

In a compact source, or a jet, the same electrons which give rise to the synchroton radiation in say the radio band can inverse Compton scatter that radiation into the X-ray band. This is known as Synchroton Self-Compton emission. The ratio of the power in the inverse Compton scattered part to the synchrotron part is roughly $\mathcal{E}_{rad}/B^2$.

3.5. *Which process?*

It may not be immediately obvious which process dominates the energy loss for a region. Generally it will be that with the shortest cooling time, but in estimating this self-absorption might need to be considered if there is much emission expected at low frequencies.

3.6. *Pair production and compactness*

The electron energies in accreting black hole systems may be mildly relativistic. They then have sufficient energy to create electron-positron pairs by particle-particle, particle-photon or photon-photon interactions. The latter process may limit the gamma-ray production in many luminous compact sources.

The probability that an outgoing photon in a hard X-ray source of radius R and luminosity L collides with another photon to produce an electrn-positron pair is roughly $\tau_{\gamma\gamma} = 0.2 n_\gamma \sigma_T R$. Now $n_\gamma > L/4\pi R^2 c m_e c^2$ so $\tau_{\gamma\gamma} > \frac{1}{20\pi} \frac{L}{R} \frac{\sigma_T}{m_e c^2}$. The last part,

$$\ell = \frac{L}{R} \frac{\sigma_{rmT}}{m_e c^2}$$

is known as the compactness, and pairs are copiously produced if $\ell > 20\pi$. Indeed if ℓ is much higher than this and the source is producing photons above an MeV, then pairs can be created at a rate faster than they can annihilate. Such catastrophic pair production will limit the temperature, and emergent hard spectrum, of such a source. Many accreting black holes appear to have $\ell \sim 10 - 100$.

Papers by Rybicki & Lightman (1980) and by Tucker (1975) give additional details of the processes discussed in this section.

4. Compact Objects – WD, NS, BH

White dwarfs (*WD*) are the collapsed cores of stars of mass less than about $8\,M_\odot$; neutron stars (*NS*) and stellar-mass black holes (*BH*) are formed in the supernova implosion of the cores of more massive stars. The formation of supermassive black holes in galactic nuclei is less certain but may be due to the collapse of a supermassive star or relativistic star cluster.

White dwarfs are supported against gravitational collapse by electron degeneracy pressure. This is essentially a consequence of the uncertainty principle (the product of positional uncertainty and momentum exceed Planck's constant divided by 2π); as the core of a star collapses under gravity the positions of all the particles are restricted more and more, so their momentum and thus (zero-point) energy must increase. The pressure due to this energy density is degeneracy pressure. More massive white dwarfs have a smaller size than less massive ones since a higher pressure is required and this can only come from restricting the particles yet further. The maximum (Chandrasekhar) mass of $\sim 1.4\,M_\odot$ occurs when the electrons reach the speed of light. When the degenerate particle is the neutron, a neutron star results. Its radius is smaller than that of a white dwarf by the factor m_n/m_e, and is about 10 km. At this point the neutrons are essentially touching and nuclear forces are also important in its structure. The object is also only a few R_s so general relativity must also be considered. There is an upper limit to the mass of a neutron star of about $3\,M_\odot$.

Black holes have an event horizon, where the redshift as seen by an external observer is infinite, of $R_S = 3M/M_\odot$ km. Matter can orbit stably about a non-spinning black hole down to $3R_s$ within which it will just fall into the hole. The horizon and corotating innermost stable orbit for a spinning, Kerr, black hole occur at smaller radii.

4.1. *Pulsars*

Spinning magnetized neutron starts were first discovered as pulsars. They have spin periods of between 1.6 ms and several seconds, and can be very good clocks.

[Rotation or pulsation of any self-gravitating object must give

$$\Omega^2 < \frac{GM}{R} \sim G\rho,$$

(*i.e.* it cannot spin faster than breakup speed or oscillate faster than the escape velocity), so values of the pulse frequency Ω limit the density ρ of the object. $P < 1\,\mathrm{s}$ implies a neutron star.]

The magnetic-field corotates with the neutron star out to a radius R_c at which the speed of a corotation particle equals c; this is the speed of light cylinder. $R_c = c/\Omega$. The field there (assumed to be dipolar) $B_c \sim B_p(R/R_c)^3$ and the energy density $\mathcal{E}_c \approx B_c^2$. This is being 'stripped off' and flows away at c as radiation so the power

$$\mathcal{P} \sim \mathcal{E}_c R_c^2 c \sim \frac{B_p^2 R^6 \Omega^4}{c^3} \sim I\Omega\dot{\Omega}.$$

The magnetic dipole model for pulsars is confirmed by observations of the Crab pulsar which lies in the Crab Nebula, the remnant of the supernova of AD1054. The blue light and X-ray emission from the Nebula is due to synchrotron radiation, for which the electron lifetimes are only a few years. There must therefore be continuous acceleration of electrons within the nebula, the energy for which is presumably the magnetic dipole radiation, which at only 33 Hz cannot propagate through the ionised gas of the Nebula. The total observed synchroton power radiated should therefore equal the rate of loss of rotational energy of the pulsar, which can be deduced from its spin down rate. They do indeed agree if the pulsar has the moment of inertia expected of a neutron star.

There is also an unpulsed, thermal component of X-ray emission from young neutron stars. On formation, the surface of a neutron star cools rapidly to a few millon K by the emission of neutrinos. It then cools more slowly over $\sim 10^5$ yr at a surface temperature of about 10^6 K emitting $\sim 10^{33}\,\mathrm{erg\,s^{-1}}$ in soft X-rays. Such emission has recently been seen from several pulsars (e.g. the Vela pulsar) and point (unpulsed) sources in supernova remnants. The results show that standard models for neutron stars with a medium-stiff equation of state are appropriate.

Further information on the processes discussed in this section may be found in Shapiro & Teukolsky (1983).

5. Accretion

Matter falling onto a compact object liberates enormous amounts of energy owing to the great depth of the potential well involved. This process of accretion at a rate $\dot{M}$ onto an object of mass M and radius R gives a luminosity of

$$L = GM\dot{M}/R,$$

which yields a mass to energy conversion efficiency of 10 per cent of the rest mass for a neutron star and a black hole. This is considerably greater than the efficiency of complete nuclear fusion which is 0.7 per cent.

The accreting matter may either be supplied by a companion star in a binary system, either as a wind or a stream from the inner Lagrangian point, or from the surrounding medium. In the latter case, known as Bondi accretion, the rate is determined by the object's mass and the density ρ and sound speed of the medium, c_s; $\dot{M} \sim 4\pi(GM)^2\rho/c_s^3$.

In most situations the accreting matter has enough angular momentum to orbit a compact object. Dissipation and friction then lead the matter to flatten and spread equatorially into an accretion disk. Here angular momentum is transported outward by friction and mass moves inward. The friction here is not simple molecular viscosity but is probably magnetic in origin (note that the disk close to a compact object will be ionised) and is usually parametrized such that the viscous stress is αP where P is the local pressure. $\alpha < 1$ and probably is about 0.1.

The accretion rate is limited in most situations by the Eddington limit. Here the outward pressure of radiation balances the inward gravitational force. Any phenomenon lasting more than a few dynamical times (thus excluding supernovae) should have $L < L_{edd}$. The basic limit is obtained by comparing the force of radiation acting on an electron (through Thomson scattering of cross-section σ_T) at the surface of the object with the force of gravity on a proton there. The electron and proton, although assumed free, are bound together by electrostatic attraction.

$$F_{grav} = \frac{GMm_p}{R^2} = F_{rad} = \frac{L}{4\pi R^2 h\nu}\sigma_T\frac{h\nu}{c},$$

where the radiative force term is the flux of photons (of typical energy $h\nu$) times the cross-section times the momentum of a photon. This gives

$$L_{Edd} = \frac{4\pi GMm_pc}{\sigma_T},$$

which is about $10^{38}\,\mathrm{erg\,s^{-1}\,M_\odot^{-1}}$.

In standard accretion disks, which probably occur in all luminous objects, the energy liberated by accretion is radiated locally from the surface of the disk, which is therefore much cooler than the local virial temperature (which exceeds 10^{11} K close to a black hole or neutron star) and is physically thin. At low accretion rates, $\dot{M} < \alpha^2\dot{M}_{Edd}$, another solution is possible. In this case the protons in the flow are close to the virial temperature so the disk is thick, toroidal in shape, and the density of the accreting matter therefore low. Provided that the energy exchange between the protons (which carry most of the mass and thus energy in the flow) and the electrons (which radiate what energy they can) is by Coulomb interactions (the rate of which depend on the square of the density), then this can be so slow that most of the energy is advected into the centre. If the object there is a black hole then the efficiency of accretion can be very low since the hole swallows the matter, energy and all. Such flows are known as Advection-Dominated Accretion Flows (*ADAF*s). They may be relevant to situations where accretion is suspected or inferred by the presence of gas surrounding a black hole, yet little luminosity is seen.

5.1. *Accretion onto magnetised neutron stars or white dwarfs*

Neutron stars and white dwarfs often have magnetic fields strong enough to channel an accretion flow when it is close to the surface. To understand what happens, we begin by assuming that the accretion flow is spherically symmetric, and that the compact object has a dipolar magnetic field. The magnetic field of the star then dominates the flow within a region of radius R_M, known as the magnetosphere, where the magnetic pressure (or stress) equals the ram pressure of the inflowing gas; $P_B > P_{flow}$, similar to the cavity in the solar wind about the Earth. $P_B \propto B^2 \propto m^2/R_M^6$ equals $P_{flow} = \rho v_{ff}^2 \propto (\dot{M}/R_M^2)\sqrt{M/R_M}$. Therefore $R_M = m^{4/7}\dot{M}^{-2/7}M^{-1/7}$, which, since $L = GM\dot{M}/R_\star$ (where $R_\star$ is the radius of the compact object) gives

$$R_M \propto L^{-2/7}m^{4/7}.$$

This scaling holds even when the magnetized object is accreting from a disk. Until the object corotates with the inner edge of the disk, it accretes angular momentum, *i.e.* the object *spins up*. The sense of $\dot{P}$ is thus opposite from that of a radio pulsar, which spins down.

Many High Mass X-ray Binary (*HMXB*) systems contain an X-ray pulsar with period ranging from 10s of ms to minutes. Observations reveal orbital periodic shifts in the spin period P, generally superimposed on a gradual spin up ($\dot{P} < 0$). These systems consist of a magnetized neutron star orbiting a massive star from which it is accreting matter, either from the star's wind or via Roche-lobe overflow. The pulses are due to the strong magnetic field funnelling the inflowing gas onto the polar caps of the neutron star, which are not aligned with the the spin axis. The pulse period is just the rotation period of the neutron star.

The specific angular momentum (*i.e.* per kg) of matter in a Keplerian orbit is $\sqrt{GMR}$, so $I\dot{\Omega} = \dot{M}\sqrt{GMR_M}$,

$$\text{so } \dot{P} = -P^2 \frac{\dot{M}\sqrt{GMR_M}}{2\pi I}.$$

Using the above scaling for R_M, we finally obtain

$$\frac{\dot{P}}{P} \propto \frac{-PL^{6/7}m^{2/7}}{I}.$$

The good agreement between data for binary X-ray pulsars and the above theory, assuming pulsar-like magnetic fields, shows that the accreting object is a neutron star, not a white dwarf. (White dwarfs have a moment of inertia about m_p/m_e times larger than that of a neutron star and so could not spin up as fast as observed.)

Most Low Mass X-ray Binaries (*LMXB*) do not pulse. Also eclipses are rare (the companion star is usually small. of mass about $0.5\,M_\odot$, and subtends a much smaller solid angle at the X-ray source than is the case for the massive stars in a *HMXB*). They appear to be undergoing relatively stable mass transfer over rather long timescales (probably $> 10^8$ yr). The lack of pulses is presumed due to a low surface magnetic field. Pulsar magnetic fields appear to decay on timescales greater than a million years, either intrinsically (as may happen with radio pulsars) or, if mass transfer has occurred, due to some effect of accretion (the details of magnetic field decay are uncertain). Continued accretion onto a neutron star with a weak magnetic field can spin it up to a ms period (the maximum period is the Keplerian period of the disk at the radius of the magnetosphere).

5.2. QPO

Many *LMXB* show complex chaotic behaviour in their X-ray flux with the appearance of complex quasi-periodic oscillations (*QPO*), perhaps due in some cases to a beat between the spin frequency of the neutron star and the orbital frequency of the inner radius of the disk (*i.e.* at R_M). The *QPO* can rapidly change frequency as the source properties (flux, X-ray colours) change.

5.3. *X-ray bursts*

Many *LMXB* show *X-ray bursts* in which the persistent X-ray flux rises by factors of 100s in a few seconds then decays back to the original level over the next few 10s of seconds. They often recur every few hours. The bursts are due to a thermonuclear flash triggered by explosive He-burning (H-burning involves the weak force and so cannot give rise to a 1 s flash) of the accreted matter. Nuclear burning gives about 6 MeV/nucleon whereas accretion gives about 200 MeV/nucleon. So we expect the ratio

$$\alpha = \frac{\text{ave. persistent flux}}{\text{time} - \text{ave burst flux}} \gtrsim \frac{200}{6} \sim 30.$$

The observed values of $\alpha \sim 50 - 200$ so the agreement is fair (especially if nuclear burning is incomplete in a burst). Bursts require that the polar field $B_p \sim 10^8$ G, otherwise matter is channelled onto the poles at high density and burns quietly and continuously.

The surface is observed to cool after a burst as a blackbody. Using $L = 4\pi R^2 \sigma T^4$ and obtaining L and T from the observed flux and spectrum enables the radius to be found. Observations show that R is constant at $\sim$10km, while L changes by a factor of about 100. (Note that General Relativistic corrections need to be applied for a precise determination of the radius.)

The peak luminosity of X-ray bursts appears to be limited by radiation pressure. This can be so high that the surface layers are blown into space. The limiting luminosity at which the force due to radiation just balances the attraction due to gravity on matter is known as the *Eddington limit*,

Pulsations are sometimes observed during the rise of X-ray bursts. This is probably due to the spread of the nuclear burning zone across the neutron star surface initially making one face hotter than the other. The pulse period is then due to the spin of the neutron star. In these cases the neutron stars are spinning at ~ 500 Hz.

5.4. *Galactic black hole candidates*

Some luminous X-ray binaries, *e.g.* Cygnus X-1, LMC X-3, probably contain black holes accreting from the companion star. The basic evidence for a black hole in these systems is the high mass for the compact X-ray emitting object (usually based on optical spectroscopic evidence of regular radial velocity variations of the companion star) and a small size, based on rapid X-ray variability (say on timescales down to ms). Unless relativistic motions are involved, an object cannot appear to vary on a timescale shorter than its light crossing time, so if it appears to vary by a large factor (say 2) on a timescale shorter than Δt it must be smaller than $c\Delta t$. The X-ray spectra of the *BHC* also often resemble those of the much more luminous.

Additional information on the processes discussed in this section may be found in Frank *et al.* (1985), Kato *et al.* (1998), Lewin *et al.* (1995).

6. Supernova Remnants

The endpoint of the evolution of a star more massive than about 8 $M_\odot$ is as a supernova. The core collapses into a neutron star (or black hole) releasing about 10^{53} erg of which about 1 per cent is in kinetic energy which ejects the envelope of the progenitor star into space at about 10^4 km s^{-1}. It interacts there with the stellar wind ejected much earlier (and at lower velocity) from the progenitor star, together with any other surrounding or interstellar gas (**Figure 4**). In doing so it causes a strong shock to occur since the ejecta is moving highly supersonically compared with the surrounding matter (*i.e.* it is moving faster than the velocity of sound in the interstellar gas). A supernova can also occur when a white dwarf accretes enough matter to reach the Chandrasekhar mass.

Across a strong shock, the gas density jumps up be a factor of 4 and, by coninuity, the velocity jumps down by the same factor. The temperature of the gas jumps to

$$T = \frac{3}{16} \frac{\mu m v^2}{k}.$$

This means that a significant fraction of the kinetic energy upstream of the shock is thermalized at the shock where it appears downstream as heat.

Much of the kinetic energy in the expanding ejecta from a supernova is thereby converted into heat. The temperatures that result exceed 10^7 K in the first few 1000 yr and expanding supernova remnants are seen as bright X-ray emitting objects (very roughly $h\nu \approx kT$ for bremsstrahlung and other radiation processes that are important). The ejecta expand more or less freely (*i.e.* $v = $ const, $r \propto t$) until the swept-up mass (*i.e.* the mass it has interacted with) has a similar mass to the ejected mass ($M_{sw} = \frac{4}{3}\pi r^3 \rho_u = M_{ej}$). After that (and assuming that radiative cooling is unimportant in the shocked gas, as is correct for typical remnants), the remnant enters the Sedov-Taylor (or Adiabatic) phase. We can understand the behaviour of the remnant here by noting that the pressure driving it

$$P_d = \frac{3}{4}\rho_u v^2 \propto \rho \dot{r}^2$$

is also $\propto E/r^3$ (pressure is energy density). Thus $r \propto \left(\frac{E}{\rho_u}\right)^{1/5} t^{2/5}$. (This also describes part of the evolution of atom bomb explosions, for which it was first applied.)

Later the remnant moves slowly, radiative cooling of the post-shock gas is important and it merges with the rest of the interstellar gas. Its size may then by about 30 pc. Supernova remnants are a major source of energy and 'metals' in the interstellar medium. Most of the elements of which the Earth is composed were forged in supernova explosions and expelled through supernova remnants.

6.1. *Non-equilibrium effects*

The hot, X-ray emitting, post-shock gas is not immediately in collisional ionisation equilibrium, but may take $\sim 10^{12}/n$ s to do so (the coefficient in this expression is very approximate and depends on the ionisation species of interest). The departure from ionisation equilibrium thus depends on the product of local density n and time since the passage through the shock.

It is also possible that the ion temperature exceeds that of the electron temperature immediately behind the shock and that Coulomb collisionial exchange is required to equilibrate them.

Shocks also accelerate cosmic rays and supernovae may be a significant source of cosmic rays in the Galaxy. Nonthermal X-ray emission from some remnants, such as parts of SN 1006, indicate the regions where acceleration is taking place.

6.2. *Compact supernova remnants*

Most supernova remnants take over 10^4 yr to reach the radiative phase, due to the low density of the surrounding gas (say $n \sim 1\,\mathrm{cm}^{-3}$). The bulk of the power radiated then emerges in the UV and the supernova remnant is never particularly luminous as an X-ray source (say $L_X < 10^{37}\,\mathrm{erg\,s}^{-1}$). Some supernovae do however occur in a dense circumstellar medium ($n > 10^7\,\mathrm{cm}^{-3}$) which results from a slow wind phase of the progenitor star. The remnant then jumps to the radiative phase within a year or

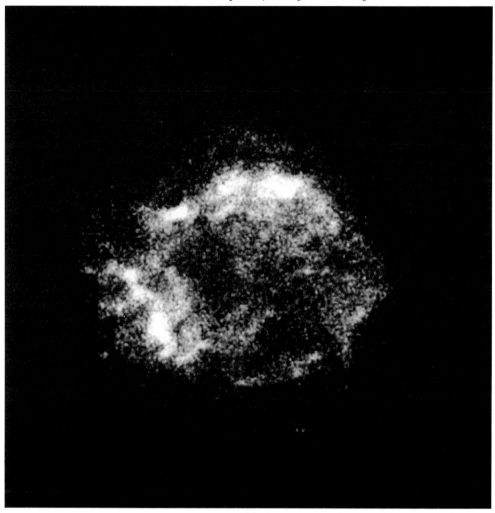

FIGURE 4. The 330 year old supernova remnant Cas A, as imaged by the *ROSAT HRI*.

two, emitting most of the kinetic energy as X-rays. The X-ray luminosity may exceed $10^{41}\,\mathrm{erg\,s^{-1}}$ for a few yr.

7. Active Galactic Nuclei

There is now good evidence from X-ray observations that Seyfert 1 galaxies are powered by disk accretion onto a massive black hole. The evidence is twofold; first, the strong qualititative similarity between the gross X-ray spectrum and variability of Seyfert 1 galaxies and Galactic Black Hole candidates such as Cygnus X-1, second, the appearance of reflection components seen in the spectrum, in particular the broad iron emission line. The breadth and skewness of this line reveals the effects of the implied strong gravity near a black hole (by large gravitational redshifts).

Optical emission-line and radio water maser studies of some low luminosity, nearby *AGN* have demonstrated the presence of central, dark massive objects in these systems. These objects are almost certainly black holes. The gas which is observed in these studies

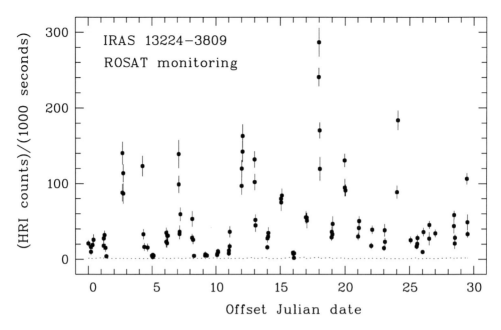

FIGURE 5. Persistent, giant variability seen in the narrow-line Seyfert galaxy IRAS 13224 with the *ROSAT HRI*. The luminosity approaches $10^{45}\,\mathrm{erg\,s^{-1}}$ at the brightest points. It is likely that amplifiaction by mild relativistic beaming, say from a matter orbiting close to a black hole, is responsible for such extreme observed variability.

is however at about 10^5 Schwarzschild radii from the black hole so does not probe the strong gravity regime.

The evidence from quasars is less clear but suggestive. Indeed, quasars show fewer X-ray spectral features than do Seyfert galaxies, possibly because the strong radiation field ionizes more of the surrounding gas.

The chaotic nature of the variability in Seyfert 1 galaxies is not understood in detail. Rapid flux changes by a factor of two do constrain the size of the emission region (which may be a flare and so a subset of the total source) and the mass-to-energy conversion efficiency of the source, provided that relativistic effects (e.g. beaming) are small (**Figure 5**). For example, the Seyfert 1 galaxy MCG–6–30–15 showed a factor of nearly 2 change in less than 100 s during an *ASCA* observation. The emission region is thus smaller than the Schwarzschild radius of a $10^7\,\mathrm{M_\odot}$ black hole.

The X-ray spectrum and its components are schematically illustrated in **Figure 6**. The underlying power-law has a photon index of about 2 over a band from less than 1 keV to above 100 keV where it exponentially cuts off with an e-folding energy of several 100 keV. It probably originates via thermal Comptonisation above the accretion disc, possibly in a flaring corona.

The 'standard' model that is emerging for the interpretation of the spectrum is of an accretion disc at a surface temperature less than 10^6 K generating about two-thirds of the total power, in the *EUV* and very soft X-rays as quasi-thermal blackbody radiation, and the other third of the power emerging via Comptonisation from hot corona flares (with $kT \sim 300\,\mathrm{keV}$ and Thomson depth $\tau_{\mathrm{T}} \sim 0.2$). The irradiation of the disc by the coronal X-ray continuum increases the soft X-ray emission and produces the reflection spectrum (**Figure 7**). This last feature consists of the reflection hump due to backscattering (it decreases at high energies due to electron recoil and at low energies due to photoelectric

Characteristic Seyfert 1 X-ray Spectrum

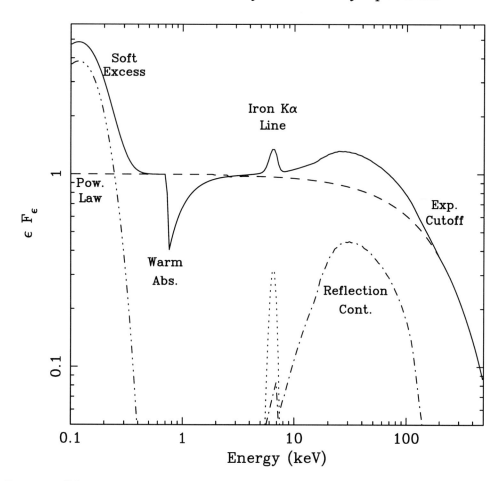

FIGURE 6. Schematic view of the broad band X-ray spectrum of a Seyfert 1 galaxy. The three known continuum components are labeled along with spectral features. We have removed the effects of absorption by neutral matter for clarity of presentation. The power law has a photon index of $\Gamma = 2$ and an exponential cutoff of energy 200 keV. The behaviour of the power law in the energy range dominated by the soft excess is not clear (and thus our plot should be not be taken literally in this regard). The solid angle of the neutral matter that generates the reflection continuum has been taken to be 2π steradians. The soft excess is taken to be a blackbody with a temperature of 0.03 keV. The warm absorber edge has a threshold energy of 0.739 keV and an absorption depth at the threshold of $\tau = 1$. The Gaussian emission line has an energy of 6.4 keV, an equivalent width of 220 eV and $\sigma = 0.5$keV.

absorption), and fluorescence, principally in the form of an iron emission line. The product of abundance and fluorescent yield (which varies as atomic number raised to the fourth power) is highest for iron. The iron in the disc of *AGN* is relatively cold and generally has M-shell electrons so the line occurs at 6.4 keV.

The quasi-blackbody emission from the disc can dominate the spectrum below 0.5 keV and is best seen in *ROSAT PSPC* spectra. Sometimes it varies in a manner different from that of the harder power-law continuum showing that it is not purely due to reprocessing. In some objects, the Narrow-Line Seyfert 1 galaxies in particular, the soft excess dominates the spectrum up to and sometimes above 1 keV.

It is also assumed in the standard model that the nucleus is surrounded at a radius of a few pc by a molecular torus. The unified model for Seyfert 1 and 2 galaxies then states that we see a Seyfert 1 when we view the nucleus directly and a Seyfert 2 when we look through the torus (or only see the nucleus via scattered light). Depending on its solid angle and Thomson depth(s) the torus should be detectable via its reflection component in Seyfert 1 galaxy spectra. In this case the iron line should be narrow. Generally, such a component is not seen and the Thomson-thick part of the torus must subtend less than about π sr. One object in which the torus reflection has, however, been detected is NGC 2992, which has faded over the last 17 yr revealing a strong narrow iron line component.

Quasars are relatively featureless in the X-ray band and show no iron component or reflection continuum.

7.1. *Warm absorbers*

The X-ray spectrum from many active galaxies shows the effects of absorption by cold material in the host galaxy. In the case of Seyfert 1 galaxies there is also often evidence for absorption by partially-ionised gas. This is presumably matter close to the central engine which has been photoionised by the intense radiation field. The main features are O VIII and O VII edges at 0.74 and 0.87 keV respectively.

In the case of MCG–6–30–15, the absorber is variable on timescales of weeks to hours. This is of great importance since this is the first time that the response of an edge to the ionizing flux is clearly seen – photoelectric absorption plays a key role in much of high energy astrophysics and here it can be tested. In detail, it is found that the O VII edge is of constant depth, even during flux changes by a factor of up to 7, while the O VIII edge increases when the flux drops.

This puzzling behaviour can be understood if there are 2 warm absorbers, one close to the central engine and one quite distant. The response of the ionisation state of gas to changes in ionizing luminosity depends on the photoionisation and recombination timescales, which in turn depend on the flux and density respectively. The data require that the recombination time of the O VII absorber exceeds about 10^6 s and that of the O VIII absorber is less than 10^4 s. The O VII absorber must then be about 100 times less dense than the O VIII absorber and so is in a different region. We identify the O VII absorber as an outer region beyond the radius of the unifying torus and in the optical Narrow Line Region; the O VIII absorber is an inner region at or within the optical Broad Line Region. The mean ionisation state of the gas depends on the ionisation parameter $\xi = L/nR^2 = (L/NR^2)(\Delta R/R)$, where L is the ionizing luminosity, n the gas density in a column density N within fractional radius ΔR at radius R. Photoionisation models produced with Ferland's *CLOUDY* code enable us to fit the *ASCA* spectra for ξ and N and so deduce $\Delta R/R$. It is generally much less than unity and approximately equals the volume filling factor (it actually equals the covering fraction, f_A times $\Delta R/R$; the common occurrence of warm absorbers suggesting that $f_A \sim 0.5$ if the Seyfert 1 aspect for warm absorbers is the same as the all-sky aspect).

The total mass in the inner warm absorber, obtained from the product of the column density, covering fraction (estimated from the frequency with which warm absorbers are seen in Seyfert 1 galaxies) and $4\pi R^2$, probably exceeds that in the observed optical/UV *BLR* clouds, and the mass in the outer warm absorber probably exceeds that in the *NLR* clouds. It therefore represents a considerable fraction of the gaseous mass around the nucleus.

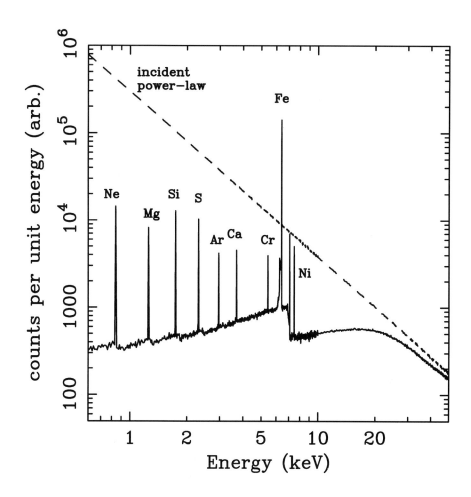

FIGURE 7. Monte-Carlo simulations of the reflection spectrum expected from a flat surface of gas of cosmic abundance. Compton scattering dominates over photoelectric absorption at the highest energies, so the albedo is relatively high. Photoelectric absorption dominates below about 30 keV with the electron-scattered continuum diminishing towards lower energies. The strongest fluorescent lines are shown.

7.2. *The broad iron line*

After some hints that the iron line, produced by reflection and commonly seen in Seyfert 1 galaxies with Ginga, is broad, definitive evidence came from *ASCA* spectra. The *FWHM* of the iron line indicates doppler velocities of tens of thousand to $100\,000\,\mathrm{km\,s^{-1}}$ in a wide range of objects.

The line in MCG–6–30–15 yields about 200 ct per *ASCA SIS* detector in 40 ks, which is the length of a typical observation. With only 200 ct it is difficult to measure more than the intensity and width of the line (it is also difficult to define the continuum at energies above that of the line). If the line originates from reflection of the inner disc about a black hole, it should show relativistic effects which skew its shape (**Figure 8**).

In particular the blue horn is brightened relative to the red horn by aberration and the whole line is broadened and shifted to the red by doppler and gravitational redshift.

A 4.5 day observation in 1994 of MCG–6–30–15 enabled the true line profile to be seen. The line is skew and consistent with that expected from within 10 Schwarzschild radii of the black hole. The disk inclination must be close to 30 deg. We have have been unable to find any other way to obtain such a broad, skew line shape. Comptonisation for instance needs a high Thomson depth (~ 5) in cold electrons ($kT < 0.25\,\mathrm{keV}$) to produce such a broad line and even then the shape is in poor agreement.

Lines from a Relativistic Disk

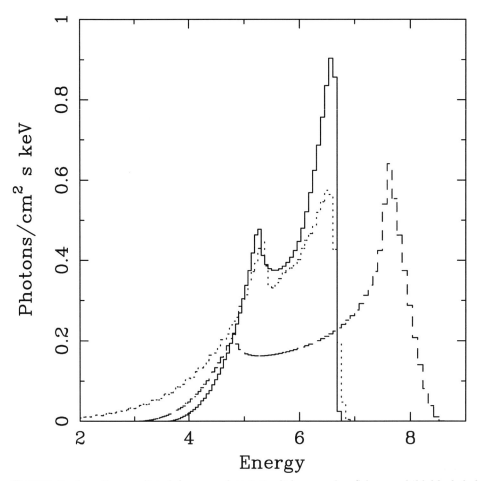

FIGURE 8. Iron line predicted from a relativistic disk around a Schwarzschild black hole at inclinations of 30 and 70 degrees (solid and dashed lines respectively) and a maximal Kerr black hole inclined at 30 degrees (dotted profile). Note that the high energy wing of the line profile is most sensitive to the disk inclination; the low energy wing is most sensitive to the inner radius which is of course smaller for the (prograde) Kerr disk.

Such cold electrons would be below the Compton temperature of the source and would downscatter the continuum, causing a break in the spectrum at 20 keV. We have also tried fitting combinations of iron K edges etc. without success. Simple reflection from the innermost disc around a massive black hole appears to give the best explanation.

Of great interest is line variability, which could be used via the reverberation method to map out the inner disc. Reverberation occurs on the light crossing time, which is is only a few thousand s for most Seyfert 1 galaxies, and so such studies are not possible with *ASCA*. During a typical light-crossing time only a few tens of ct are detected. Nevertheless, longer term changes may occur and we have looked at the line profile of MCG–6–30–15 in segments of the 4 day observation. Distinct changes are seen showing that the line variability is more complex than simple reverberation.

When the source was brightest the line is dominated by a narrow peak at 6.4 keV, whereas when it is weakest, the narrow component is small and we see a large red wing. These effects can be explained if we adopt a model in which a only few flares take place above the disc at any given time (as suggested by the large rapid variability). Then if the bright phase is due to a few flares on the approaching side of the disc, both the blue horn of the line and the continuum are relativistically boosted. The converse occurs if the flares are on the receding side of the disk, with the red horn being boosted. The magnitude of the intensity and redshift of the red horn when the source is weakest may however require the very innermost parts of a disc in Kerr geometry (i.e. the hole is spinning). A detailed comparison of the observed line shape when the flux was a minimum with model profiles of lines expected for different values of black hole spin shows that this interpretation requires that the spin parameter $a/m > 0.95$. The reason that the minimum corresponds to the largest redshift seen may simply be due to the flares then happening closer to the black hole. Gravitational lightbending then causes us to see fewer photons whilst more impact the disk and promote fluorescence.

The broad iron line has been seen in several dozen Seyfert 1 galaxies, in some Seyfert 2 galaxies (the ones which are not Compton thick), and in some radio galaxies.

7.3. *Ionised disks*

The above discussion assumes that the disk is X-ray cold, which means that all the iron in the disk has all its L-shell electrons. This is plausible for a Seyfert galaxy with a black hole of say $10^7 - 10^8 \, M_\odot$ running at a few per cent of the Eddington limit. At higher accretion rates iron may be more ionised which can cause the iron line to occur at higher energies ($6.7 - 6.9$ keV) and change in amplitude. When an incomplete L-shell is present then resonance scattering of the line followed by an Auger de-excitation can destroy the line rapidly. When there are just the K-shell electrons (i.e. H– and He–like iron), no Auger ejection can occur and the line is strong.

This does not fit simply with the disappearance of the line in quasars, which plausibly are accreting at a higher Eddington fraction (i.e. closer to the Eddington limit). No very strong lined intermediate class is known. Perhaps the surface (i.e. outer few Thomson depths) of the disk changes too such that the disk jumps to being very highly ionised (iron completely stripped of electrons).

The blackbody emission from disks around *BHC* is likely to make their disks partially ionised, which may explain the large smeared edges seen in spectra of some of them. If the main absorber below the iron edge, oxygen, is completely ionised then the disk becomes substantially reflective below the edge, and by contrast makes the edge apparent in the spectrum. When it is not completely ionised the reflection continuum below the iron edge is low and the edge is not apparent in the observed spectrum.

7.4. *Jets and blazars*

Compact radio-loud quasars and BL Lac objects are often luminous X-ray sources. The main reason is the Synchrotron self-Compton process where the same electrons that make the radio emission, inverse Compton scatter those photons (and any other photons external to the emission region) to X-ray and Gamma-ray energies. The level of X-ray emission is very sensitive to the size of the emission region and other parameters. They are consequently often highly variable.

Most of this emission is beamed in our direction by a jet. Limits on the bulk Lorentz factor of the jet can be established when gamma-rays are detected, by requiring that the compactness, ℓ, in the jet rest frame is less than say 100 so that photon-photon collisions have not turned most of the gamma-rays into electron-positron pairs.

Further details of the processes discussed in this section may be found in Blandford *et al.* (1990) & Mushotzky *et al.* (1993).

8. The X-ray Background

There is considerable background emission at all X-ray energies which is generally considered to be due to many unresolved faint sources. This X-ray background (*XRB*) has two main 'handles' by which it can be studied – spectrum and isotropy – and a secondary handle in the nature of resolved sources, which may be a guide to what the faint sources might or might not be.

8.1. *The spectrum of the XRB*

The spectrum of the X-ray background (or backgrounds) is soft (i.e. steep) below one keV and hard (i.e. flat) above, with most of the energy density (which is about 10^{-4} of that in the microwave background) residing at about 30 keV. The soft part can be divided into several components, one being Galactic in origin and due in part to our surrounding hot Local Bubble, another being due to unresolved quasars and other unobscured *AGN*. Other components could be a Galactic halo and contributions from starburst galaxies and groups of galaxies. The soft X-ray background has been well studied with *ROSAT* and at least 80 per cent of it has been resolved.

The spectrum of the harder X-ray background above 1 keV is well fit by 30–40 keV bremsstrahlung. The *COBE* satellite limits on Compton distortion of the spectrum of the microwave background ruled out an origin in bremsstrahlung in a hot intergalactic medium. The source of the background is still debated. The spectrum in the 1-8 keV band is close to a power-law with a photon index of 1.4, which is flatter than that of any major known class of extragalactic source. This gives rise to Boldt's spectral paradox: removal of the residual contribution of known classes of source makes the remaining spectrum steeper and even more unlike known sources. Some new class of source, or a new appearance of a known class is required. The simplest and most widely discussed possibility is that it is due to a population of absorbed sources, probably *AGN*. If this is the correct explanation then a comparison with the X-ray spectra of quasars indicates that most *AGN* are obscured with only 10–30 per cent unobscured.

8.2. *The isotropy of the XRB*

The X-ray background fluctuates in intensity across the Sky due to variations in the number of unresolved sources in the telescope beam. The major contribution to these fluctuations (called $P(D)$ from radio astronomy: where P is the probability of observing a deflection, or intensity, D) is due to sources at the flux level where there is about one source per beam. Studies of these fluctuations enables the number counts of faint sources to be determined at levels much below that where source confusion sets in (which is at the flux level where there is about 1 source per 40 beams, or pixels; source confusion means that unresolved sources in the beam or pixel of the source of interest make the determination of its flux to a precision of better than the 10 per cent level impossible).

Fluctuations in excess of those expected from $P(D)$ estimates can be attributed to source clustering or large scale structure. This can in principle be determined if a) measurements from an imaging system allow the source counts to be known accurately and b) the region is scanned with a wide beam instrument which measures the XRB intensity accurately (note that the signal from point sources becomes clearer the smaller the beam, whereas that from the XRB increases with beamsize). It may then be possible to constrain large scale structure at redshifts of about 2 (probably where most of the XRB originates) directly from the XRB smoothness.

Source counts and fluctuation studies have been carried out in both the soft (0.5 − 2 keV) and harder (1 − 7 keV) X-ray bands, allowing the integrated contribution of sources down to the one per beam limit to be determined (about 60 per cent in the soft band from *ROSAT* and about 30 per cent in the harder band from *ASCA*).

8.3. *Known source contributions*

Large numbers of sources have been identified to faint flux levels in *ROSAT* images. Quasars (optically broad-line *AGN*) dominate down to a knee in the source counts at a flux level of about $10^{-14} \, \mathrm{erg \, cm^{-2} \, s^{-1}}$. Below that a population of optically narrow-line objects becomes increasingly important. It has been suggested that these objects are starbursts alone, but the latest data suggest that the X-ray emission is due to *AGN*, possibly with a large starburst component dominating the optical spectra. High excitation lines or a broad base to some lines indicates that an *AGN* is present. Note that the typical spectrum of an unabsorbed quasar is much steeper than that of the XRB, except below 1 keV.

In the harder band, flat spectrum sources are beginning to emerge at the fainter flux levels. These are probably the dominant class to contribute to the XRB. It is likely that they are obscured *AGN*, but clear signatures from a large population have not yet emerged.

There is still a possibility of a new population source which has a power-law(ish) spectrum fitting the XRB spectral shape. The current standard model in which many absorbed sources are summed together, with a range of absorption column densities and redshifts, is certainly the plausible explanation but it may have a difficulty in accounting for the smooth featureless appearance of the power-law spectrum of the XRB in the 1–10 keV band.

More information about the processes discussed in this section may be found in Fabian & Barcons (1992), Hasinger (1997).

9. X-rays from Clusters of Galaxies

The X-ray emission from clusters of galaxies is dominated by diffuse radiation from the intracluster medium (**Figure 9**). This is composed of gas at roughly the virial temperature of the cluster ($\sim 2 \times 10^7 - 10^8 \, \mathrm{K}$). The mass of gas is about 20 per cent, and the stars in the galaxies another 5 per cent, of the total cluster mass; the remainder is in the form of dark matter. The gas has a metallicity which is about 30 to 40 per cent of Solar abundance in iron, with the other elements agreeing well with enrichment by Type II supernovae (i.e. massive progenitors), at least in the rich clusters.

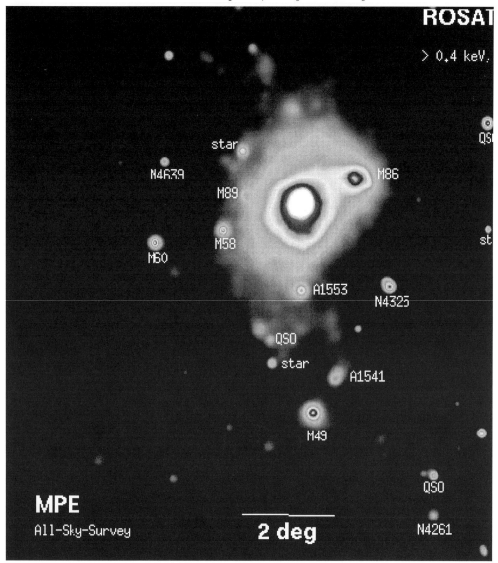

FIGURE 9. The Virgo cluster in X-rays as mapped by the *PSPC* during the *ROSAT* All Sky Survey. The central peak is around the bright elliptical galaxy M87. Other galaxies in the Virgo cluster, as well as background quasars and clusters, are indicated.

Bremsstrahlung is the dominant emission process for the intracluster medium, with line radiation from the metals adding to the spectrum.

9.1. *The gas distribution*

The intracluster gas acts as a fluid on galactic scales. The electron-electron coupling time, $t_{e-e} \approx 2\,10^5 T_8^{\frac{3}{2}} n_{-3}^{-1}\,\mathrm{yr}$, where $T = 10^8 T_8\,\mathrm{K}$ and $n = 10^{-3} n_{-3}\,\mathrm{cm}^{-3}$, and the electron-ion coupling time is 1840 times larger. The crossing time of the cluster exceeds this,

$$t_{cross} = \frac{R}{\langle v^2 \rangle^{1/2}}$$

$$\approx 10^9 R_{\rm Mpc} v_8^{-1} \,{\rm yr}.$$

where $v = v_8 10^8 \,{\rm cm\,s^{-1}}$ is the velocity dispersion of the cluster galaxies.

The sound crossing time of the gas,

$$t_s = \frac{R}{c_s} \approx t_{cross} = 10^9 R_{\rm Mpc} T_8^{-\frac{1}{2}} \,{\rm yr}$$

where the size of the region is $R_{\rm Mpc}$ Mpc. This is much less than the age of the cluster (say 5×10^9 yr) provided that $R \sim 5 T_8^{1/2}$ Mpc. Any radial flows of the gas must therefore be subsonic otherwise impossibly large mass flow rates are implied ($\gg 10^4 \,{\rm M_\odot \, yr^{-1}}$). Intracluster gas is then close to hydrostatic support so that

$$\frac{dP_{gas}}{dr} = -\rho_{gas}\frac{d\phi}{dr} = -\rho_{gas}g.$$

This means that measurements of P_{gas} and ρ_{gas}, the gas pressure and density as a function of radius r (i.e. $n_e(r)$ and $T_e(r)$), allows $\phi(r)$ and so $M_{cluster}(r)$ to be determined. Much of the mass of a cluster, $M_{cluster}$ is dark matter and X-ray observations can probe its distribution. The gas will arrange itself (convect) so that isobaric surfaces are on equipotentials which will be roughly spherical, even for quite a lumpy or flattened mass distribution. Analysis therefore generally assumes spherical equipotentials.

We now ignore these cautionary remarks and proceed under the assumption that the gas is completely hydrostatic and uniform. This is a useful approximation for the bulk of the gas in an average sense. If the gas is isothermal then hydrostatic equilibrium implies

$$kT\frac{dn_e}{dr} = -n_e \mu m \frac{GM(r)}{r^2},$$

and

$$\rho_{gas} \propto \exp\left(-\frac{\phi(r)}{(kT/\mu m)}\right),$$

where μm is the mean molecular mass. Also, if the galaxies have an isothermal velocity distribution (also unlikely) then

$$\rho_{gal} \propto \exp\left(-\frac{\phi(r)}{\sigma_{los}^2}\right).$$

where σ_{los} is the line-of-sight velocity disperion of the cluster. Then

$$\rho_{gas} \propto (\rho_{gal})^\beta,$$

where

$$\beta = \frac{\mu m \sigma_{los}^2}{kT}.$$

We might expect $\beta \sim 1$ so $c_s^2 \approx \sigma_{los}^2$. However, its application indicates otherwise. The above isothermal-isothermal model suggests the use of King's approximation of an isothermal distribution where

$$\rho_{gal}(r) = \rho_{gal}(0)\left(1 + \left(\frac{r}{a}\right)^2\right)^{-\frac{3}{2}},$$

so

$$\rho_{gas}(r) = \rho_{gas}(0)\left(1 + \left(\frac{r}{a}\right)^2\right)^{-\frac{3\beta}{2}},$$

where a is the core radius. This can then be fitted to X-ray images of clusters (emissivity$\propto \rho_{gas}^2$). Often

$$\langle\beta_{image}\rangle = 0.65,$$

so

$$\rho_{gas} \approx \rho_{gas}(0)\left(1 + \left(\frac{r}{a}\right)^2\right)^{-1}.$$

On the other hand, X-ray spectral and optical velocity dispersion measurements of T and σ_{los} sometimes give

$$\langle\beta_{spec}\rangle \approx 1 \approx 2\langle\beta_{image}\rangle.$$

The reasons for this discrepancy are not yet clear, although it is unlikely that the gas and galaxies are precisely isothermal; and the King law assumed for the mass distribution is probably too simple. Use of the Navarro, Frenk and White dark matter profile instead of the King one eliminates much of the mismatch.

It is possible to obtain gas density and temperature profiles without assuming an equation of state. The steeply rising surface brightness profile in clusters means that the X-ray surface brightness profile can be deprojected, assuming some geometry (e.g. spherical) and a distance to the cluster, to yield count emissivities as a function of radius. The emissivity depends upon n_e and T_e as well as the detector response and the effects of intervening photoelectric absorption, the last two of which are assumed known. A further relationship between n_e and T_e is obtained from the pressure via the equation of hydrostatic equilibrium. The densities obtained in this way are usually determined to better than 10 per cent, whereas the temperatures are somewhat dependent upon the assumed value of g (usually estimated from σ_{los}) that is used in the hydrostatic equilibrium equation. One pressure, typically the outer pressure, is required to start the solution and this is usually adjusted so that most of the cluster gas has a temperature consistent with X-ray spectral measurements.

9.2. *The Sunyaev–Zeldovich effect*

X-ray emission is not the only means of observing the hot intracluster medium. The low energy photons of the cosmic microwave background can be Compton-scattered on passing through the gas and experience an energy shift, $\Delta\epsilon$, given by

$$\left\langle\frac{\Delta\epsilon}{\epsilon}\right\rangle \approx \frac{4kT_e}{m_e c^2},$$

where ϵ is the initial photon energy. The proportion of photons scattered is given by the Thomson depth,

$$\tau_T = \sigma_T \int n_e dl.$$

The result is a microwave dip in the direction of the cluster in the Rayleigh-Jeans part of the cosmic blackbody spectrum

$$\frac{\Delta T}{T} = -\frac{2k\sigma_T}{m_e c^2}\int n_e T_e dl.$$

This is the Sunyaev-Zeldovich effect and measures the integral of the pressure along the line of sight. It has been detected in many clusters.

Since the X-ray luminosity is proportional to $\int n_e^2 T_e^{\frac{1}{2}} dl$, a combination of X-ray and microwave measurements can lead to an independent estimate of cluster distance. To understand this, consider a spherical region of gas of radius R at distance D, subtending an angle θ at the observer. The X-ray luminosity

$$L_X \propto n_e^2 T_e^{\frac{1}{2}} R^3$$

and

$$\frac{\Delta T}{T} \propto n_e T_e R.$$

The observer measures the flux

$$F_X \propto \frac{L_X}{D^2} \propto n_e^2 T_e^{\frac{1}{2}} \theta^3 D,$$

so, substituting for n_e,

$$F_X \propto \left(\frac{\Delta T}{T}\right)^2 \frac{\theta}{T_e^{\frac{3}{2}} D}.$$

Measurements of F_X, $\frac{\Delta T}{T}$, and T_e then lead to an estimate of D, which from $cz = HD$ gives the Hubble constant. Of course in practice, the cluster is not a uniform isothermal sphere of gas and both X-ray and microwave decrement maps must be used.

9.3. *Cooling flows*

The gas density in the core of many clusters is sufficiently high that the radiative cooling time is less than the age of the cluster (i.e $t_{cool} < H_0^{-1}$). The gas density then rises in order to maintain the pressure needed to support the weight of the overlying gas. This establishes a cooling flow. The inflow velocities are highly subsonic. X-ray observations of this phenomenon indicate that the mass flow rates can be substantial and in the range of hundreds of solar masses per year. We may be observing the continued formation of central cluster galaxies.

The cooling time of the gas can be obtained from density and temperature profiles estimated by the method given at the end of Section 9.2. Where $t_{cool} < H_0^{-1}$, the rate at which mass is deposited through cooling, $\dot{M}$, is given by

$$L_{cool} \approx \frac{5}{2} \frac{kT}{\mu m} \dot{M},$$

where L_{cool} is the total luminosity within that region. The factor of 5/2 represents the enthalpy of the gas (i.e. the sum of thermal energy and PdV work done). Short cooling times are associated with high gas densities and bright regions and so X-ray images of cooling flows show a highly peaked surface brightness profile. This is common in 30 - 50 per cent of rich clusters and in elliptical galaxies. Most nearby clusters (Virgo, Centaurus, Hydra, Perseus etc.) contain cooling flows and it is only those with two dominant central galaxies such as Coma that do not show a sharp X-ray peak.

Corroborative evidence for cooling is provided by observed low X-ray temperature spectral components in the cooling flow region (i.e. where the radiative cooling time is less than say 10^{10} yr). The emission measures ($\int n_e^2 dV$) of gas at the lower temperatures agree well with a cooling interpretation.

As the surface brightness profile gives $L(r)$ we can determine $\dot{M}(r)$, the mass deposition profile. The X-ray images are not as peaked as they could be if all the matter flowed to the centre. We generally find that

$$\dot{M}(r) \propto r.$$

Distributed mass deposition is presumably due to a range of densities present at all radii; the gas is inhomogeneous.

When the flow is inhomogeneous we can estimate $\dot{M}(r)$ by assuming that the gas is composed of a number of phases, the densest of which cools out of the flow at the radius under consideration, or by model fitting. In the first approach, the cooling region is divided into a number of concentric shells of size compatible with the instrumental resolution. The luminosity, δL_i of the ith shell can then be considered to be the sum of the cooling luminosity of the gas cooling out at that radius from the mean temperature T_i at rate $\delta\dot{M}$ and the luminosity of gas flowing across the shell experiencing temperature and potential changes ΔT_i and $\Delta\phi_i$;

$$L_i = \delta\dot{M}_i \frac{5}{2}\frac{kT_i}{\mu m} + \left(\frac{5}{2}\frac{k\Delta T_i}{\mu m} + \Delta\phi_i\right).$$

In many cases, the rate of mass deposition is large, with $50 - 500\,M_\odot\,\mathrm{yr}^{-1}$ being common for a cluster of galaxies. An outstanding problem now presents itself, namely what does the cooled gas form? If stars, then the IMF must be non-standard or the central galaxies would be much bluer and brighter than is observed.

The problem of what forms from a cooling flow has caused the assumptions behind the phenomena of cooling flows to be questioned. A heat source, for example, could offset the cooling so that little gas is actually deposited. Heat sources, or fluxes, that have been considered are cosmic rays, conduction, galaxy motions and supernovae. None of these proposals successfully confronts the evidence from spectroscopic X-ray line measurements of low temperature components that are entirely consistent with a cooling interpretation. It would indeed be a strange heat source that allowed the gas to lose 90 per cent of its thermal energy in cooling from 80 million K down to 8 million K but then prevented it cooling further. There is, of course, no evidence for large stockpiles of gas at some intermediate temperature. The X-ray spectroscopic data also indicate that cooling flows are long lived since gases with a wide range of cooling times give the same mass cooling rate.

X-ray spectra of cooling flows show evidence for excess photoelectric absorption affecting the emission cooling gas. What is doing the absorbing is unclear. Cold dusty gas clouds is a possible absorbing medium, formed perhaps from the cooled gas. There is however no evidence for such distributed cold gas at other wavelengths. Optical line nebulosities and massive star formation are often found over the innermost few kpc of the dominant galaxy at the centre of a cooling flow. This is presumably the fate of the cooled gas at the centre, but it usually accounts for at most 10 per cent of the X-ray inferred mass deposition rate.

9.4. *The hot interstellar medium of early-type galaxies*

X-ray emission from early-type galaxies was detected in the first Einstein Observatory images of the Virgo cluster. These showed diffuse patches of emission centred on the larger galaxies, together with more widespread and intense emission on M87. A plume of X-ray emission was found to the NW of M86, which is plunging through the Virgo cluster at $1500\,\mathrm{km\,s}^{-1}$. This plume provides strong evidence for a gaseous origin for the X-ray emission. Neither M dwarfs nor binary X-ray sources (possible sources that might appear diffuse in great numbers) would be preferentially swept out of a galaxy.

Generally,

$$L_X \propto L_v^\alpha; \quad 1.5 < \alpha < 2.3.$$

The total observed mass of X-ray emitting gas is $\sim 10^9$ to $10^{10}\,M_\odot$ with the gas density rising as $r^{-\gamma}$ ($1.2 < \gamma < 1.6$). It is debatable as to whether the emission follows the stellar profile well. Any outer radius of the X-ray emitting gas is certainly $> 30\,\mathrm{kpc}$ and poorly determined. It is not clear that one could be defined. The influence of surrounding gas and stripping etc. is only clear in a few cases. For many elliptical galaxies the gas temperature is $\sim 0.7 - 1.5 \times 10^7\,\mathrm{K}$.

The thermal pressure of the gas is much higher (a factor ~ 100) than that in spiral galaxies and the cooling time of much of the gas (most within 30 kpc radius)$\leq H_0^{-1}$. The immediate implications of the X-ray observations are that there are no steady-state winds and so no large heat input from supernovae. Furthermore early-type galaxies must be very massive and so possess extensive dark haloes. There is widespread cooling and, presumably, some continuous star formation. The interstellar medium in elliptical and S0 galaxies is continuously being recycled and redistributed.

There are strong similarities between the hot *ISM* in elliptical galaxies and the cooling flows in cluster cores, except that in ellipticals there is the added complication of the continuous injection of stellar mass loss at rates similar to those by which cooling depletes the atmosphere.

X-ray measurements of the hot gas in elliptical galaxies and clusters provides a powerful means for determining their gravitational mass profiles. The equation of hydrostatic equilibrium may be written as

$$\frac{d\phi}{dr} = - \left(\frac{kT_{gas}}{\mu m} \right) \left(\frac{d\ln\rho_{gas}}{dr} + \frac{d\ln T_{gas}}{dr} \right).$$

$$\phi = \int \frac{GM}{r^2} dr$$

is thus obtained from the gas pressure and density profiles. Although the density profiles measured so far are robust, the temperature profiles are somewhat uncertain.

The isophotes of X-ray emission are generally much rounder than the optical isophotes, which is evidence that the X-ray gas is responding to a dark matter potential, not that due to the stars alone. In summary, the X-ray data on elliptical galaxies demonstrates that they have massive dark haloes and a low supernova rate.

9.5. *The baryon fraction*

The gas fraction in clusters is at least 15 per cent and probably 25 per cent. The total baryon fraction is then about 30 per cent when the contribution of stars in the member galaxies is included. Assuming that there is no mechanism for clusters to have a gas-to-dark-matter ratio higher than anywhere else in the Universe (there is not time enough for any plausible segregation to work), then this implies that the baryon fraction of the Universe is 30 per cent. Now the fraction of the closure density of the Universe in baryons is determined from a comparison of cosmic nucleosynthesis calculations with helium and deuterium abundance measurements to be only about 8 per cent (for $H_0 = 50 \, \text{km} \, \text{s}^{-1} \, \text{Mpc}^{-1}$), i.e $\Omega_b = 0.08 \pm 0.02$. This, together with the above 30 per cent, means that the Universe must have a matter density parameter $\Omega_m \sim 0.3$, where "m" refers to the matter (much of it dark) which clusters now have. The Universe is therefore open. A cosmological constant is needed if the total density parameter $\Omega_0 = 1$.

9.6. *Clusters, the cosmic power spectrum and evolution*

The number density of clusters in the Universe depends on the extent of fluctuations in the dark matter distribution set down in the early universe. They provide a good measure of the fluctuation spectrum – the cosmic power spectrum – on scales around about 16 Mpc. The important issue here is that X-ray measurements provide a good way to parametrize clusters, particularly using the temperature of the intracluster gas. The temperature function $N(> T)$, the number of clusters per unit volume with temperatures exceeding temperature T, is commonly used for this purpose.

The evolution of this function with redshift can then help determine the density parameter Ω_m. If $\Omega_m = 1$, as in standard cold dark matter models, then hierarchical clustering, in which smaller clusters merge to form large clusters, proceeds rapidly up to the present day. If Ω_m is much smaller, as the high baryon fraction indicates, then evolution too is much smaller at the present time. The data show little change in the temperature function with redshift, corroborating the low Ω_m result. Moreover, studies of the amount of substructure in X-ray images of clusters support this conclusion.

Further estimates can be made from the luminosity function, which is easier to measure but relies on knowledge of the luminosity to temperature relation. Simple gravity collapses predict that $L_x \propto T^2$, whereas observations suggest $L_x \propto T^3$. There is also significant non-statistical scatter seen when these quantities are plotted against each other. Cooling flow clusters, for example tend to be more luminous than non-cooling-flow clusters at the same temperature.

The X-ray emission from clusters is therefore useful for cosmological studies. It indicates that Ω_m is small. The precision with which Ω can be defined does however require an appreciation and understanding of the non-gravitational processes (e.g supernova heating) which have shaped the intracluster medium.

Additional discussion the processes discussed in this section may be found in Sarazin (1988), Fabian (1994), & Fabian (1992)).

REFERENCES

BLANDFORD R.D., NETZER H., WOLTJER L., 1990 *Active Galactic Nuclei, Springer-Verlag*

FABIAN A.C., 1992 *Clusters and Superclusters of Galaxies, Kluwer*

FABIAN A.C., 1994 *ARAA* **32**, 277

FABIAN A.C., BARCONS X., 1992 *ARAA* **30**, 429

FRANK J., KING A., RAINE D., 1985 *Accretion Power in Astrophysics, CUP*

FRASER G.W., 1989 *X-ray detectors in astronomy, CUP*

HASINGER G., 1997 *A&AS* **120**, 607

KATO S., FUKUE J., MINESHIGE S., 1998 *Black-Hole Accretion Disks, Kyoto Univ. Press*

LEWIN W.H.G., VAN PARADIJS J., VAN DEN HEUVEL E.P.J., 1995 *X-ray Binaries, CUP*

MUSHOTZKY R.F., DONE C., POUNDS K.A., 1993 *ARAA* **31**, 717

RYBICKI G.B., LIGHTMAN A.P., 1980 *Radiative Processes in Astrophysics. Wiley-Interscience*

SARAZIN C.L., 1988 *X-ray emission from clusters of galaxies, CUP*

SHAPIRO S.L., TEUKOLSKY S.A., 1983 *White Dwarfs, Neutron Stars and Black Holes*

TUCKER W., 1975 *Radiation Processes in Astrophysics, MIT press*

High Energy Data Analysis Techniques and Archive Access

By HERMANN BRUNNER

Astrophysikalisches Institut Potsdam, An der Sternwarte 16, D-14482 Potsdam, Germany

In the following article we will go, step by step, through some basic facts about X-ray detectors, general X-ray data analysis techniques, and through the content and functionality of some of the mayor X-ray data archives and software packages. Finally, we will examine a number of typical X-ray data analysis tasks in greater detail, both in the lectures and the tutorials, enabling the reader to download X-ray data sets from the archives, experiment with them, and perform some of his or her favourite data analysis tasks.

1. Introduction

In these practical lectures which are to follow, the readers will be presented with the data analysis tools and data archives required to tackle by themselves some of the scientific issues, presented in the lectures on "High Energy Astrophysics" by Andy Fabian. All that is required will be access to the Internet, a moderately powerful computer, and last, but not least, your own creativity.

The practical lectures will be concerned with X-ray astronomy in the hard and soft X-ray range, covering the spectral range from roughly 0.1 to several tens of keV. I.e., we will not discuss gamma ray astronomy and cosmic rays which, according to the usual definition, also fall into the realm of high-energy astrophysics. Apart from my lack of experience in these fields it will probably also be to the benefit of the reader, to have the opportunity study a more focused field to greater depth.

Since Andy Fabian has discussed X-ray telescopes and instruments from a "semi-theoretical" point of view, it is probably in order to take the more traditional approach and begin the practical lectures which a brief historic review of X-ray astronomy instrumentation which will then lead us on to an initial review of X-ray data analysis and calibration issues.

1.1. *A brief review of X-ray detectors and instruments*

X-ray astronomy is probably not much older than many of the students attending this Winter School (or at any rate, this statement used to be true when I was a student). Those interested in the complete story may want to look into such books as *The invisible Sky* by B. Aschenbach and J. Trümper. X-ray astronomy began with sounding rocket flights in the early sixties. At this time and shortly thereafter it was established for the first time that there are X-ray emitters beyond the earth and in fact as far away as clusters of galaxies and quasi-stellar objects.

1.1.1. *Proportional counters*

The type of X-ray detector used in these first experiments, the proportional counter, is even today one of the workhorses of X-ray astronomy. And in fact, one of the data analysis tasks, which we will tackle together will be the analysis of data from present day proportional counters. In its simplest form, proportional counters consist of two sets of wires in a box filled with a suitable counter gas (typically Argon or Xenon are used) to which a high voltage is applied. X-ray photons entering the box through an entrance window will ionise some of the atoms of the counter gas and the resulting primary electrons will then be accelerated towards the anode wires resulting in a cascade of secondary electrons which are picked up by the anode wires. Other than in the similarly designed Geiger counter, the proportional counter is set up in such a way that the signal at the anode wire is proportional to the energy of the incoming X-ray photon, allowing the detector to measure the energy of each X-ray photon. Speaking of Geiger counters you are probably aware that the main purpose of Geiger counters is in the realm of radioactivity, i.e. they measure both particles and photons. Since proportional counters (and other types of X-ray detectors) are also quite sensitive to particle radiation, the suppression of particle background radiation is one of the central issues in observational X-ray astronomy.

There are several possible approaches that can be applied to suppress particle background radiation. Fortunately, the duration and shape of the signal at the anode wires is different for photons and particle events, such that, by means of clever electronics, already at this stage, some discrimination between particle and photon events is possible. Since particle events tend to leave a long ionisation trail in the detector such that several anode wires will sense the resulting charge, so-called anti-coincidence between the signals measured by different anode wires can also be used to tell photons from particle events. Finally, particularly in the case of satellite based X-ray instruments, one way to avoid unacceptable high particle background is to avoid environments altogether where high doses of particle radiation are to be expected. As you know, the magnetic field of the earth captures charged particles from the solar wind and directs them mostly towards the poles, giving rise to complicatedly shaped radiation belts. High particle background can be avoided by shutting down the detector at times where the satellite comes close to the poles or close to the so-called south-Atlantic anomaly, a region of enhanced particle radiation. Some satellites have highly eccentric orbits where they spend most of the time at large distances from the Earth completely outside of the radiation belts.

As is obvious from the design of the proportional counter, in its simplest form it does, by itself, not provide any directional information for the incoming photons. Traditionally, this limitation is overcome by the use of collimators. These are sets of parallel pipes (of honeycomb or rectangular shape) set up in front of the detector entrance window, which block off the path of photons that enter the collimator at an angle with respect to the collimator axis. Collimators are usually designed to have opening angles in the range from half a degree to several degrees, thereby providing a rough positional information on the observed X-ray sources.

1.1.2. *Imaging X-ray detectors*

More sophisticated directional information can be obtained by substituting the collimator with an "optical" telescope which focuses the incoming X-ray photons onto the detector. Since reflection of photons at X-ray energies by and large is only possible at small angles, grazing incidence telescopes, also known as "Wolter telescopes", are used. These consist of nested paraboloid and hyperboloid mirrors, where the incoming photons are in turn reflected from a paraboloid and hyperboloid mirror shell, before being focused onto the focal plane. For technical reasons some Wolter telescopes have been designed to use conical approximations of the hyperboloid and paraboloid mirror surfaces. While resulting in a degradation of the imaging capabilities of the telescope this allows for a cheaper design and also permits the nesting of a large number of mirror shells, thereby increasing the effective area of the instrument. In addition to the optical mirror a position sensitive X-ray detector is needed in the focal plane. There are several different approaches:

(*a*) Let's again begin with the proportional counter which we have already studied in some detail. If a proportional counter is fitted with a set of many, closely spaced, parallel anode wires and a similar set of cathode wires, at right angles to the anode wires, then the charge released will be restricted to the anode and cathode wires closest to the location where the incoming photon hit the focal plane, thereby providing energy and at the same time 2-dimensional positional information. This technique has been employed in the *PSPC* (for "Position Sensitive Proportional Counter"), one of the two focal plane detectors of the *ROSAT* satellite, which is most well known for its All-Sky Survey of the soft X-ray sky.

(*b*) The other focal plane detector of the *ROSAT* X-ray telescope, the *HRI* (for "High Resolution Imager") which is based on a different technique uses a so-called micro channel plate to provide positional information. This channel plate consists of a large number of small pipes, each of which functions as a tiny photo-multiplier tube when high voltage is applied. When an X-ray photon enters the front of a pipe and interacts with its wall, an electron is released which, due to the high voltage applied to the pipe, gets accelerated along its length, resulting in the release of many secondary electrons. When leaving the pipe, these electrons hit a large resistive disk anode, which is read out by four wires attached to its corners. The location of the event is derived from the difference in voltage which each of the four readout wires senses. While providing high spatial resolution, micro channel plate detectors do not provide any spectral information on the incoming X-ray photons.

(*c*) The most widely used positionally sensitive X-ray detector for energies up to 10 keV at the present moment and in the years ahead will be the CCD detector. Since CCD detectors are not peculiar to X-ray astronomy, but rather are very well known to anyone who has had some experience in optical astronomy, we probably don't have to discuss the properties of CCD detectors here. Note, however, that other than in optical astronomy, CCD detectors are usually used in a photon counting mode in X-ray astronomy, i.e., the CCDs are read out at sufficiently short intervals such that, generally, only one photon (or no photon at all) will be contained in one frame. Since the year 1979, when the HEAO 2 satellite, also known under the name *Einstein* Observatory was launched, imaging X-ray telescopes have become the state of the art in soft X-ray astronomy.

1.2. *Recent X-ray satellite missions*

Having discussed some of the basic principles of X-ray detectors, we will now briefly go through the major X-ray satellite missions which are presently in their operational phase. A number of important upcoming satellite missions to be launched in the next few years will be discussed further on in the context of the future of X-ray astronomy.

1.3. *ASCA*

The Japanese X-ray satellite *ASCA* which was launched in February 1993 has four bi-conical, grazing incidence X-ray telescopes with 120 mirror shells each. Two of the telescopes have CCD detectors, which in the case of *ASCA* are called *SIS* (for Solid-state Imaging Spectrometer) and the other two use gas scintillation proportional counters (called *GIS* for Gas Imaging Spectrometer). Both detectors provide sensitivity in the energy range from about 0.4 to 10 keV. The *SIS* cover a field of view of 22 by 22 arcminutes and provide an energy resolution of 2% at 5.9 keV. The *GIS* have a larger field of view (50 arcminute diameter) but provide a somewhat lower spectral resolution (8% at 5.9 keV). Due to the bi-conical mirror design the spatial resolution is quite poor with a half-power diameter of the on-axis point spread function of 2.9 arcminute. The combined effective area of the four telescopes is about 1300 cm^2 at 1 keV and 600 cm^2 at

7 keV. *ASCA* was the first X-ray satellite mission that combined an imaging capability in the energy range up to 10 keV and the first to use CCD detectors. It has been particularly successful in studying spectral features in Active Galactic Nuclei, in studying the hot gas in clusters of galaxies and in investigating the X-ray background radiation.

1.4. BeppoSAX

This is an Italian X-ray satellite with contributions from the Space Research Organisation of the Netherlands (*SRON*), from *ESA*, and from the Max-Planck-Institut für extraterrestrische Physik, Germany, which was launched in April 1996. It includes a number of different, mostly non-imaging X-ray instruments, which cover the large energy range from about 0.1 to 300 keV. The low energy range up to 10 keV is again covered by four bi-conical grazing incidence X-ray telescopes which have two slightly different types of gas scintillation proportional counters. Three of the telescopes have *MECS* detectors (Medium Energy Concentrator Spectrometer) and the remaining telescope uses the *LECS* detector (Low Energy Concentrator Spectrometer) which has a thinner entrance window, but is otherwise identical. At an energy of 6 keV, the collecting area and energy resolution are about 150 cm^2 and 8%, respectively.

Higher energies are covered by the *HPGSPC*, a gas scintillation counter which is sensitive in the 4–120 keV energy range, and the *PDS*, a phoswich detector sensitive in the 15–300 keV energy range. Both of these have collimators with fields of view of 0.5° *FWHM*. Their effective areas are, respectively, 240 cm^2 at 30 keV for the *HPGSPC* and 600 keV at 80 keV for the *PDS*. At 60 keV, the energy resolution of the *HPSPC* is at 4% and the *PD*s has a energy resolution of 15%. The lateral shields of the *PDS* may also be used as an omni-directional Gamma Ray Burst monitor, which is sensitive in the 60–600 keV range. Finally there are the so-called Wide Field Cameras (*WFC*), sensitive between 2 and 30 keV, which have 20 degree diameter fields of view in opposite directions.

BeppoSAX has lately become famous through its role in pinpointing the locations and identifying the enigmatic Gamma Ray Bursts. But it has also contributed to many other fields of X-ray astronomy.

1.4.1. The Compton Gamma Ray Observatory (CGRO)

CGRO, the second of *NASA*'s great observatories, launched in April 1991, is a large gamma ray observatory, with several different instruments providing sensitivity over an energy range covering six orders of magnitude, from 30 keV to 30 GeV. Since these lectures are mainly intended to cover the field of X-ray astronomy, as opposed to the higher energy gamma rays, we will only briefly touch on the *CGRO* instrumentation. The *BATSE* instrument, a Sodium Iodide scintillation detector, is an all-sky monitor in the energy range from 20 keV to 1 MeV. *OSSE*, a collimated NaI scintillation detector and the only other instrument onboard the *CGRO* providing sensitivity at X-ray energies covers the energy range from 50 keV to 10 MeV and has a field of view of 3.8° × 11.4°. Finally, there are the *COMPTEL* and *EGRET* instruments which cover the energy range from 0.8 to 30 MeV and from 30 MeV to 10 GeV, respectively.

1.5. ROSAT

The *ROSAT* satellite, designed and operated by Max-Planck Institut für extraterrestrische Physik, Germany (with significant contributions from the USA and UK) is most well known for its first imaging All-Sky Survey of the soft X-ray sky, in the energy range from 0.1 to 2.4 keV. The All-Sky Survey was performed in the first half year of the satellite's operation, after its launch in June 1990. More than 7 years of observations of individual astronomical objects, following the All-Sky Survey performed in the first half year of operation of *ROSAT* so far, have provided a wealth of data on all fields of X-ray astronomy. Its principal instrument is a Wolter telescope with 4 nested mirror shells and a total collecting area of 1200 cm^2 and an on-axis point spread function of 15′ *FWHM*. The focal plane instrumentation consists of two redundant Position Sensitive Proportional Counters (*PSPC*) which provide positional and also moderate (43% at 0.93 keV) energy resolution and the High Resolution Imager (*HRI*) which is a micro channel plate detector. Providing no energy resolution and less sensitivity than the *PSPC* detectors, its positional resolution of about 1.7″ *FWHM* is far superior. The counter gas of the *PSPC* detectors was exhausted after about 4 years of operation. After a failure of the remaining *HRI* detector, *ROSAT*'s guest observer programme was terminated in September 1998.

In addition to the X-ray telescope, *ROSAT* also has a small co-mounted Wolter telescope, provided by the UK, which is sensitive in the XUV spectral range (20 – 180 eV).

1.5.1. *The Rossi X-ray Timing Explorer (RXTE)*

Launched in December 1995, *RXTE* is a mission that was designed to perform detailed time variability studies while providing only moderate spectral resolution and no imaging capabilities. The *RXTE* detectors cover the energy range from 2 to 250 keV. The *PCA* (for Proportional Counter Array) is a large area (0.56 m^2) proportional counter, sensitive in the energy range from 2 to 60 keV. The *HEXTE* High Energy X-ray Timing Experiment has an effective area of 2 × 800 cm^2 and covers the energy range from 15 – 250 keV. Both instruments are non-imaging instruments employing collimators. In addition there is an All-Sky Monitor (*ASM*), sensitive in the 2 to 10 keV energy range.

1.6. *General data analysis and calibration techniques*

X-ray data analysis is the art of making the most out of generally relative small numbers of photons, such that proper treatment of count statistics plays an important role. From a practical point of view, it is probably useful to discuss spatial and spectral analysis of X-ray data in an independent fashion. Let's begin with the spectral analysis. The general approach in X-ray astronomy is to start out with the photons as they enter the instrument and follow them through the detection process, i.e, in the usual lingo, the photons are 'folded' through the detector. Mathematically, this is achieved by multiplying a vector containing the source fluxes in each of a number of different energy bands (given, e.g., in units of photons/cm^2 s keV), first with the effective areas of the instrument in each energy band and with the width of each energy band (such that the unit is now photons/s intercepted by the instrument in each energy band), and finally with a dimensionless matrix which describes the detailed properties of the detection process. The result of the matrix multiplication is source counts per second measured in each energy band. Starting with a physical model spectrum incident on the detector, we thus have predicted the response of the detector, which we can now compare with the actual observed source counts. This comparison is usually performed by means of the χ^2 statistics or some similar statistical approach (such as like, e.g., Maximum Likelihood statistics) and the parameters of the model spectrum are adjusted such that the model predicted count rate spectrum best describes the data.

At this point usually the question is raised whether the reverse approach, i.e., starting out with the observed count rate spectrum and 'folding' it backwards through the detector wouldn't be more appropriate. I think the response to this question is related to the generally relatively poor energy resolution (as compared to, e.g., optical spectroscopy) and to unavoidable uncertainties in the calibration of the detector response matrix, which make the inversion of the response matrix rather poorly defined. Forward folding at any rate is the approach which, with only few exceptions, has been used almost exclusively in X-ray astronomy.

Having discussed the analysis of X-ray spectra let's now move on to the spatial analysis of X-ray data. Prominent examples of spatially extended X-ray emitters are, for example, supernova remnants, or clusters of galaxies. This discussion is naturally limited to the analysis of data from imaging X-ray instruments. Spatial quantum efficiency variations over the face of the detector as well as geometric and reflectivity effects due to the telescope are usually described in the form of an exposure image which gives the effective area of the instrument (telescope plus detector) at each image pixel. The exposure map which is dependent on energy needs to be calculated for each observation individually.

Another important issue when studying extended X-ray emission is the point spread function which is dependent both on the location of the source in the field of view and on energy. One undesired effect of grazing incidence telescopes, is their rather strong degradation of the point spread function at large off-axis angles. Apart from geometrical optics, scattering due to the micro-roughness of the mirrors also needs to be taken into account, which results in rather wide scattering wings of the point spread function. Depending on the type of X-ray detector, the intrinsic spatial resolution of the detector also has to be considered.

Finally, one effect that needs to be considered, both for spectral and spatial analysis is dead time (i.e., time periods when the detector is not able to detect or process incoming photons) which may either be due to the physics of the detection process, or to telemetry limitations.

2. X-ray Data & Software in Public Archives

Having acquainted ourselves with X-ray instrumentation and some basic data analysis techniques we will now have a closer look into the data archives and software packages available through the Internet. For each of the ongoing X-ray satellite missions which we discussed in the first section, access to the data archives, data analysis software, and to extensive documentation is available through the World Wide Web. Web sites are usually maintained by the respective data centres, and the hardware or instrument teams of each satellite. We will first discuss the principal structure of the X-ray data and then look at the contents and access methods of some important archives.

2.1. *Data formats and data access*

Note that public access to the data in most cases is only possible after the proprietary period of exclusive data rights of the principal investigator (*PI*) of an observation have expired. In most cases the data enter the public domain one year after they were made available to the *PI*. Depending on the type of instrument and observing mode, the data are usually available in the form of raw or calibrated event lists, or binned energy channel spectra or time series. In the case of imaging X-ray detectors event lists will contain raw or linearised X and Y detector co-ordinates and typically also the corresponding position, projected onto the sky (given in right ascension and declination, or in local X and Y sky co-ordinates), using the attitude information of the satellite at the time the event was detected. The energy of each detected X-ray photon is provided in the form of the raw channel number as determined by the instrument electronics and additionally as a calibrated energy or energy channel. It has become customary to refer to the raw energy channel number as *PHA* channel (for *pulse height amplitude*) and to the calibrated energy channels as *PI* channels (for *photon intensity*). Pre-processed data products, such as calibrated source or background spectra, time series, or images are also available in most cases.

Access to individual data sets is made possible through the use of database systems which allow to search the archive for data sets which cover a particular area of the sky, or which were performed in a particular time period. More advanced database systems will also allow a search for data sets of particular astronomical object (using a large range of naming schemes) or object class. Of course, the most straight forward way to access the archive, i.e. if an individual data set is already known, is through its unique sequence number.

2.1.1. *HEASARC*

Archived data of many previous and ongoing X-ray satellite missions are accessible through the High Energy Astrophysics Archive (*HEASARC*) at *NASA*'s Goddard Space Flight Center. Access to the data is possible through the Browse database system originally developed at *ESA*'s Space Science Department, for the European *EXOSAT* X-ray satellite mission. From the *HEASARC* Web site the browse database system can be accessed through its forms based Web interface `W3browse`. The data are provided in *FITS* format, using the *OGIP* header keyword convention which we will discuss in a later lecture. The *FITS* data format, which is described in full detail in a series of papers in Astronomy & Astrophysics Supplements was originally developed as a platform-independent data exchange format in radio astronomy, but today may be regarded as a quasi-standard in all branches of astronomy. Details on the *OGIP* convention may be found on the Web pages of *HEASARC*'s *FITS* Working Group at `http://heasarc.gsfc.nasa.gov/docs/heasarc/ofwg/ofwg_intro.html`.

Beginning which the presently active X-ray missions, the *HEASARC* archive contains the complete archived data of the *ASCA*, *ROSAT*, and *RXTE* missions. Public *CGRO* and *BeppoSAX* data are to be added in early 1998. The *HEASARC* archive also contains data from the *BBXRT*, *Copernicus*, *COS B*, *DXS*, *EXOSAT*, *Einstein*, *GINGA*, *GRANAT*, *HEAO-1*, *HEAO-3*, *OSO-7*, *OSO-8*, *SAS-2*, *SAS-3*, *TENMA*, *VELA5B*, and *UHURU* satellites. We will not discuss these missions here. Brief outlines and links to more detailed information may be found on the 'missions page' of the *HEASARC* Web site.

While the *HEASARC* archive is a good starting point to find data on many X-ray satellite missions, first hand information may be obtained from many of the data centres and hardware groups which built and operate each satellite. We will now have a look at some of the data centres of current satellite missions.

2.2. *ROSAT archive at* MPE, *Garching*

The *ROSAT* data centre at the Max-Planck Institut für extraterrestrische Physik (*MPE*) in Garching, Germany, accessible at `http://wave.xray.mpe.mpg.de/rosat/` features both an online archive of all public *ROSAT* data and source catalogues of the *ROSAT* All-Sky Survey and pointed phase. The *ROSAT* data archive is accessible through a Web forms based interface (The *ROSAT* Archive Browser) which allows one to search for individual data sets using a large selection of search criteria. Except for the full *FITS* format data, preview images, both in *GIF* and *FITS* format, and some information on each observation, including the abstract of the original proposal, are available. The *FITS* format data are provided in the same format as in the *HEASARC* database and are compatible with all available *ROSAT* data analysis software packages (i.e., *FTOOLS* or *MIDAS/EXSAS*; see next chapter). In addition to the observational data, the archive also contains a full set of *ROSAT* calibration data files, as well as observation logs and timelines.

2.3. *BeppoSAX data centre*

The *BeppoSAX* mission web page of the Science Data Centre (http://www.sdc.asi.it/) at present contains documentation on the *BeppoSAX* instruments, timelines and observation logs.

2.4. *Getting acquainted with X-ray data analysis packages*

There are a number of different data analysis software packages available for the analysis of astronomical X-ray data. Probably the most widely used are the *FTOOLS* and *XANADU* packages, which can be used for the analysis of data from a number of different X-ray satellite missions. At this moment there are *FTOOLS* sub-packages which permit the analysis of data from the *ASCA*, *Einstein* Observatory, *CGRO*, *ROSAT*, *BeppoSAX*, and *RXTE* satellite missions.

The *IRAF* based *PROS* package was developed for *ROSAT* data analysis, but may also be applied for X-ray data from other missions. The *EXSAS* package was developed as a sub-package of the all-purpose astronomical data analysis package *MIDAS* for the analysis of *ROSAT* data and is presently being ported to the *FTOOLS* environment.

Below, we will discuss the *FTOOLS*, *XANADU*, and *EXSAS* packages in some detail and in the next section will follow through some example data analysis sessions in detail. Obviously, the best way to get acquainted with these software packages will be to read the respective manuals and to get some hands-on experience and play around with the data.

In addition to these general-purpose X-ray data analysis packages, there are a number of utilities and tools useful both for studying and interpreting the X-ray data and for planing new observations. These are the flux-conversion program *PIMMS* (the WWW version is called W3Pimms), which permits to predict source count rates in each of the different X-ray instruments, the Galactic NH program, *W3NH*, which gives the galactic neutral hydrogen column density at each celestial location, the program *Viewing* which provides the dates when an object or co-ordinate position is observable from a specific mission, and the WebSpec program, which provides a WWW, forms interface to the *XSPEC* program (see below) to permit the modelling of X-ray spectra as observed by specific X-ray instruments. In addition, a number of utilities for data, time, and co-ordinate conversion are available. These tools can either be downloaded, to be installed on one's local machine, or can be run directly from the *HEASARC* Web site via a Web forms interface.

2.5. *FTOOLS*

FTOOLS is a general-purpose package for processing and analysing data in the *FITS* (Flexible Image Transportation System) data format. It consists of over 200 individual utilities which are organised into a number of different sub-packages, which handle generic data manipulation and display tasks and the calibration and analysis of data from different X-ray satellite missions. *FTOOLS* calibration tools automatically interface with the instrument specific files in the caldb calibration database.

The user interface and the communication between the individual *FTOOLS* is based on the *IRAF* parameter system. *FTOOLS* can either be run from within the *IRAF* environment or as a stand-alone package from the operation system command line prompt. The *FTOOLS* programme code and documentation is available from the *HEASARC*

Web site, both in source code form and pre-compiled for most widely used operating systems and architectures.

Available *FTOOLS* sub-packages for general data analysis tasks are *CALTOOLS* containing calibration-related tasks, *HEASARC* for X-ray analysis-related tasks, and *FIMAGE* and *TIME* for image and time-related tasks. The *FUTILS* package contains tasks for general *FITS* file manipulation. The *FTOOLS* packages for specific X-ray satellite missions are *ASCA*, *EINSTEIN*, *GRO*, *ROSAT*, *SAX*, *VELAB5*, and *XTE*, respectively.

The *FTOOLS* tasks are implemented as Fortran source code contained in a C wrapper. The *FTOOLS* package also makes extensive use of Perl scripts. An *FTOOLS* script central to X-ray analysis is *XSELECT* which links together many individual *FTOOLS* needed for data selection and data preparation in X-ray astronomy.

All of these packages are included in the standard *FTOOLS* release which can be downloaded from the *HEASARC* Web site. At the time of the writing of this document the latest version of *FTOOLS* is release 4.0. To install and run the *FTOOLS* for X-ray data analysis the user should obtain the latest versions of the User's guide to *FTOOLS*, The *FTOOLS* Installation Guide, and the *Xselect User's Guide*, all of which are available as Postscript files, and in part also as *HTML* documents at the *HEASARC* Web site.

2.5.1. *XANADU*

The *XANADU* packages contains the three sub-packages *XSPEC*, *XRONOS*, and *XIMAGE* for spectral, temporal, and image data analysis, respectively. In particular, the spectral fitting and modelling program *XSPEC* is very widely used and may be regarded as a quasi-standard for the spectral modelling of X-ray data. All three of these packages seamlessly work with the *FTOOLS* package and the `caldb` calibration database. They can be downloaded in total or individually from the *HAESARC* Web site.

To find out more about the origin of the name of the *XANADU* package, look up the poem by *Samuel Tayler Coleridge*, or follow the link `http://lheawww.gsfc.nasa.gov/~kaa` `/poem/poem.html`.

You may also want to read the novel by Douglas Adams *Dirk Gentey's Holistic Detective Agency*.

Here we will look at the *XSPEC* package in some detail, discussing its main commands and features and refer the reader to the documentation available, both in Postscript and *HTML* form at the *HEASARC* Web site.

2.6. *XSPEC*

XSPEC is versatile and easy to use interactive program for the modelling and fitting of X-ray spectra. One of its strong points is the large number of physical model spectra which are built into the program. These range from simple power law and exponential spectra, to physical models of hot thermal plasmas. The models are grouped into additive and multiplicative models, which can be combined in a straightforward fashion. It is also possible to link additional spectral models with the programme, or to supply additional spectral models in table form. *XSPEC* also provides an easy way to combine the data from different X-ray instruments.

2.6.1. *SAOimage*

SAOimage is a quite user friendly and versatile image display program written at the Smithsonian Astrophysical Observatory for displaying *FITS* format images on X11 window screens. The program which was developed particularly with X-ray astronomical applications in mind is available from `http://tdc-www.harvard.edu/software/saoimage`, the *SAO* Web site.

2.6.2. *EXSAS*

The *EXSAS* package is a comprehensive data analysis package for the analysis of *ROSAT* data. It was developed at the Max-Planck Institut für extraterrestrische Physik (*MPE*) in Garching, Germany which also built and operated the *ROSAT* satellite.

EXSAS was designed as a sub-package (which in *MIDAS* are called *contexts*) of the general-purpose astronomical data analysis package *MIDAS* (Munich Image Display and Analysis System). *MIDAS*, which was developed by the European Southern Observatory (ESO), which also has its headquarters in Garching, in scope and functionality resembles the *IRAF* package and may be regarded as its European counterpart. Presently a version of *EXSAS* is being developed which will comply with the *FTOOLS* standard, the first release of which should be available at the time of the printing of this book, making *EXSAS* much more accessible for users who are not acquainted with the *MIDAS* package. The *EXSAS* package may be downloaded from the *ROSAT* Web site at *MPE* (`http://www.xray.mpe.mpg.de/exsas`). It is available both in source code form and pre-compiled for most widely-used architectures. The *MIDAS* package, which at the present moment is required to run *EXSAS*, is freely available from the ESO Web site (`http://www.eso.org/midas/`). The main documents which the prospective user should obtain from the *MPE ROSAT* Web site are the *EXSAS* Command Summary which gives a one page description of each *EXSAS* command and the comprehensive *EXSAS* User's Guide, which gives a full description of the commands and of typical data analysis sessions. Both documents are available in Postscript and *HTML* form.

3. Some Sample Data Analysis Sessions

On the following pages we will go through a number of typical X-ray data analysis sessions, starting from the identification of the appropriate data sets using X-ray source catalogues, and observation logs from different X-ray satellite missions, downloading the data, and finally performing some basic X-ray data analysis tasks. We will look at the extraction of source and background photons, the binning of images and spectra, and the fitting of model spectra using the *FTOOLS*, *XANADU*, and *MIDAS* packages. Note however, that this document is not meant as a substitute for the proper user's documentation which is available for each of these data analysis packages.

3.1. *Retrieving data from the archives*

Before we will be able to perform some data analysis tasks ourselves, we need to retrieve the appropriate data sets from on of the data archives available through the Internet. Here, we will, as an example, retrieve a data set observed by the *ROSAT* satellite both from the archive at *ROSAT*'s home base, the Max-Planck-Institute für extraterrestrische Physik in Garching, Germany, as well as from the *HEASARC* archive located at the Goddard Space Flight Center.

The *ROSAT* archive at *MPE*, Garching may be accessed through the Web page `http://wave.xray.mpe.mpg.de/rosat/`.

Click on `Archive Browser` to be presented with several different options to access the archive. An efficient way to search for a particular observation is through the `Master Interface`, which allows you to search for observations by co-ordinates, time of observation, target name, or observation sequence number. Choosing the target name option, entering the string `Puppis` will result in a list of observations that have been performed of this beautiful supernova remnant. Note that this interface requires a strict match between the input string and catalogue entry and you run the risk that your target name query will not give you all observations of the desired target. To make sure that you don't miss any observations it is probably better to query by target co-ordinates. We may display a quick look image of the observation in *GIF* format by clicking on the letter g. Auxiliary information on the observation is available by clicking on the letters a which will display an abstract of the original science proposal to observe this source or p which will display the main parameters of the observation as listed in the header keywords of the *FITS* data files. Finally, clicking on the letter d displays links to the individual data files of each observation. You will need to download all of these compressed *FITS* data files to perform *ROSAT* data analysis of your own. The format of the files is such that they may be fed immediately into the data analysis packages which we have discussed (see the examples below). You may retrieve the files by clicking on the links (don't forget to hold the right mouse button if you are using the Netscape Navigator as Web browser) or by using the ftp command at the operating system command line prompt if this is more convenient.

We will now retrieve the same observations from the *HEASARC* archive. The data are best accessed using the *W3Browse* interface at http://heasarc.gsfc.nasa.gov/W3Browse/.

Using, e.g., the Advanced interface, we need to select the ROSAT mission, select a catalogue (mark the ROSPUBLIC ROSAT Archival Data entry in the first input window) and again specify the target name string in the second input window. Note that, in this case, the input target name will be resolved using the *SIMBAD* database, such that typing Puppis A in the input window will give us a list of all public *ROSAT* observations of the supernova remnant. We now need to mark the box in front of the desired data sets, click on the list all data products button, and finally click on the submit button to access a list of data files for the selected observations which are available for downloading. The files which are identical to the files in the archive at *MPE* may be downloaded by marking the boxes in front of each file and by clicking on the retrieve button. Since this is only a small example of the many uses to which the *HEASAC* archive can be put, the reader is encouraged to explore the many options of the astronomical database access tools (*W3Browse, Argus, Astrobrowse, Sky View*) which are available through the *HEASARC* home page.

3.2. *Data selection*

As a first example we will perform simple selection and extraction tasks on data from the *ROSAT* satellite, using the *XSELECT* program which is part of the *FTOOLS* package and then move on to repeat an equivalent *ROSAT* data preparation sequence, using the *MIDAS/EXSAS* package. As already mentioned in the previous section, the *FTOOLS* package may be used in conjunction with the data from many different X-ray missions and instruments. It will be beyond the scope of this document, however, to discuss the details of the data selection process relating to the requirements for each different instrument and we will only briefly mention instrument-dependent aspects of the data selection.

3.2.1. *Data selection with XSELECT*

XSELECT is a command line interface to the *FTOOLS*, originally developed for *ASCA* which allows to apply intensity, phase, region, spectral, and timing filters. Based on the selection criteria, good-time intervals (*GTI*) are created. The input data can be organised through an observation catalogue. *XSELECT* is invoked by typing *XSELECT* at the command prompt. A typical session will include the commands read events to specify the event file on which one is going to work, the command `filter region` to select, e.g. events from an area on which a spectral analysis is to be performed, the command `select` which will actually perform a selection, and the command `extract` which will create binned light curves, spectra, and images from the filtered event lists. Filter regions may be defined interactively using the image display program *SAOimage*. See the *Xselect User's Guide* by J. Ingham, which is available from the *HEASARC* Web site at `http://heasarc.gsfc.nasa.gov/docs/software/ftools/xselect/xselect.html` for a full description of *XSELECT*. A short sample *XSELECT* session, adapted from the Xselect User's guide, is given below. The *XSELECT* session opens a *ROSAT PSPC* event file, extracts an image, selects an area from the image using the image display program *SAOIMAGE*, and finally extracts a spectrum for further analysis with the spectral fitting program *XSPEC*.

```
> xselect
 ** xselect V1.1a **
> Enter session name >[xsel] session1
> session1: ASCA > read events
> Enter Event file list >[] rp7000000.fits
 Got new mission: ROSAT and new instrument: PSPC
> Reset the mission and instrument? >[yes]
 This is ROSAT
 Notes: xselect set up for         ROSAT
 Keywords for time and pha are     TIME        PI
 Units of time are                 s
 Keywords for image are            X           Y

 ...

session1:ROSAT-PSPC > extract image
 Extractor 1.0q
 XPI 2.1t
 Doing file: rp700000.fits
 Getting FITS WCS Keywords
100% completed

 ...

session1:ROSAT-PSPC > saoimage
session1:ROSAT-PSPC > filter region target.reg
session1:ROSAT-PSPC > extract spectrum
 Extractor 1.0q
 XPI 2.1t
 Doing file: rp700000.fits
 Getting FITS WCS Keywords
100% completed

 ...

session1:ROSAT-PSPC > save spectrum
> Give output file name >[] spec1
```

The data will be grouped using the following:

```
Bad 1-11, 212-256&group 12-211 10&show group&show quality&exit
```

```
> Group ( or rebin ) the spectrum before outputting? >[yes] no
```

...

```
Wrote spectrum to spec1.pha
```

```
session1: ROSAT-PSPC > exit
```

3.2.2. *ROSAT data preparation using the MIDAS/EXSAS package*

In the *EXSAS* package data selection is performed using the so-called *projection package*, which may either be controlled by command line interface, or through a command file. The main commands for performing selections are `select` to perform spatial, temporal, and energy selections, and `bin` to create binned images, spectra, and time series. As a first step the data need to be converted into the *MIDAS/EXSAS* data format using command `intape/rdf`. The command `set/correction correct` optionally corrects the output of the following projection commands for telescope vignetting. Fitting model spectra can either be performed within the *EXSAS* package or the *XSPEC* spectral fitting program may be used, in which case the command `outtape/xspec` will convert the spectra into the format required by *XSPEC*. A simple sequence of commands which performs the same tasks as the *XSELECT* example, above, is given below. Please refer to the *EXSAS* Command Summary and the *EXSAS* User's Guide for a detailed description of the calling sequence.

```
> inmidas
Midas 001> set/context exsas
Midas 002> intape/rdf disk rp
Midas 003> bin/ima events 30 i=events o=image
Midas 004> create/display
Midas 005> load/image image
Midas 006> set/correction correct
Midas 007> select/ring cursor bin ampl 10,12,211 i=events o=spec1
Midas 008> outtape/xspec
Midas 009> bye
```

3.3. *Model fitting using the XSPEC package*

Below we will go through an example session of the spectral fitting program *XSPEC*. Typing *XSPEC* at the command prompt starts an *XSPEC* session. Basic *XSPEC* commands needed for a simple spectral fitting session are: The command `data` to load spectral data files which have been prepared by the *XSELECT* package (see above), the command `backgrnd` to load the background spectrum, the command `model` to define a model spectrum, and the command `fit` to adjust the parameters of the model spectrum to best represent the data. The detector response matrix may be specified using the command `response` and the energy channel range on which the spectral fit is to be performed can be specified using the command `ignore`. The command `newpar` lets one set the spectral model parameters to appropriate starting values. Both the observed model and the model fit may be displayed using one of the may options of the `plot` command. Access to the extensive built in help function is available through the `help` command. Note that this is only a small fraction of the many commands which *XSPEC* offers to analyse the data. Both the data and model commands feature a very flexible and versatile syntax which lets the user specify arbitrary combinations of model spectra and data sets. The example below loads the source spectrum *spec1.pha*, the background spectrum *back1.pha*, and the response file *resp*. A power law model with absorption (wabs + po) is set and the first model parameter is set to the value 0.05 and fixed at that value. Finally a model fit is performed and the result is displayed:

```
xspec> data spec1.pha
xspec> bckgrnd back1.pha
xspec> response resp
xpsec> model wabs + po
xspec> newpar 1 0.05
xspec> freeze 1
xspec> fit
xspec> plot ufspec
```

4. Future X-ray Satellite Missions

The remaining years of the century will see a number of highly interesting X-ray satellite missions, which will advance X-ray astronomy a long way. In the order of expected launch date these are: the *Chandra* Observatory, now expected to be launched probably early in 1999; *ESA*'s cornerstone mission, *XMM*, to be launched in mid 1999; the Japanese *Astro-E* mission scheduled for launch in early 2000; and the Russian Spectrum-X-Gamma mission with an expected launch date in 1999, or later. In addition, the German small satellite project *ABRIXAS*, to be launched in early 1999, will perform the first imaging survey of the complete X-ray sky in the energy range up to 10 keV.

4.1. *The Chandra X-ray Observatory (AXAF)*

Following the Hubble Space Telescope and the Compton Gamma Ray Observatory, *Chandra* will be the third of *NASA*'s "Great Observatories" astronomy satellites. Its main strongpoint will be the sub-arcsecond spatial resolution of its X-ray telescope which, for the first time, will provide X-ray astronomers with imaging capabilities at the level of ground-based optical telescopes. *Chandra* is equipped with a 4-fold nested mirror system with a geometric area of 1200 cm^2. Its two focal plane instruments, the *ACIS* CCD detector array and the *HRC* channel plate detector both provide sensitivity in the 0.1 to 10 keV range. The *HRC* is designed to achieve the full sub-arcsecond resolution of the telescope, while the *ACIS* CCD detector, providing somewhat lower spatial resolution, also has some spectral capability. Higher resolution spectroscopy can be performed using the *LEGT* and *HEGT* grating spectrometers which provide spectral resolutions in the range from $E/\Delta E = 200 - 2000$. A full set of documents and links on the *AXAF* instrumentation as well as documents and tools for preparing proposals for *Chandra* observing time are available from the Web site of the *Chandra* Science Centre, located at the Center for Astrophysics in Cambridge, MA at `http://asc.harvard.edu/`.

4.2. *Abrixas*

In April 1999 the *ABRIXAS* satellite will be launched into low earth orbit, where it will begin its three-year scanning operation of the whole X-ray sky (energy range 0.5 to 10 keV). It will thus extend the X-ray All Sky Survey performed by *ROSAT* to harder energies. Abrixas is expected to detect of the order of 10 000 new X-ray sources, many of which will be absorbed Seyfert galaxies which, due to their intrinsic low energy absorption, were not visible in the *ROSAT* All Sky Survey. *ABRIXAS* is a German small satellite project, which is based on the mirror technology and CCD detectors developed for *XMM*. Its mirror assembly consists of seven grazing incidence telescopes with 27 mirror shells each of which share a common CCD detector. The design is optimised for the all-sky scanning operation of *ABRIXAS*. Additional details on the Abrixas mission are available on the Abrixas Web pages at *MPE*, Garching and Astrophysikalisches Institut Potsdam: `http://wave.xray.mpe.mpg.de/abrixas` and `http://www.aip.de:8080/~friedric/abrixas/abrixas_start.html`.

4.3. *XMM*

Launched less than a year later than *Chandra*, *ESA*'s cornerstone mission, *XMM*, will surpass *Chandra* in terms of collecting area by a factor of 4 (at 0.1 keV) to 10 (at 8 keV). The large collecting area of *XMM* is achieved by three identical X-ray telescopes, each consisting of 58 closely spaced thin mirror shells, manufactured by an innovative casting technique. Its spatial resolution of better than 15 arcsec (*FWHM*) is somewhat worse than that of *Chandra*, but still compares favourably with many of the previous and presently operational X-ray telescopes. All three X-ray telescopes have CCD detector arrays as focal plane instruments. Two of these are conventional *MOS* CCD detectors while one telescope is fitted with a pn CCD detector array especially developed for use in X-ray astronomy. Two of the telescopes also include the *RGS* grating spectrometers providing spectral resolution in the range $E/\Delta E = 100 - 600$. In addition to the X-ray telescopes, *XMM* will also carry an optical monitor telescope, allowing optical filter spectroscopy to be performed contemporaneously with the X-ray observations. A routine pipeline processing of all *XMM* data will be performed by the *XMM* Survey Science Centre consortium, which will also provide optical cross-identifications of serendipitous X-ray sources. An archive of all *XMM* data and data products will be kept at *ESA*'s Villafranca tracking station, near Madrid, Spain. Up-to-date information on the *XMM* mission is available from *ESA*'s *XMM* Web page at `http://astro.estec.esa.nl/XMM`.

4.4. *Astro-E*

Japan's next generation X-ray satellite mission will include five identical conical thin foil mirrors, similar to the ones already employed by *ASCA*. Four of the telescopes will have the *XIS* CCD detectors as focal plane instrumentation, while the fifth telescope will be the first X-ray calorimeter ever to be flown on an X-ray mission which will provide unprecedented spectral resolution to be achieved in a non-dispersive instrument. The combined effective area of the four telescopes used for imaging is 1800 cm^2, at 1 keV while, the fifth telescope used in conjunction with the calorimeter, has an effective area of 450 cm^2 at 1 keV. Both the X-ray telescopes and instrumentation are contributed by the United States. In addition, *Astro-E* also includes a collimated high energy X-ray detector, providing sensitivity in the 0.4 to 700 keV energy range. For additional information, consult the *Astro-E* Web pages of the *Astro-E* Guest Observer Facility at the Laboratory for High Energy Astrophysics (`http://legacy.gsfc.nasa.gov/docs/astroe/astroegof.html`), or at `http://www.astro.isas.ac.jp/xray/mission/astroe/astroeE.html`, the *Astro-E* Web page of ISAS.

4.5. *Spectrum-X-Gamma*

Built under the leadership of the Russian Space Research Institute *IKI*, with substantial contributions from a large number of European institutions, Spectrum-X-Gamma combines a multitude of instruments covering the spectral range from the UV to hard X-rays. Some of its main experiments are the two *SODART* conical foil mirror telescopes with the *HEPC* (2 – 24 keV) and *LEPC* (0.5 – 8 keV) proportional counters in the focal plane and the Jet-X instrument, consisting of two X-ray telescopes employing CCD detectors as focal plane instruments. The *OXS* Objective Crystal Spectrometer may be used in conjunction with the *SODART* telescope to perform high-resolution spectroscopy. High X-ray energies up to 150 keV are covered by the *MAR-LIME* coded mask telescope, which uses a proportional counter. The *FUVITA* and *TAUVEX* UV telescopes cover the spectral range from 90 to 120nm and 140 to 300nm, respectively. Consult the Web page of the US Spectrum-X-Gamma Coordination Facility at `http://hea-www.harvard.edu/SXG/sxg.html` for a description of all experiments to be flown on Spectrum-X-Gamma. This site also gives a list of links to the contributing institutes. The Spectrum-X-Gamma Web page of the Russian Space Research Institute, *IKI*, which still is under construction, is available at `http://hea.iki.rssi.ru/SXG/SXG-home.html`.

REFERENCES

ASCHENBACH, B. AND TRÜMPER, J. 1998 The Invisible Sky: ROSAT and the Age of X-ray Astronomy Springer, 1998

BLACKBURN, J.K. 1994 FTOOLS: A FITS data processingand analysis software package ADASS IV, ASP Conf. Series, Vol. 77, 1995, p. 367

GROSBOL, P., HARTEN, R.H., GREISEN, E.W., AND WELLS, D.C. 1988 Generalized Extensions and Blocking Factors for FITS. Astron. Astrophys. Suppl.. 73, 359

GREISEN, E.W., AND HARTEN, R.H. 1981 An Extension of FITS for Groups of Small Arrays of Data. Astron. Astrophys. Suppl., 44, 371

HARTEN, R.H., GROSBOL, P., GREISEN, E.W., AND WELLS, D.C. 1988 The FITS Table Extension. Astron. Astrophys. Suppl., 73, 365

WELLS, D.C., GREISEN, E.W., AND HARTEN, R.H. 1981 FITS: A Flexible Image Transport System. Astron. Astrophys. Suppl., 44, 363

ZIMMERMANN, H.U., BECKER, W., BELLONI, T., DÖBEREINER, S., IZZO, C., KAHABKA, P., SCHWENTKER, O. 1994 Exsas User's Guide Edition 4, ROSAT Science Data Center, MPE Report 257